P9-CCC-504

The Germination of Seeds

FOURTH EDITION

Titles of Related Interest

FAHN
Plant Anatomy, 3rd Edition

GOODWIN & MERCER
Introduction to Plant Productivity, 3rd Edition

GREEN *et al.*
Chemicals for Crop Improvement & Pest Management, 3rd Edition

KENT
Technology of Cereals, 3rd Edition

LOCKHART & WISEMAN
Introduction to Crop Husbandry, 6th Edition

NASH
Crop Conservation & Storage, 2nd Edition

WAREING & PHILLIPS
Growth & Differentiation in Plants, 3rd Edition

WIDDOWSON
Towards Holistic Agriculture: A Scientific Approach

Journals of Related Interest

Biochemical Systematics and Ecology

Current Advances in Ecological Sciences

Current Advances in Plant Sciences

Outlook on Agriculture

Phytochemistry

Soil Biology & Biochemistry

(Sample copy gladly sent on request)

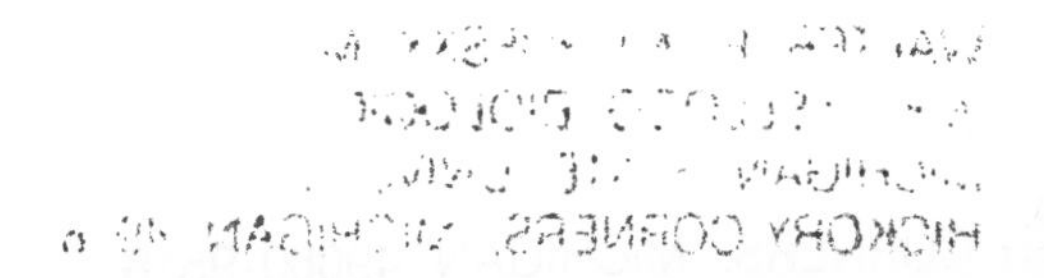

The Germination of Seeds

FOURTH EDITION

by

A. M. MAYER
Professor of Botany,
The Hebrew University of Jerusalem, Israel

and

A. POLJAKOFF-MAYBER
Professor of Botany,
The Hebrew University of Jerusalem, Israel

PERGAMON PRESS

OXFORD · NEW YORK · BEIJING · FRANKFURT
SÃO PAULO · SYDNEY · TOKYO · TORONTO

U.K.	Pergamon Press plc, Headington Hill Hall, Oxford OX3 0BW, England
U.S.A.	Pergamon Press, Inc., Maxwell House, Fairview Park, Elmsford, New York 10523, U.S.A.
PEOPLE'S REPUBLIC OF CHINA	Pergamon Press, Room 4037, Qianmen Hotel, Beijing, People's Republic of China
FEDERAL REPUBLIC OF GERMANY	Pergamon Press GmbH, Hammerweg 6, D-6242 Kronberg, Federal Republic of Germany
BRAZIL	Pergamon Editora Ltda, Rua Eça de Queiros, 346, CEP 04011, Paraiso, São Paulo, Brazil
AUSTRALIA	Pergamon Press Australia Pty Ltd., P. O. Box 544, Potts Point, N.S.W. 2011, Australia
JAPAN	Pergamon Press, 5th Floor, Matsuoka Central Building, 1-7-1 Nishishinjuku, Shinjuku-ku, Tokyo 160, Japan
CANADA	Pergamon Press Canada Ltd., Suite No. 271, 253 College Street, Toronto, Ontario, Canada M5T 1R5

First edition 1963

Second edition 1975

Reprinted 1978; 1979

Third edition 1982

Reprinted 1986

Fourth edition 1989

Library of Congress Cataloging-in-Publication Data

Mayer, A. M.
The germination of seeds/by A. M. Mayer, and
A. Poljakoff-Mayber.—4th ed.
p. cm.
Bibliography: p.
Includes indexes.
1. Germination. I. Poljakoff-Mayber, Alexandra, *1915-*
II. Title.
QK740.M38 1989 581.3'33—dc19 88-39999

British Library Cataloguing in Publication Data

Mayer, A. M. (Alfred Max), *1926-*
The germination of seeds.-4th ed.
1. Seeds. Germination I. Title II. Poljakoff-
Mayber, A. (Alexandra), *1915-*
582.'0333

ISBN 0-08-035723-7 Hardcover
ISBN 0-08-035722-9 Flexicover

Printed in Great Britain by BPCC Wheatons Ltd, Exeter

Contents

Preface to the Fourth Edition

Seven years have elapsed since the third edition of this book was prepared. In preparing this fourth edition we have retained the basic structure of "Germination of Seeds", but wide ranging revisions have been made. In the period between the last and the present edition there has been a continued steady publication of papers on germination. The techniques of molecular biology and of membrane research were increasingly applied to studies of the early stages of germination. As a result of this increase in information more than 30% of the text has been rewritten and well over 300 new references have been added. We have deleted whole sections from the third edition, but have retained most of the older references because we feel they are still useful and important and should not be forgotten or rediscovered.

We have attempted to cover, as far as possible, most new developments in research on germination, but no attempt has been made to cover or to cite all the literature. We have continued to be selective and to be led by our personal predilections. A new short chapter dealing with seed technology and unconventional ways of propagating plants has been added.

We would like to thank Drs Leopold, Huang, Roberts, Fahn, Werker, Karssen, Agami and Gutterman for permission to reproduce figures from their published data.

Lastly we would like to thank the departments whose hospitality we enjoyed, while we were on sabbatical leave, during which much of the preparatory work on this edition was done: A. M. M. at the Department of Plant Sciences, University of Oxford and the Boyce Thompson Institute for Plant Research at Cornell and A. P-M. at the Department of Applied Plant Physiology, Pierre and Marie Curie University, Paris and the Department of Post-harvest Physiology, CNRS, Meudon.

Jerusalem
July, 1988

A. M. M.
A. P-M.

Preface to the Third Edition

Although only six years have elapsed since the 2nd edition appeared, there has been an outburst of publications on germination. Two important books have appeared in the interval — The physiology and biochemistry of seed dormancy and germination, edited by A. A. Khan (North Holland, 1977) and The Physiology and Biochemistry of Seeds, by J. D. Bewley and M. Black (Springer, 1978) and a summary of a Bath-Sheva de Rothschild seminar on Control Mechanisms in Germination was published by the Israel Journal of Botany (Vol. 29, 1980/81). Readers of this volume are certainly advised to consult them for further information. Inevitably we had to be extremely selective in the additions made to this volume, and some tables and figures had to be removed, to permit space for new data. Nevertheless some of the older classical data have been retained, because we felt that they were the basis for much modern research. These older data are too easily forgotten.

We have maintained the overall structure of the second edition but have revised every section of the book. In some cases details, in others conception has been altered. As much as possible new information has been integrated into the structure. Choice has been subjective and as before we did not try to cover everything. Our aim remains to provide an accurate introduction to the subject for undergraduate and graduate students and to provide the research worker not familiar with the field with an adequate bibliography to allow him to follow up his interests. The addition of new references will be found to be very considerable.

Our thanks are due to a number of scientists for permission to use data: Dr. A. Hadas for Fig. 3.2, Dr. E. Werker for preparing Fig. 1.3m and Fig. 4.1b and permission to use Fig. 1.3k and 1.3l, Dr. Y. Gutterman for Table 3.9, Dr. Cavanagh and the Royal Society of Victoria for Fig. 4.1a, Professor A. A. Khan for Fig. 4.9, Professor H. Beevers for Fig. 5.22 and Dr. M. Coccuci for Fig. 6.4.

Since the first edition of this book appeared an enormous amount of new information has become available on germination. Although at least an outline of the events occurring is beginning to become clear, many details are missing and especially how germination is regulated as an integrated process of development is still unclear. It could serve as a model of development and as such should continue to interest all biologists, and this quite apart from the importance of germination in agriculture.

Jerusalem
May, 1981

A. M. M.
A. P-M.

Preface to the Second Edition

In preparing the new edition of this book we were faced with the difficulty of bringing it up to date, in view of the very large number of papers published in the last 12 years, and yet remaining within a reasonable framework as far as size is concerned. The original structure of this book has been retained, but the last chapter dealing with cryptobiotic states in general has been omitted. This helped to compensate partly for the increase in size of this book.

The material in all the chapters has been brought up to date. This will be marked especially in Chapters 4, 5 and 6 to which new subsections have been added, dealing with dormancy-inducing hormones, metabolism and especially protein and nucleic acid metabolism and information on the metabolic effects of growth promoters and inhibitors such as gibberellic acid, cytokinins and abscisic acid. A section on seed establishment has been added in Chapter 7.

In order to increase the usefulness of the book for those who wish to extend their reading the literature citations have been increased. We felt that the reader should be able to go as far as possible directly to the newer literature, which is dispersed among very many journals in different disciplines.

As in the first edition we have been selective in our choice of topics and the data cited and the selections made were necessarily very subjective. Obviously the field of germination has not been completely covered. Nevertheless, we hope that a reasonable balance has been maintained.

We are grateful to all those who have granted us permission to reproduce or otherwise utilize data from their published work.

Our thanks are due to especially Dr. D. E. Briggs and Academic Press for the use of Fig. 6.3; to Drs Gutterman and Shain for the electron micrographs in Fig. 1.3i and Fig. 5.15b and Mrs Marbach for that in Fig. 1.3j; and to Drs D. E. Briggs, W. R. Briggs, M. J. Chrispeels, E. E. Dekker, L. S. Dure, M. M. Edwards, L. Fowden, R. H. Hageman, B. Juliano, D. L. Laidman, C. Kolloffel, A. Marcus, E. Marre, L. G. Paleg, P. Rollin and P. F. Wareing for material as cited in the text.

We hope that this new edition will encourage more biologists to enter the area of research on seed germination. It is obvious that very much is still to be learned on this subject. Seed germination is of great importance in biology, agriculture and food production for the human race. Its study still presents a great challenge to the inquiring mind.

Jerusalem
November, 1974

A. M. M.
A. P-M.

Preface to the First Edition

Although this volume on germination is one of a series of monographs on subjects in plant physiology, we have not attempted to cover the entire field of germination. The number of papers is vast and goes back 50 to 100 years. For this reason we have cited those papers which appeared to us as of importance. No doubt important papers have been omitted and differences of opinion are possible both as regards the selection and the arrangement of the material. An attempt has been made to treat the available information critically and to arrange it in an integrated form. At the same time it is impossible to avoid personal predilections and this has resulted in a more extensive treatment of those chapters dealing with subjects of special interest to the authors.

Our thanks are due to many scientists who have given us permission to use figures, tables and data from their published works, as follows: Professor P. Maheshwari and the McGraw-Hill Book Co. for Fig. 1.1, Professor W. Troll and the Gustav Fisher Verlag for Fig. 1.3(a) and (b), Professor R. M. McLean and Longmans Green and Co. Ltd. for Fig. 1.3(g), MacMillan and Co. Ltd. for Figs. 1.3(b), (f) and (h), and Fig. 1.4(a) from Hayward: *Structure of Economic Plants*, Dr. E. E. Conn and the Journal of Biological Chemistry for Table 5.9, as well as to Drs Y. Oota, M. Yamada, J. E. Varner, Y. Tazakawa, H. Albaum, S. P. Spragg, E. W. Yemm, H. Halvorson, A. Marcus, S. B. Hendricks, H. B. Sifton, S. Isikawa, P. F. Wareing, M. Holden, S. Brohult and A. M. McLeod for the material as cited in the text.

Our thanks are also due to others who have helped us in the preparation of the material.

Special thanks are due to Professor A. Fahn for reading and commenting on Chapter 1, and to the late Professor Hestrin and Dr. A. Lees for reading and commenting on Chapter 8, which is an attempt to demonstrate that the behaviour of seeds in their natural environment is not an isolated phenomenon in biology.

Lastly we are greatly indebted to our editor, Professor P. F. Wareing, for his critical reading of the entire manuscript and for his many comments and suggestions, which helped to give the book its present form. Nevertheless, it goes without saying that the views expressed and the attitude taken are entirely our responsibility.

Jerusalem

A. M. M.
A. P-M.

1

The Development of the Seed and the Structure of Seeds and Seedlings

The term germination is used to refer to a fairly large number of processes, including the germination of seeds, and of spores of bacteria, fungi and ferns, as well as the processes occurring in the pollen grain when the pollen tube is produced. The common denominator in all these processes is the resumption of interrupted growth and development. Although all these processes are germination, in the following we will confine ourselves to the use of the term germination to the seeds of higher plants, the Angiosperms. Extension beyond this would lead to discussions which are well outside the scope of this monograph.

I. Seed Development

The seed of Angiosperms develops from a fertilized ovule, while the ovary itself develops into a fruit. The ovules develop from the placenta of the ovary. The ovule consists of the nucellus surrounded by one or two *integuments* and is connected to the placenta by a small stalk — the *funiculus*. The integuments do not cover the nucellus completely — at the free end of the ovule a gap remains — the *micropyle*, through which the pollen tube will have access to the ovule. The area where the integuments fuse with the funiculus is designated the *chalaza*.

The form of the ovule, and hence eventually the position of the embryo in the seed, is different in different plants. The nucellar apex and the micropyle may be in a straight line with the funiculus, or the ovule may be bent forming an angle with the funiculus, or the apex of the nucellus may face backwards so that the chalaza is along most of the length of the outer integument (Fig. 1.1.a).

The nucellus is considered to be the megasporangium. A megaspore mother cell develops in the ovule and undergoes a meiotic division to form four cells, homologous to megaspores. Three of these cells degenerate, while the remaining one develops into an embryo sac. The haploid nucleus of the embryo sac undergoes three successive divisions resulting in a structure with eight nuclei. Three nuclei, at the micropylar end, differentiate into cells, one of which becomes the actual egg cell — the female gamete. Three nuclei are situated at the opposite side of the embryo sac and two at the centre. At fertilization, two generative nuclei from the pollen tube — the male gametes — enter the embryo sac; one gamete fuses with the nucleus of the egg cell, while the other fuses with the two nuclei in the centre of the ovule (Fig 1.1. b, c, d). Successive divisions of the triploid nucleus will form the storage

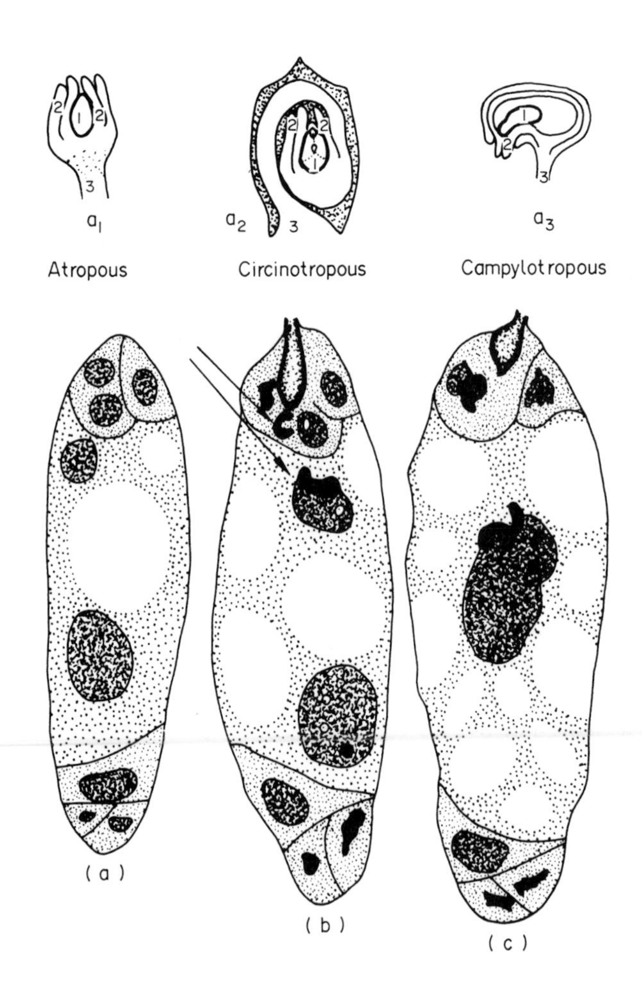

FIG. 1.1. Ovule and embryo sac before and after fertilization. (a) General appearance of ovule in different positions: (a_1) Atropous, (a_2) Circinotropous, (a_3) Campylotropous. (1 — Embryo sac; 2 — Integuments; 3 — Funiculus). (b), (c), (d) Embryo sac of *Lilium martagon* before and after fertilization. (b) Mature embryo sac. (c) Discharge of pollen tube into embryo sac (male nuclei) see arrows. (d) Contact of one male nucleus with egg nucleus and the other with two polar nuclei (after Maheshwari, 1950).

tissue of the seed — the endosperm. In the early stages of development of the seed the endosperm surrounds the embryo which develops from the fertilized egg cell. The integuments of the ovule develop into the seed coat, or testa — terms that are often used as synonyms. However, according to Corner (1976), "testa" is applicable only to the constituents of the seed coat which are formed from the outer integument. The elements of the seed coat developing from the inner integument are referred to as "tegment".

Although this pattern of development is very common, there are many variations, concerning the number of integuments, the number of nuclei in the embryo sac and the pattern of their arrangement. The anatomy and morphology of seed development is described in greater detail by Fahn (1982), Boesewinkel and Bouman, (1984) and Evenari (1984).

The three basic structures — embryo, endosperm and seed coat — differ in their genetic origin. While the embryo and the endosperm are derived from both maternal and paternal genetic material, the seed coat is derived from diploid maternal cells.

Frequently the endosperm tissue is triploid, but its ploidy can be very high and may differ in various regions of the endosperm. This polyploidy arises from secondary processes after division of the primary nucleus. The endosperm may remain coenocytic throughout or it may show cellular organization. In some cases the coenocytic endosperm remains completely liquid, for example in *Cocos nucifera* (Bhatnagar and Johri, 1972).

The endosperm may persist as a storage tissue; alternatively it may degenerate partially or fully, remaining as a rudimentary tissue particularly in those cases in which the cotyledons serve as storage organs. In some cases it may become fused with the seed or fruit coat.

In addition to the three basic structures, other tissues may occasionally participate in the make up of the seed, which is by no means constant. The seed coat is sometimes made up of tissues other than the integuments, such as the nucellus or the endosperm. The seed coat itself varies greatly in form; it may be soft, gelatinous or hairy, although a hard seed coat is the form most commonly encountered.

Relatively little is known about the ultrastructure of the dry seed. Normal fixation procedures involve the introduction of at least some water, which immediately changes ultrastructure. Careful studies of seeds in the air-dry state reveal that both cells and organelles show shrinkage and folding, especially of the cell wall. The convolution of the cellular membranes and walls appears to be due to water loss by the cytoplasm, since no corresponding loss occurs in the walls and membranes themselves (Webb and Arnot, 1982; Opik, 1985). Despite these changes, many cell organelles have been identified in dry seeds, with the exception of Golgi bodies and in many cases ribosomes. Endoplasmic reticulum is abundant.

Although cells of seeds during desiccation undergo considerable changes, these do not apparently cause disorganization of the cells and tissues. Changes may occur, however, at the level of the molecular structure of membranes, as is perhaps indicated by the leakiness of seeds immediately after having been placed in water.

II. Structure of the Seed Coat

The seed coat is a structure of considerable importance because it forms the barrier between the embryo and its immediate environment. The seed coat is formed from the integuments, but in most seeds many layers of tissue forming the integuments are destroyed and only the residual parts make up the seed coat. In some cases the seed coat is formed mainly from the outer integument, such as in *Pancratium maritimum* (Werker and Fahn, 1975); in other cases it is formed mainly from the layers of the inner integument as in the Malvaceae (Netolitzky, 1926). Greater detail of seed coat structure can be found in Fahn (1982).

A most important property of the seed coat is its permeability to water and sometimes to gases. Impermeable seed coats confer seed dormancy. As long as water entry is blocked, no germination can occur. It is not yet fully understood what makes the seed coat impermeable to water. This problem was studied in some detail in the genus *Pisum* and in *Sida spinosa*. In *Pisum* seeds (Werker *et al.*, 1979) a continuous layer of palisade cells is present. The caps of these cells are very hard and are impregnated with pectinaceous material. In these cells, or in the layer of cells immediately below the palisade, the osteosclereid cells, quinones are laid down. It is the combination of these two features which renders the seeds impermeable to water.

In *Sida spinosa* the onset of coat impermeability was correlated with the accumulation of insoluble lignin-like polymers (Egley *et al.*, 1983; 1986). These polymers were synthesized by peroxidase catalysed oxidation of phenolics, a process which occurred in the seed during development on the mother plant. In addition to the activity of enzymes in inducing seed coat impermeability (such as polyphenol oxidase or catalase) in the formation of water impermeable substances which impregnate the cell walls of the palisade layer, deposition of callose was also suggested as a cause of impermeability (Egley and Paul, 1982; Egley *et al.*, 1983; Bhalla and Slattery, 1984). Although it is known that the changes responsible for the impermeability of the seed to water are often localized in definite tissues or cells, the basic process is not yet fully understood (Rolston, 1978).

In the seeds of *Pancratium maritimum* the seed coat develops mainly from the outer integument (testa). Two or three outer layers of the testa are compressed and form a dark, water-impermeable protective layer. The rest of the testa is composed of large, dead cells with numerous airspaces (Fig. 1.2). All cell walls are, apparently, impregnated with quinones. The air-filled testa renders the seeds buoyant and this, together with the impermeability of the outer layer to water, permits them to float, for example in sea water (Werker and Fahn, 1975). The tegment (developing from the inner integument) has almost deteriorated.

A frequently observed feature of seed coats, for example in *Linum, Plantago* or *Lepidium*, is the presence of mucilagenous materials in one of the layers (Fahn, 1982). In parasitic plants, such as mistletoe (*Arceuthobium*), a mucilage which contains oligogalacturonic acid together with cellulose elements covers the seed. This mucilagenous cover, called viscin, is sticky and may aid adherence of the seeds to their substrate (Paquet *et al.*, 1986).

Appendices developing from other parts of the mother plant may be attached to the seed coat, for example the elaiosome in the seed of *Pancratium parviflorum* (Fig. 1.4.k). The elaiosome usually contains large amounts of lipids and serves as an attractant to insects, such as ants (see Chapter 7). The elaiosome is considered by some anatomists as a rudiment of a third cotyledon (Fahn, 1982).

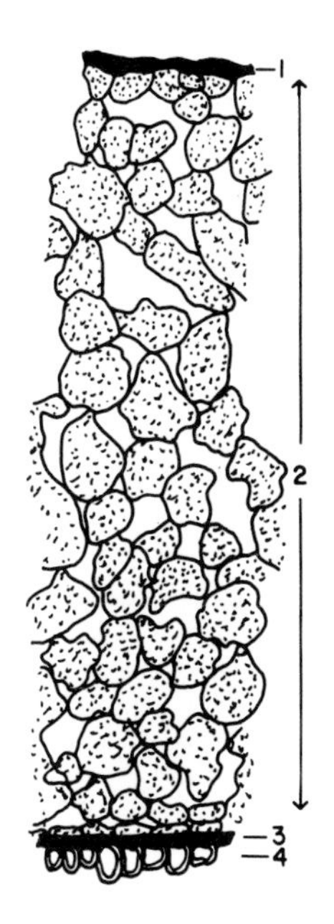

FIG. 1.2. Structure of seed coat of *Pancratium maritimum* (adapted from Werker and Fahn, 1975). 1 — Protective layer; 2 — outer integument; 3 — inner integument and nucellus; 4 — endosperm. Note the protective layer impregnated with apparently water impermeable dark brown material. The shaded cells of the outer integument are dead and together with the extensive intercellular spaces form a sort of air cushion which gives the seed its buoyancy.

III. Variability Between Seeds

In some plants seeds arise by apomixis or nonsexual processes, the seed usually being formed from a diploid cell. Nonsexual embryo formation also occurs in tissue culture, a technique which is well known today. In such embryos there is no desiccation or quiescent stage. They may be stored by cryopreservation techniques (Vasil and Vasil, 1980).

In some cases more than one embryo develops in the seed. Cases of polyembryony may arise due to division of the zygotic cell, existence of more than one embryo sac in the ovule, or other aberrations during seed formation have been reported. Polyembryony is common in some *Citrus* species, in *Linum usitatissimum*, in *Ponalipina* and in *Opuntia*. In some plants, *Linum* for example, the polyembryony is of haploid embryos.

The size and shape of seeds is extremely variable. It depends on the form of the ovary, the condition under which the parent plant is growing during seed formation and on the species (see Chapter 7). Other factors which determine the size and shape of seeds are the size of the embryo, the amount of endosperm present and to what extent other tissues participate in the seed structure. Some of the variety in seed form is illustrated in Fig. 1.3. Variability in seed shape also exists within a given species and is then referred to as seed polymorphism. Characteristic of polymorphic

FIG. 1.3. Variation in seed form (all scales in mm). (a) *Xanthium strumarium*, (b) *Pisum sativum*, (c) *Ricinus communis*, (d) *Pancratium maritimum*, (e) *Citrullus colocynthis*, (f) *Phaseolus coccineus*, (g) *Raphanus sativus*, (h) *Phaseolus vulgaris*, (i) *Papaver orientalis*.

seeds is that they differ not only in shape or colour, but also in their germination behaviour and dormancy. Among the better known examples are *Xanthium pennsylvanicum*, in which there are differences in germination requirements of the upper and lower seeds (Esashi and Leopold, 1968), *Salsola volkensii*, in which chlorophyllous and achlorophyllous embryos occur (Negbi and Tamari, 1963), and *Suaeda* in which the endosperm may be either present or absent (Zohary, 1937; Harper *et al.*, 1970; Roberts, 1972).

The normal seed contains materials which it utilizes during the process of germination. These are frequently present in the endosperm, as in *Ricinus*, tomato, maize and wheat (Fig. 1.4.a,b,c,d). The endosperm may contain a variety of storage materials such as starch, lipids, proteins or hemicelluloses (Fig. 1.4l). But an

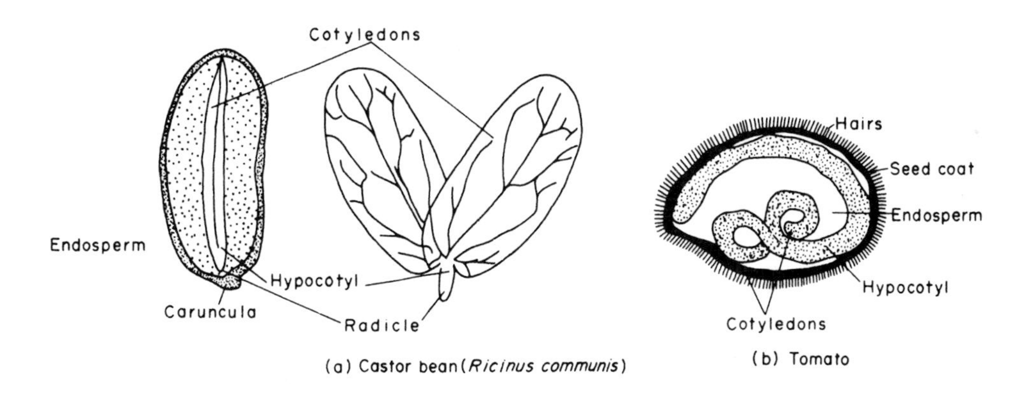

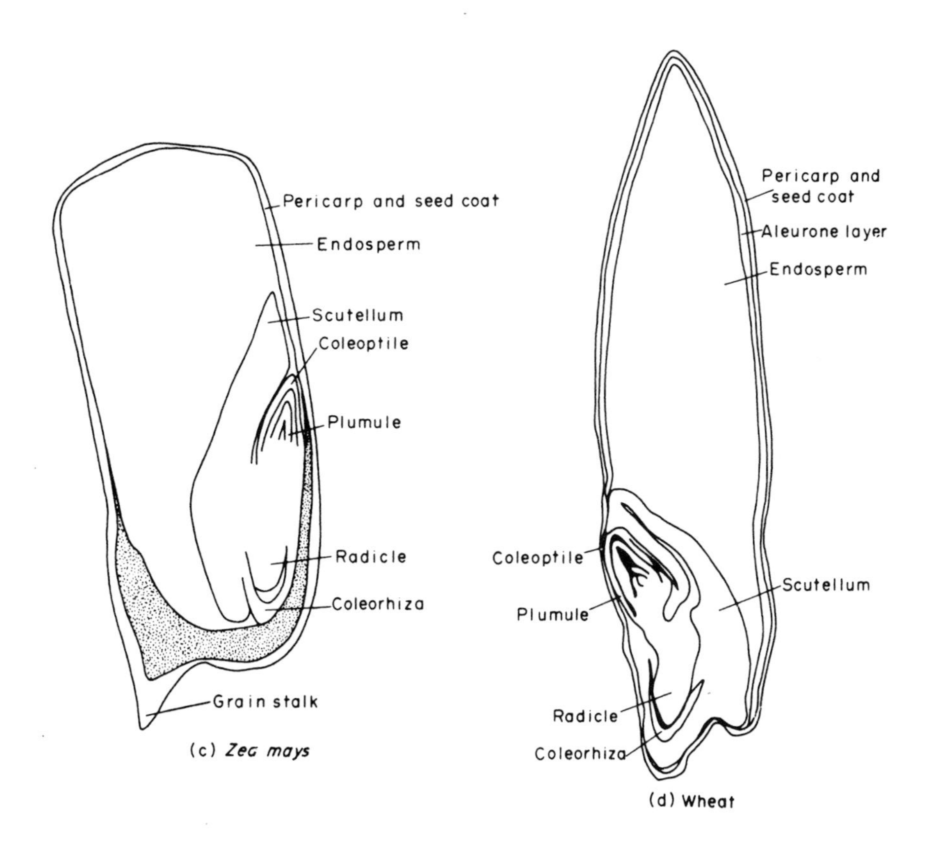

FIG. 1.4. Structure of various seeds and of one-seeded fruit in which seed and fruit coats are fused. (a) Castor bean (Troll, 1954). (b) Tomato. (c) *Zea mays*. (d) Wheat.

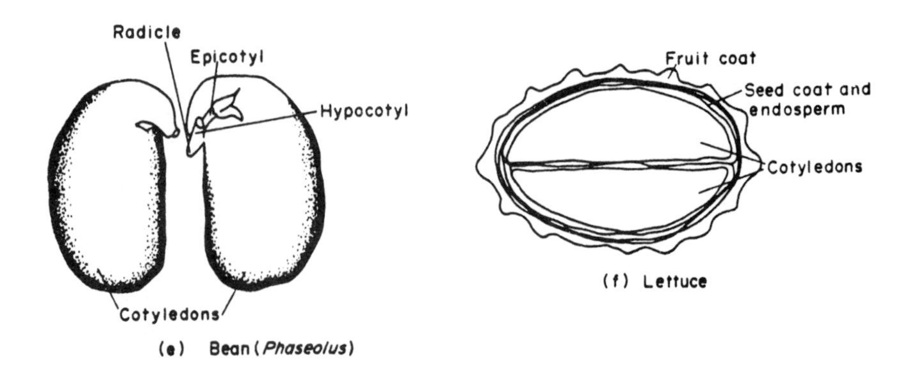

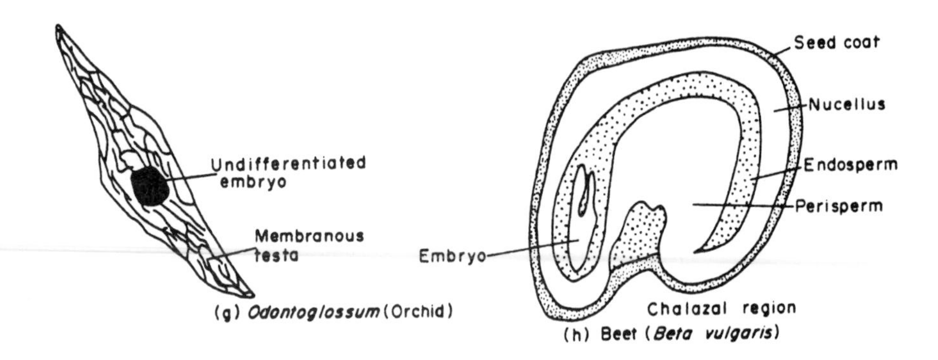

FIG. 1.4. (e) Bean (Troll, 1954). (f) Lettuce.
(g) Orchid (McLean & Ivimey-Cook, 1956). (h) Beet.

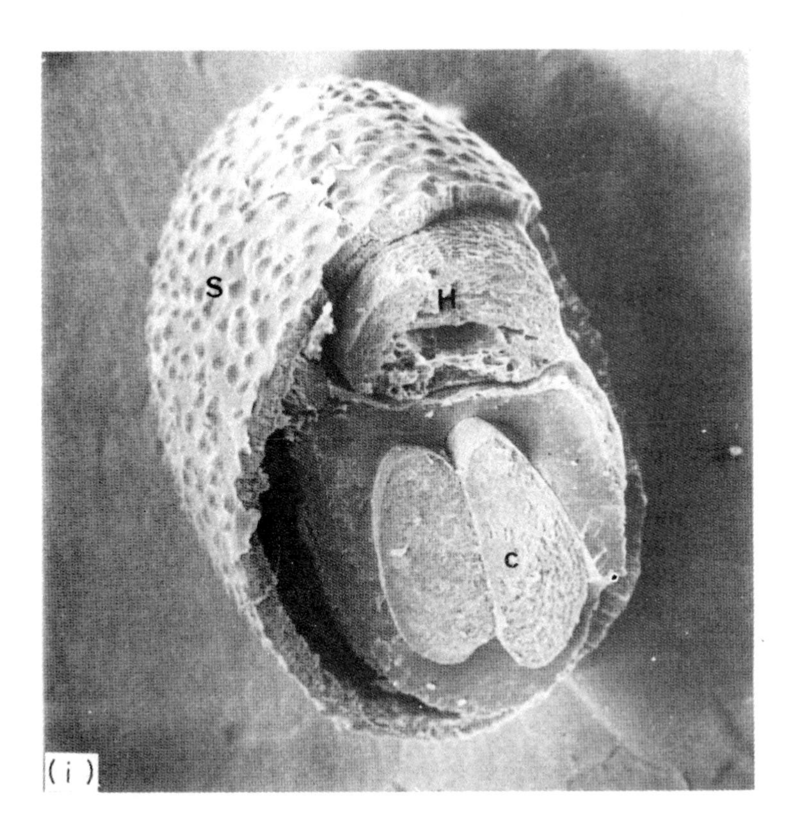

FIG. 1.4. (i) Seed structure of *Trigonella arabica*. Seed cut. Magnification ×67. S — seedcoat, C — cotyledons, H — hypocotyl (root axis) (courtesy D. Y. Gutterman) scanning electron micrograph.

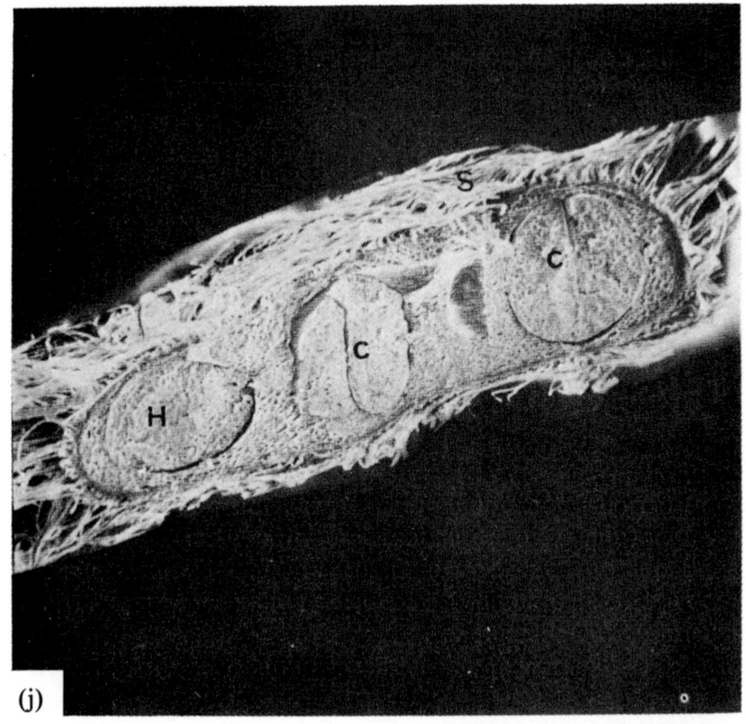

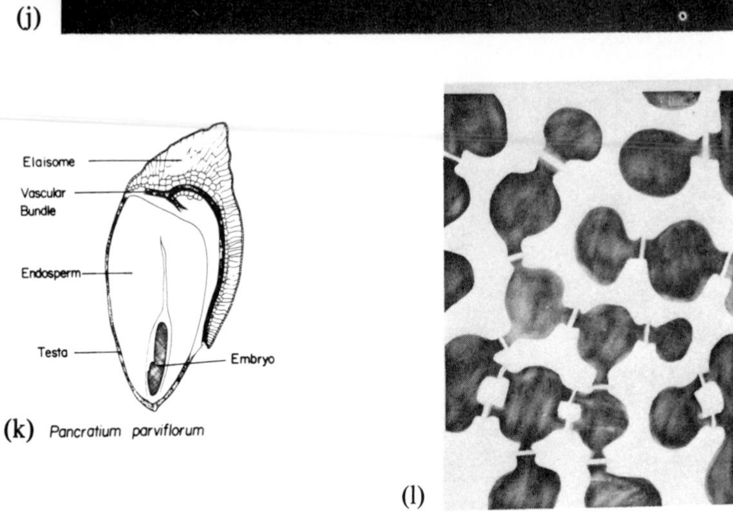

FIG. 1.4. (j) Seed structure of tomato. Seed cut. Magnification × 40. Note that the cotyledons have been cut twice due to the curled position of the embryo. S — seedcoat, C — cotyledons, H — hypocotyl (root axis) (courtesy Dr I. Marbach) scanning electron micrograph. (k) *Pancratium parviflorum* (After Werker and Fahn, 1975). (l) Section through endosperm of date *Phoenix dactylifera*, showing heavily thickened cell walls containing storage materials. Dark areas — cell lumen (courtesy Dr E. Werker).

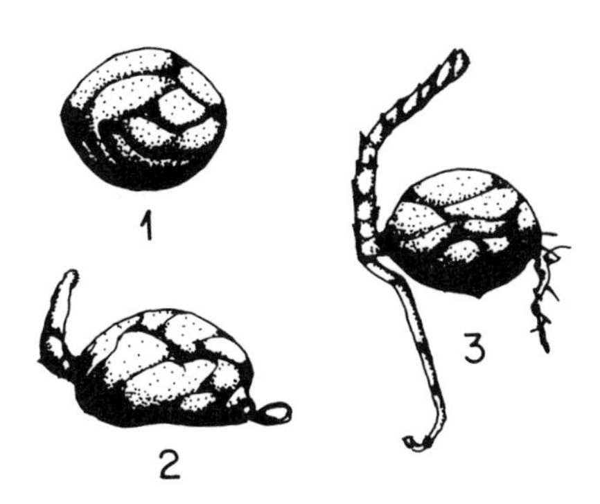

FIG. 1.4.(m) Seed and seedling of *Symphonia globulifera* (after Corbineau and Come, 1986). 1 — seed, 2 — germinated seed, 3 — seedling (note atrophied embryonic root and adventitious root at the base of the plumule).

an endosperm is by no means invariably present nor is it always the chief location of reserve materials. In many plants the endosperm is greatly reduced, as in the Cruciferae, and in the orchids its formation is entirely suppressed. Reserve materials are present elsewhere, for example in the cotyledons of the embryo, as in beans and lettuce (Fig. 1.4.e, f). Orchids contain virtually no reserve materials and therefore represent a very special case of seed structure. In some plants the storage materials are contained in the perisperm e.g. *Beta vulgaris* (Fig. 1.4.h). The perisperm originates from the nucellus, e.g. in the Caryophyllaceae and in *Coffea*, and not from the embryo sac; genetically the perisperm is therefore of maternal origin. Examples of the structure of seeds as seen with the scanning electron microscope are shown in Fig. 1.4.i, j.

Seeds are formed in the ovary and this develops into the fruit. Fruits arising primarily from the ovary are known as "true" fruits, while fruits in which other structures or several ovaries and their related structures participate are often termed "false". Both types of fruits may be dry or fleshy. In many plants the integuments and the ovary wall are completely fused, so that the seed and fruit are in fact one entity, for example in the grains of grasses and in lettuce (Fig. 1.5). In other cases additional parts of the plants, such as glumes, bracts and neighbouring sterile florets, remain attached to the fruit and form a bigger dispersal unit. The exact nature and classification of these organs need not concern us here. A few of the very variable seed-bearing structures are illustrated in Fig. 1.6.

The embryo consists of a radicle, a plumule or epicotyl, one or more cotyledons, and a hypocotyl which connects the radicle and the plumule. The embryo may be variously located within the seed and may either fill the seed almost completely, as in the Rosaceae and Cruciferae, or be almost rudimentary, as in the Ranunculaceae. The classification of seeds is often based on the ratio of the size of the embryo to that of the storage tissue.

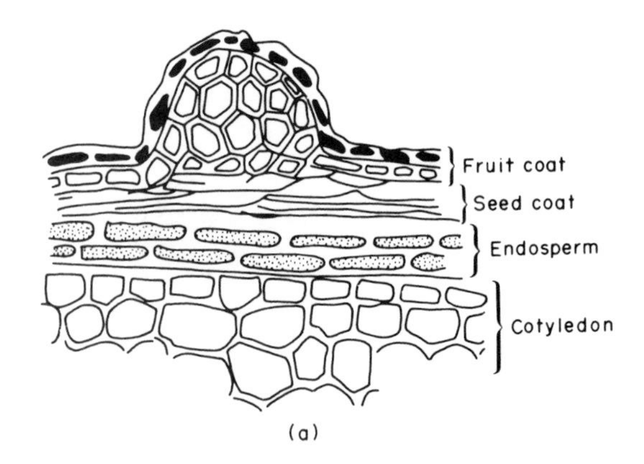

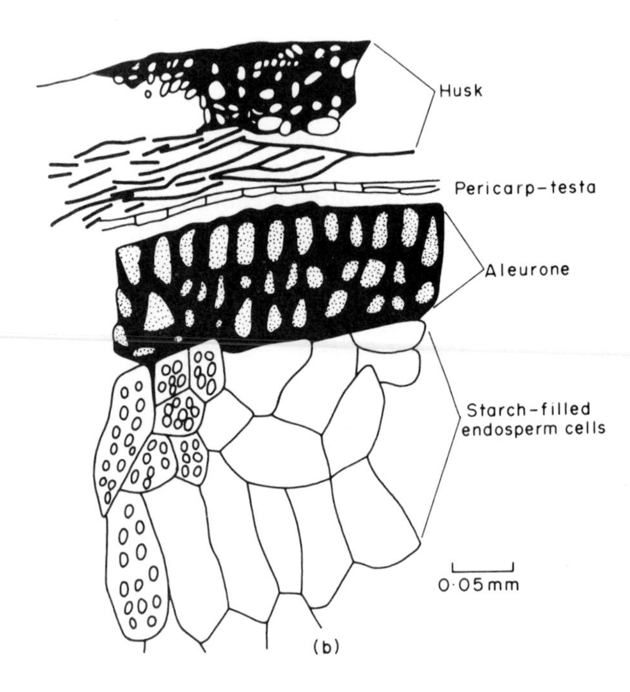

FIG. 1.5. Fusion of seed and fruit coat in lettuce and barley seeds. (a) Lettuce (after Hayward, 1938). (b) Barley (after McLeod, 1960).

FIG. 1.6. Forms of dry fruits and dispersal units (all scales in mm). (a) *Quercus ithaburensis*, (b) *Atriplex halimus*, (c) *Avena sativa*, (d) *Arachis hypocaea*, (e) *Papaver orientalis*, (f) *Ricinus communis*, (g) *Helianthus annuus*, (h) *Lactuca sativa* var. Grand Rapids, (i) *Zea mays*, (j) *Medicago polymorpha*, (k) *Triticum vulgare*, (l) *Xanthium strumarium*, (m) *Sinapis alba*, (n) *Tipuana tipu*, (o) *Erodium maximum*, (p) *Rumex rosea*, (q) *Scabiosa prolifera*.

IV. Accumulation of Reserve Materials

The seed, either alone or as part of a dispersal unit, is in a way a self-contained unit. It contains an embryo and storage tissues which supply the embryo while it develops into a seedling during the initial stages of germination.

The main storage materials accumulating in the seed storage tissue are carbohydrates (sugars and oligo- and polysaccharides), proteins and lipids. Seeds also contain mineral reserves, especially phosphates.

These storage substances are synthesized from assimilates transported into the seed from other parts of the plant, while the developing seed (and the fruit) serves as a "sink". The seed coat has a special role in the transport of assimilates from the parent plant to the embryo. It apparently constitutes the chief interface through which sucrose, glutamine and other compounds are unloaded into the seed. Since there is no vascular connection between the seed coat and the developing seed, the transport is a complex process (Murray, 1987). There is a certain sequence in accumulation of the various reserve material as can be seen for instance in the development of the almond seed (Hawker and Buttrose, 1980).

The almond embryo sac contains two ovules, but usually only one develops into a seed. Twelve weeks after flowering the fruit reaches its full size, but the space enclosed by the pericarp is still filled by the nucellus. The endosperm develops between the 16th and the 20th week, but at the same time the embryo also begins to increase in size at the expense of the endosperm. The developing embryo accumulates proteins and lipids, which are apparently synthesized utilizing the starch and sucrose present in the young embryo (Fig. 1.7).

Weeks after flowering	5 10 15 20 25 30 35
Fresh weight (g)	0.1 1.0 1.8 2.2
Protein % F.W.	0.06 0.08 0.1
Lipids % F.W.	0.06 0.15 0.35
Starch % F.W.	0.3 0.7 0.5
Sucrose % F.W.	1.0 3.2 3.4 1.0

FIG. 1.7. Changes and sequence of accumulation of reserve materials in the developing almond embryo (compiled from data of Hawker and Buttrose, 1980).

In field peas (*Pisum arvense*) if $^{14}CO_2$ was fed to the plants in the early stage of their development (40 days after sowing), only 2% of the ^{14}C could be located, ultimately, in the seeds. However, 7.5% of the ^{14}C was found in the seeds when $^{14}CO_2$ was fed to the plants 120 days after sowing (Pate and Flinn, 1973). On the other hand 51% of ^{15}N, which was fed as $^{15}NO_3^-$ in the early stages of seedling development, found its way into the seed. The ^{15}N was translocated continuously from the older into younger organs and eventually to seed protein, while the ^{14}C fed to the plant in its early growth stages was either utilized in respiration or was trapped in cell wall materials which do not undergo intensive turnover. The photosynthetic capacity of the pod does not contribute much to the reserve materials of the seed. Its main contribution is apparently through the refixation of the respiratory carbon dioxide of the developing seeds and the pod tissue itself. This is also valid for other legumes (Pate and Flinn, 1973; Oliker *et al.*, 1978; Pate and Herridge, 1978).

Carbohydrates are stored either as starch in grains, which are formed in the leucoplasts of cells of the storage tissue (endosperm in seeds of Gramineae or cotyledons in the dicotyledonous plants), or as components of the cell wall, such as hemicellulose and other oligosaccharides as in dates (Fig. 1.4.l) or in *Nasturtium* seeds (see Chapter 2).

Storage proteins are synthesized on the endoplasmic reticulum (ER) mainly on RER. They are apparently directed by the dictyosomes into the vacuoles, where they accumulate in the protein bodies (Fig. 1.8.b, c, d) (Buttrose, 1963; Mollenhauer and Moore, 1980).

In numerous seeds carbon skeletons are stored chiefly in the form of lipid. This form of storage is economically worthwhile for energy conservation, because of the high caloric value of lipids. In nature, in most small seeds, the main storage material is indeed lipid. The lipids are synthesized in plastids (Yamada and Usami, 1975), usually from sucrose transported to the seeds, but many points concerning lipid accumulation in the developing seed are as yet unclear. The available knowledge has been recently summarized by Slack and Browse (1984), see also Chapter 2.

Many other substances, some of them toxic to animals and man, such as alkaloids, cyanogenic glycosides, saponins and others are also accumulated in various seeds. For greater detail see Chapter 2 and a review by Bell (1984).

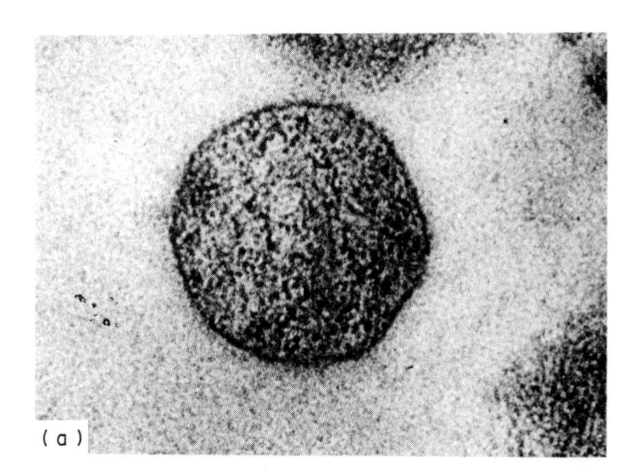

FIG. 1.8. Subcellular particles characteristic of seeds. (a) Electron micrography of a glyoxysome from castor beans × 60,000 (courtesy of Dr Y. Shain).

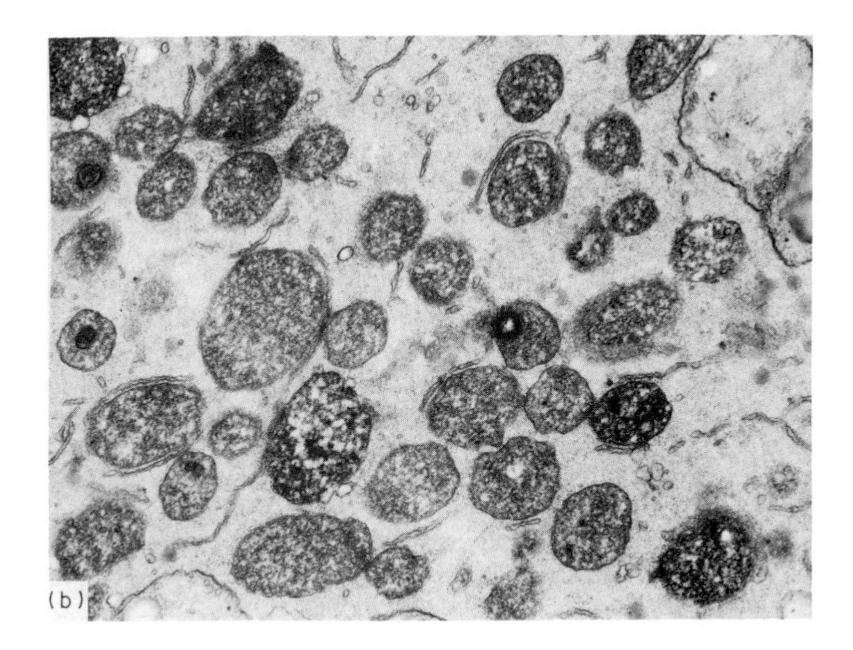

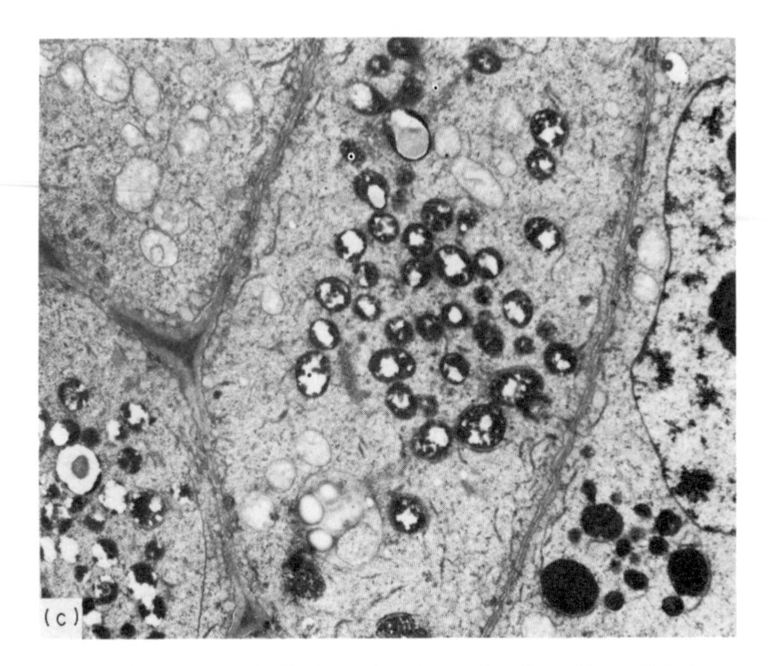

FIG. 1.8.b & c. Protein bodies from the sub-apical region of the radicle from a germinating pea seed. (b) Intact protein bodies. (c) Protein bodies, at the beginning of dissolution of the protein (from data of Hodson, Di Nola & Mayer).

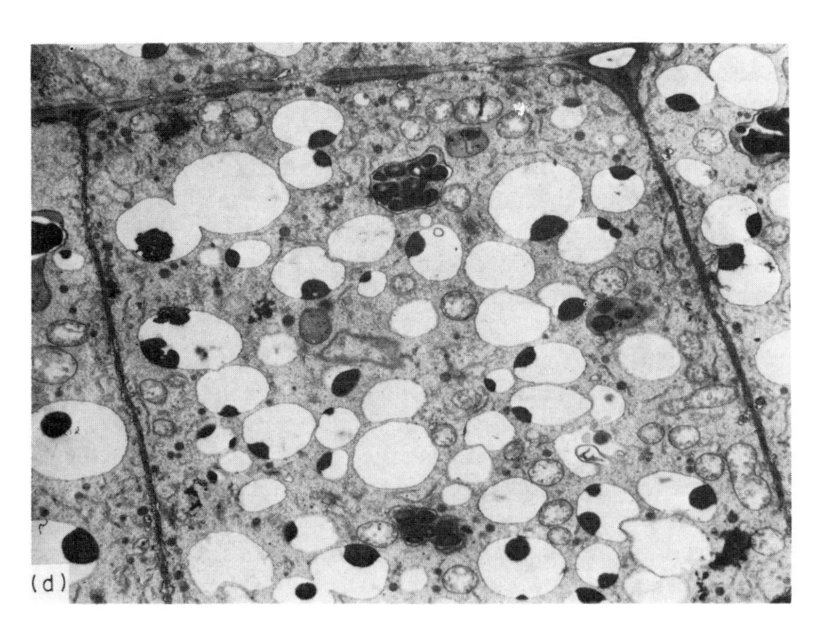

FIG. 1.8.d. An advanced stage in the dissolution of the protein, with appearance of vacuoles (from data of Hodson, Di Nola and Mayer).

V. Seeds and Seedlings

Maturation of the developed seed is closely connected with desiccation. Most mature seeds contain only 5-10% water and the embryo is apparently desiccation tolerant. Moreover, the desiccation period seems to be, in many seeds, an essential stage of development in which a switch-over occurs from a pattern associated with development to one characteristic of germination (Kermode and Bewley, 1986; Kermode *et al.*, 1986; Quatrano, 1987). Four stages may be distinguished in seed development: (1) histodifferentiation — which is actually seed formation; (2) maturation, during which the embryo reaches its maximal size and most of the storage materials are accumulated; (3) desiccation and (4) quiescence.

A mature seed (at the end of the second stage), will not germinate in the "moist" stage. It will, however, germinate if it is artificially dried before rehydration for germination. This change in the pattern of behaviour due to drying is accompanied by a change in the nature of the synthesized proteins: before drying storage proteins accumulate in the cotyledons (*Phaseolus*) or endosperm and cotyledons (*Ricinus*); drying stops their production and the proteins synthesized during the subsequent rehydration of the seeds are identical to those synthesized in the seed during normal germination (Kermode *et al.*, 1986). However, in some plants the intermediary drying is not necessary. Moreover, in *Ricinus* excised embryos will germinate even if the seeds were not dried. It appears therefore that the "environment" of the embryo in the seed actually prevents it from germinating. The change occurring during artificial or natural desiccation may be connected in some way with the decrease in the ABA concentration in the seed (Kermode and Bewley, 1986). Some parts of the developmental process appear to be controlled by ABA (Quatrano, 1987; see also Chapter 6). However, ABA does not seem to provide the single critical signal which switches off the genes responsible for development. The signals which cause the switching from the developmental stage to that of germination are still unknown. Moreover, the transition is not abrupt. The desiccation stage is associated with many other changes, for example the ability of the aleurone of cereals to respond to gibberellins (Armstrong *et al.*, 1982). Some proteins, whose production is clearly a part of development, still continue to be produced early on in germination. Later they, and the messages coding for them, disappear (see also Chapters 5 and 6).

The process of germination leads eventually to the development of the embryo into a seedling. Seedlings are classified as *epigeal*, in which the cotyledons are above the ground and are usually photosynthetic, and *hypogeal*, in which the cotyledons remain below ground. In this latter case the cotyledons are either themselves the source of reserves for the seedling or receive materials from the endosperm. The size, shape and relative importance of the various tissues of seedlings are as variable as those of the seeds and fruits, and we will not attempt to classify them here. A few typical seedling forms are shown in Fig. 1.9. These indicate the large morphological variability which is met during germination.

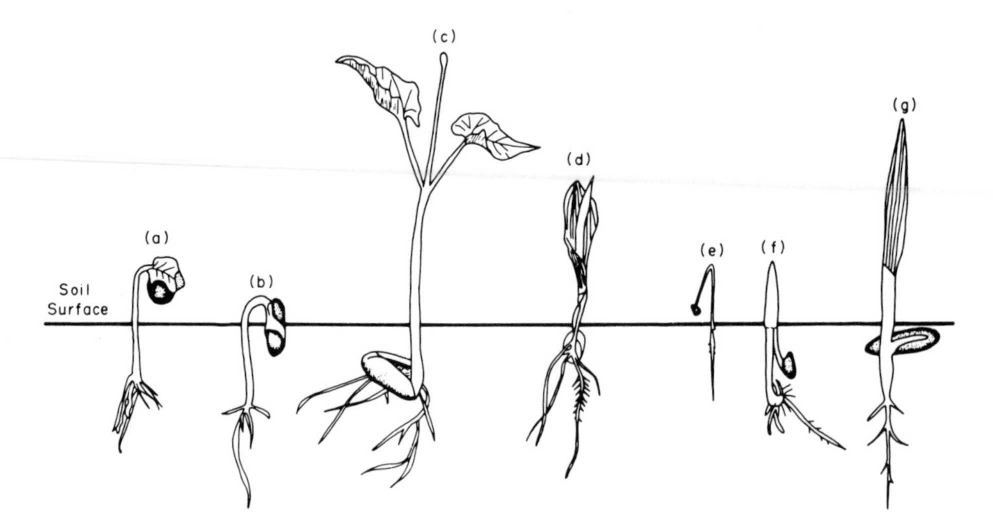

FIG. 1.9. Seedling morphology. (a) *Ricinus communis* (epigeal), (b) *Phaseolus vulgaris* (epigeal), (c) *Phaseolus multiflorus* (hypogeal), (d) *Zea mays* (Hypogeal), (e) Onion (epigeal), (f) *Tradescantia virginica* (hypogeal), (g) *Phoenix dactylifera* (hypogeal).

In many plants special absorbing tissues develop which take up the reserve materials from the endosperm. In the onion the tip of the cotyledon is actually embedded in the endosperm and withdraws the materials from it (Fig. 1.9.e). The cotyledon with the seed attached is eventually raised above the ground and becomes green. In *Tradescantia* also the cotyledon is embedded in the endosperm, but the emerging organ is the coleoptile which is partly joined to the cotyledon and partly to the hypocotyl. In the Gramineae specialization has proceeded further and the scutellum fulfils the special function of an haustorium, which withdraws substances from the endosperm.

In some species the scutellum does not develop during germination, as in rice and wheat. In other species, for example *Zea*, it shows some development, while in oats it grows into the endosperm and develops markedly during germination (Negbi, 1982).

In various palms the haustorium, formed from the tip of the cotyledon, has become greatly enlarged. In a few cases the tip of the primary root rather than the scutellum (or the cotyledon) becomes the haustorium.

An extreme case of pattern of development in seed and seedling structure is that of *Symphonia globulifera* — a tropical dicotyledonous tree belonging to the family Guttiferae. The seed has a thin seed coat which ruptures easily (Corbineau and Come, 1986). Inside the seed coat no cotyledons or endosperm can be distinguished. The whole seed has become reduced to a hypertrophied axis. The radicular pole consists of a small group of undifferentiated meristematic cells which at germination produce one or more seminal roots. The gemmular pole consists of a meristematic dome with two undeveloped scales. At germination, a scale-covered stem grows from this pole. At the base of the stem adventitious roots develop, which will form the root system of the seedling. The seminal root atrophies at an early stage of development (Fig. 1.4.m). The development of the seed from the fertilized ovule has not yet been described in detail.

Such complications are absent in most of the dicotyledonous plants. Some straightforward examples are provided by the castor bean, *Ricinus communis*, which has epigeal germination with an endosperm, and *Phaseolus vulgaris* (French bean) which is epigeal without endospermic tissue. In contrast, *Phaseolus multiflorus* is hypogeal and non-endospermic, while examples of hypogeal endospermic germination are provided by *Hevea* among the dicotyledons and *Zea mays* among the monocotyledons (Fig. 1.9).

VI. Germination

After this brief discussion of forms of seeds, their development and their structure, we are now in a position to try and define germination and to begin a discussion of the processes which lead up to it. We may regard germination of the seed of the higher plant as those consecutive events which cause a dry quiescent seed, in response to water uptake, to show a rise in its general metabolic activity and to initiate the formation of a seedling from the embryo. The exact stage at which germination ends and growth begins is extremely difficult to define. This is particularly difficult because we identify germination by the protrusion of some part of the embryo from the seed coat, which in itself is already a result of growth. There is no general rule as to which part of the embryo first pierces the seed coat. In many seeds this is the radicle and therefore germination is frequently equated with root protrusion. However, in some seeds it is the shoot which protrudes first, for example in *Salsola*.

The piercing of the seed coat by part of the embryo is caused by cell division, cell elongation or both. In other words, the protrusion of part of the embryo through the seed coat is in fact the result of growth. Not all definitions would include cell division as such in the term growth. However, cell division is usually followed by an enlargement of the daughter cells and therefore constitutes growth. It is uncertain whether the process of germination can be equated with the process of growth which leads to protrusion of part of the embryo through the seed coat. Probably the fundamental processes which cause germination are different from those of growth.

A further question which arises is whether cell division or cell elongation is the first process which occurs when the embryo pierces the seed coat. In some seeds it has been shown that cell division occurs first and is followed by cell elongation, while in other cases the reverse has been observed. Thus in lettuce seeds which are germinated at 26°C, cell division begins after some 12-14 hr and at about the same time cell elongation in root cells can also be observed (Evenari *et al.*, 1957).

In cherry seeds the changes in cell number and length of embryonic organs were followed during after-ripening under conditions of stratification at 5°C (Pollock and Olney, 1959). The length of the embryonic axis increased by some 7% during stratification at 5°C. This increase in length was attributed to a combination of cell division and elongation. Changes at 5°C and 25°C were similar for the first 4 weeks of stratification but following this, axis length increased much more at 5°C than at 25°C. At the latter temperature no germination was obtained. Thus cherry and lettuce seem to be similar with regard to the more or less simultaneous occurrence of elongation and division. In germinating seeds of *Pinus thunbergii*, in contrast, it seems that cell division precedes cell elongation, but the evidence is not entirely clear (Goo, 1952). Cell division occurs during a period when there is very little water uptake and before germination. In *Pinus lambertiana*, division and elongation were observed visually to occur simultaneously, but thymidine incorporation occurred much earlier than cell elongation, indicating that cell division might precede elongation (Berlyn, 1972).

In the germination of *Zea mays*, the first change is cell enlargement in the coleorhiza. Cell division in the radicle occurs later, when the latter breaks through the coleorhiza and the seed coat. It is the coleorhiza which first pierces the seed coat (Toole, 1924). A similar situation apparently exists in barley.

A further attempt to differentiate between division and elongation during germination of lettuce has been made by Haber and Luippold (1960). Gamma-radiation and low temperatures were used to delay cell division and hypertonic solutions of mannitol used to arrest cell elongation. Despite the fact that irradiation and low temperatures prevented mitosis, germination as observed by root protrusion occurred. When, however, cell elongation was prevented by hypertonic solutions, germination was prevented, although cell division could clearly be noted. Haber and Luippold concluded that division and cell elongation are affected by different factors and that root protrusion is primarily the result of cell elongation. Cell division only serves to increase the number of cells which can elongate. It is by no means clear whether these observations also apply to other seeds.

The attempts to differentiate between germination proper and growth have also led to discussions as to the extent of reversibility of the early stages of germination. In most seeds initial water uptake is reversible, i.e. the seeds can be dried again without damage. In later stages when water uptake has proceeded some way, such reversibility is not always the case. This has at times been taken to be the stage at which germination ended. However, some seeds, whose seedlings are very resistant to desiccation, may be dried without damage when the seedling root has clearly protruded. Thus the occurrence of reversibility is also not a satisfactory way of differentiating between germination and growth. In some cases at least the stage of germination itself has a different temperature optimum than subsequent growth. In apple embryos radicle growth proceeds at 30°C, while the initial steps of germination are prevented by this temperature (Perino and Come, 1977).

Cell division and cell elongation only occur once the cells of the seed have been activated in some way, so as to permit the control of these processes by various factors. Although it appears that for visible germination only one of these processes is necessary, there can be no doubt that for normal development and growth of the seedling both cell division and cell elongation are essential. It is thus possible to differentiate between these growth processes and the process of activation, which precedes growth and which may be termed germination.

However, we will not during the discussion of germination, attempt to make such a precise differentiation but rather will try to describe all those processes which take place up to seedling formation.

Bibliography

Armstrong C., Black M., Chapman J. M., Norman H. A. and Angold R. (1982) *Planta* **154**, 573.

Bell E. A. (1984) in *Seed Physiology*, p. 245 (Ed. D. R. Murray), Academic Press, Sydney.

Berlyn G. P. (1972) in *Seed Biology*, vol. 1, p. 223 (Ed. T. T. Kozlowski), Academic Press, New York.

Bhalla P. L. and Slattery H. D. (1984) *Ann. Bot.* **53**, 125.

Bhatnagar S. P. and Johri B. M. (1972) in *Seed Biology*, vol. 1, p. 78 (Ed. T. T. Kozlowski), Academic Press, New York.

Boesewinkel F. D. and Bouman F. (1984) in *Embryology of Angiosperms*, p. 567 (Ed. B. M. Johri), Springer Verlag.

Buttrose M. S. (1963) *Aust. J. Biol. Sci.* **16**, 305.

Corbineau F. and Come D. (1986) *Seed Sci. Technol.* **14**, 585.

Corner E. J. H. (1976) *The Seeds of Dicotyledons*, Cambridge University Press, Cambridge.

Egley G. H. and Paul R. N. (1982) *Am. J. Bot.* **69**, 1402.

Egley G. H., Paul R. N. and Lax A. R. (1986) *Physiol. Plant.* **67**, 320.

Egley G. H., Paul R. N., Vaughn K. C. and Duke S. O. (1983) *Planta* **157**, 224.

Evenari M. (1984) *Botanical Review* **50**, 153.

Evenari M., Klein S., Anchori H. and Feinbrunn N. (1957) *Bull. Res. Council, Israel* **6D**, 33.

Esashi Y. and Leopold A. C. (1968) *Plant Physiol.* **43**, 871.

Fahn A. (1982) *Plant Anatomy*, 3rd ed. Pergamon Press, Oxford.

Goo M. (1952) *J. Jap. Forestry Soc.* **34**, 3.

Haber A. H. and Luippold H. J. (1960) *Plant Physiol.* **35**, 168.

Harper J. L., Lovell P. H. and Moore K. G. (1970) *Ann. Rev. Ecol. Syst.* **1**, 327.

Hayward H. E. (1938) *The Structure of Economic Plants*, MacMillan, New York.

Hawker J.S. and Buttrose M. S. (1980) *Ann. Bot.* **46**, 313.

Hodson, M. J., Di Nola, L. and Mayer, A. M. (1987) *J. Exp. Bot.* **38**, 525.

Kermode A. R., Bewley J. D., Dasgupta J. and Misra S. (1986) *Hort. Science* **21**, 1113.

Kermode A. R. and Bewley J. D. (1986) *J. Exp. Bot.* **37**, 1887.

McLean R. M. and Ivimey-Cook W. R. (1956) *Textbook of Theoretical Botany*, vol. 2, Longmans Green, London.

McLeod A. M. (1960) *Wallerstein Lab. Comm.* **23**, 87.

Mahershwari P. (1950) *An Introduction to the Embryology of Angiosperms*, McGraw-Hill, New York.

Mollenhauer H. H. and Morre D. J. (1980) in *The Biochemistry of Plants*, vol. 1 *The Plant Cell*, p. 437 (Ed. N. E. Tolbert), Academic Press, New York.

Murray D. R. (1987) *Am. J. Bot.* **74**, 1120.

Negbi M. (1982) *Bot. J. Linnean Soc.* **88**, 205.
Negbi M. and Tamari B. (1963) *Is. J. Bot.* **12**, 124.
Netolitzky F. (1926) *Anatomie der Angiospermen Samen* in *Linsbauer* 12th ed. *Handbuch der Pflanzenanatomie*, Borntraeger, Berlin.
Oliker M., Poljakoff-Mayber A. and Mayer A. M. (1978) *Am. J. Botany* **65**, 366.
Opik H. (1985) *Ann. Bot.* **56**, 453.
Paquet P. J., Knutson D. M., Tinnin R. O. and Tocher R. D. (1986) *Bot. Gaz.* **147**, 156.
Pate J. S. and Flinn A. M. (1973) *J. Exp. Bot.* **24**, 1090.
Pate J. S. and Herridge D. F. (1978) *J. Exp. Bot.* **29**, 401.
Perino C. and Come D. (1977) *Physiol. Veg.* **15**, 469.
Pollock B. M. and Olney H. O. (1959) *Plant Physiol.* **34**, 131.
Quatrano R. S. (1987) in *Plant Hormones and their Role in Plant and Growth and Development*, p. 494 (Ed. P. J. Davies), Nijhoff.
Roberts E. H. (1972) in *Viability of Seeds*, p. 321 (Ed. E. H. Roberts), Chapman and Hall, London.
Rolston M. P. (1978) *Bot. Rev.* **114**, 365.
Slack C. R. and Browse J. A. (1984) in *Seed Physiology*, p. 206 (Ed. D. R. Murray), Academic Press, Sydney.
Toole E. H. (1924) *Am. J. Bot.* **11**, 325.
Troll W. (1954) *Praktische Einfuhrung in die Pflanzenmorphologie*, Gustav Fisher, Jena.
Vasil I. K. and Vasil V. (1980) *Int. Rev. Cytology*, Suppl. **11 A**, p. 145, Academic Press.
Webb M. A. and Arnot H. J. (1982) *Am. J. Bot.* **69**, 1657.
Werker E. (1980/81) *Is. J. Bot.* **29**, 22.
Werker E. and Fahn A. (1975) *Bot. Gaz.* **136**, 396.
Werker E., Marbach I. and Mayer A. M. (1979) *Ann. Bot.* **43**, 765.
Yamada M. and Usami Q. (1975) *Plant Cell Physiol.* **16**, 879.
Zohary M. (1937) *Sond. Bei. Bot. Cent.* **56A**.

2

Chemical Composition of Seeds

The chemical composition of seeds shows the same variability as is seen in other plant characteristics. Quantitatively dominant are the storage materials. These, as already mentioned, are primarily confined to special seed tissues, the cotyledons or the endosperm, or in some cases the perisperm. These storage materials, to a considerable extent, characterize seeds and are the economically most significant part of the seed.

Seeds can be divided into those whose main storage material is carbohydrate and those whose main storage material is lipid. Lipid-containing seeds are by far the more common of these two, although among economically important seeds this preponderance does not occur to the same extent. Seeds containing proteins can belong to either group. Almost no seeds are known in which the predominant storage material is protein, although there are exceptions, such as soybean or *Machaerium acutifolium* that has been reported to contain 66% protein (Coutinho and Struffaldi, 1972). Precise information about seed composition is available chiefly for seeds which are used either for food or in industry. The chemical composition of seeds of various vegetable crops is less well known, while the information about the composition of seeds of wild plants is extremely scant.

The constituents of seeds are determined genetically, but the relative amounts of these constituents are sometimes dependent on environmental factors such as mineral nutrition and climate. Thus, deficiency in sulphur, phosphorus or potassium induces changes in the relative amounts of the different storage proteins in pea cotyledons (Randall *et al.*, 1979). Sulphur deficiency induces a relative decrease of legumin, while the effect on vicilin varies. Vicilin can be resolved into four main fractions. Sulphur deficiency induces a relative decrease of one of these and an increase of another. The most marked effect of sulphur deficiency is the decrease in a 22 KD protein which is located in the cytoplasm — not in the protein bodies. Potassium and phosphorus deficiencies induce a relative increase in the legumin and vicilin (fraction 4). Increases in the external potassium concentration in the range of 0.2-200 ppm induce a decrease of soluble protein in the cytoplasm and a decrease in the percentage of protein located in the protein bodies (Randall *et al.*, 1979). In soybeans, the level of sulphur supply affects the relative amounts of glycinin and β-conglycinin (Gayler and Sykes, 1985). The amount of phytic acid phosphorus in these seeds can be regulated by changing the external phosphorus supply to the plant (Raboy and Dickinson, 1984; Raboy *et al.*, 1985). Nitrogen supply does not appear to have any major effect on the composition of storage proteins.

It appears that such compositional changes, within quite a wide range, have little effect on the germination of the seeds. Such changes in seed composition are probably a fairly normal occurrence and can be tolerated by the germinating seed.

However, changes in the enzyme proteins, or the structural proteins of the organelles, can be tolerated to a much lesser extent.

Climatic effects on seed composition were demonstrated by Iwanoff (1927a, b) who showed that the protein content of wheat varied in various parts of Russia while that of peas was constant (Table 2.1). In different populations of the genus *Vaccinium*, latitude affected size, thickness of seed coat and also germination behaviour (Crouch and Vander Kloet, 1980). Modern plant breeding practice has permitted selection for quantitative differences in seed constituents. Thus soybeans have been bred for high protein content, flax for oil content and wheat for both protein and starch content. In some breeding experiments, Woodworth *et al.* (1952) selected maize for various contents of protein and oil. Starting with maize containing 4.7% oil and 10.9% protein they obtained, after 50 generations of selection, four varieties having 15.4 or 1.9% oil and 19.5 or 4.9% protein. By selective breeding the composition of proteins can also be changed. Thus mutants of maize have been obtained in which the lysine and tryptophan content of the endosperm was almost doubled (Mertz *et al.*, 1966). High lysine lines of barley (Munck *et al.*, 1970) and of sorghum have been developed in recent years. This has been achieved either by changing the relative amount of different storage proteins or by altering the morphology of the seed by inducing changes in the size of the embryo size or the number of layers of cells in the aleurone of cereals (for review see Payne, 1983). Such changes hold out hope for improving the nutritional value of these cereals.

TABLE 2.1. *Protein content of wheat (Triticum vulgare) and peas (Pisum sativum) grown in various parts of Russia and harvested in 1924 (after Iwanoff, 1927a, b)*

			Protein content % dry weight	
Place	Longitude	Latitude	Wheat	Peas
Severo Dvinsk	46°18′	60°46′	11.92	27.00
Moscow	37°20′	55°48′	14.30	26.56
Kiev	30°28′	50°27′	19.32	30.37
Saratov	45°45′	51°37′	21.01	30.37
Omsk	70°32′	55°11′	18.69	28.12
Krasnojarsk	92°52′	56°01′	19.03	26.06
Vladivistok	131°57′	43°05′	11.86	26.87

Tables 2.2, 2.3, 2.4 and 2.5 illustrate some of the variability in seed composition. A more detailed example of the composition of the seeds of a wild *Amaranthus*, and an *Amaranthus* species used as food grain, is shown in Table 2.6. From these tables it will be seen that in addition to carbohydrates, proteins and lipids, seeds contain minerals, tannins, phosphorus compounds and many other substances.

TABLE 2.2. *Chemical composition of seeds (after Wehmer, 1929, Anderson and Kulp, 1921; Gutlin Schmitz, 1957; and Czapek, 1905)*

	Percentage of air-dry seeds				
	Carbohydrates				
	Starch	Sugar	Proteins	Fats	Lecithin
Zea mays	50 – 70	1 – 4	10·0	5	
Pisum sativum	30 – 40	4 – 6	20·0	2	
Arachis hypogea	8 – 21	4 – 12	20 – 30·0	40 – 50	1·2
Helianthus annuus	0	2	25·0	45 – 50	
Ricinus communis	0	0	18·0	64	
Acer saccharinum	42	20	27·5	4	
Triticum	60 – 75		13·3	2·0	0·7
Fagopyrum esculentum		72·0	10·0	2·0	0·5
Chenopodium quinoa		48·0	19·0	5·0	
Aesculus hippocastanum		68·0	7·0	5·0	
Castanea vesca		42·0	4·0	3·0	
Quercus penduncułata		47·0	3·0	3·0	
Linum usitatissimum		23·0	23·0	34·0	
Brassica rapa		25·0	20·0	34·0	
Papaver somniferum		19·0	20·0	41·0	
Cannabis sativa		21·0	18·0	33·0	0·4
Amygdalus communis		8·0	24·0	53·0	0·9
Aleurites moluccana		5·0	21·0	62·0	

TABLE 2.3. *Chemical composition of lettuce seeds (original)*

	Air-dry seeds (mg/g)
Total dry weight	960.0
Ash	46.0
Phytic Acid	20.0
Sucrose	30.0
Glucose	2.0
Fat	370.0
Total nitrogen	40.0
Protein nitrogen	37.0
Soluble nitrogen	1.0
Riboflavin	0.012
Ascorbic acid	0.29
Carotene	0.004
Total P (free and bound)	8.5 – 14

TABLE 2.4. *Chemical composition of soybeans (compiled from data of Morse, 1950 and Smith and Circle, 1978)*

A. Major constituents	% of moisture-free basis
Moisture	7·0 – 8·0
Ash	3·3 – 6·4
Fat	16·0 – 18·0
Fibre	3·5
Protein	38 – 46
Pentosans	4·4
Sugars	7 – 10
Phosphorus	0·5 – 1·1
Potassium	0·8 – 2·4
Calcium	0·2 – 0·3
B. Mineral constituents	**% air-dry weight of seeds**
Magnesium	0·25 – 0·35
Sulphur	0·1 – 0·5
Chlorine	0·03 – 0·04
Iodine	trace
Sodium	0·15 – 0·6
Copper	0·0012
Manganese	0·002 – 0·004
Iron	0·006 – 0·013
C. Vitamins	**μg/g**
β-carotene	0·2 – 2·4
Ascorbic acid	200
Thiamine	11 – 17·5
Pyridoxine	6·4
Nicotinic acid	20 – 26
Pantothenic acid	12
Inositol	1900 – 2600
Biotin	0·6

TABLE 2.5. *Composition of Vicia faba seeds (from data of White, 1966)*

Total Dry matter % Fresh weight	90·03%
Dry matter in Embryo + Cotyledons	86·28%
Composition of Embryo + Cotyledons	
Carbohydrates	56·67%
Total Nitrogen	35·8 %
(Insoluble N_2	31·69%)
Fat	2·03%
Organic Acid	1·00%
Ash	9·29%
Composition of carbohydrate fraction	
Hexose	0·16%
Sucrose	4·02%
Starch	42·48%
Pectin	1·69%
Hemicellulose	6·66%
Cellulose	1·66%

TABLE 2.6 *Composition of seeds of Amaranthus retroflexus and of A. cruentus (from data of Woo, 1919 and Becker et al., 1981)*

% Air dry seeds	*A. retroflexus*	*A. cruentus*
Water	8·6	6·5
Lipids	7·8	7·7
Polysaccharides	47·2	62·0
Reducing sugars	none	trace
Non-reducing sugars (after hydrolysis)	1·2	3·0
Nitrogen	2·5	3·0
Protein	15·0	17·8
Soluble	3·0	
Insoluble	12·0	
Ash	4·2	3·5

I. Carbohydrates

Starch is the main storage material in the most important field crops — wheat, rice, maize and sorghum. These crops, between them, account for most of the grain used by the human population the world over. Starch is a polysaccharide composed of α-1,4 glucan with α-1,6 glucan side chains and serves as a source of glucose for the germinating seed. But starch is not the only storage carbohydrate, other poly- and oligosaccharides are often found in many seeds. Some seeds like those of date palm (Keusch, 1968) or ivory nut (Aspinall *et al.*, 1958) have as the main storage material polysaccharides with a $(1 \longrightarrow 4)\beta$ mannan core. Other seeds contain compounds referred to as hemicelluloses which are heteropolysaccharides in nature. Thus galactomannans are present as the main reserve carbohydrate in the endosperm of Leguminoseae, for example *Trigonella* (Reid and Meier, 1970) and in *Ceratonia* account for 35% of the seed weight. The cotyledons of lupins contain a hexosan, while hemicelluloses are present in the cotyledons of *Impatiens* and *Primula*. In some species of the Liliaceae and Iridaceae, galactoglucomannans are present, and even more complex carbohydrates have been reported which contain galactose, mannose, glucose and xylose. These complex carbohydrates are stored both in the cell walls of the endosperm and the aleurone layer of various seed species. The composition of the galactomannans is quite variable, both in the ratio between the constituent sugars and their sequence (Reid, 1984; 1985).

The most common oligosaccharide in seeds is sucrose but trisaccharides and higher oligosaccharides — penta and hexasaccharides — are present in seeds of many species (Matheson, 1984).

The amount of sugars, such as raffinose may be quite high in the embryonic axes of seeds (Koster and Leopold, personal communication). In pea axes sucrose accounts for 5% of the dry weight and stachyose for 8 – 9% and in addition there is almost 3% verbascose. In soybean axes there are almost 10% stachyose and 3% sucrose, while in corn the dominant sugar is sucrose — (10%) — with about 3% raffinose.

In the sequence of utilization of the storage carbohydrates, during germination the smaller oligosaccharides are utilized first, before the main storage material (starch or lipid) are hydrolysed.

Many other carbohydrates occur in seeds, not necessarily as storage materials. Thus many seeds contain various mucilages. Chemically the mucilages are polyuronides, mainly galacturonides, or galactomannans or complex compounds containing both galactomannans and uronic acids (Tookey *et al.*, 1962). In addition, various sugars, hexoses and pentoses, have been isolated from mucilages. The polyuronides are frequently associated with proteins. Occasionally cellulose fibres are found in mucilages.

Mucilages occur either on the seed surface, as in flax, or in special cells in the seed coat, as in *Brassica alba*. *Plantago* seeds constitute a commercial source for mucilages.

Mucilages are also found in the cotyledons, as in some seeds of the Malvaceae (*Kosteletzkya* for instance). The mucilages may, sometimes, account for over 35% of the weight of the seed. The role of the mucilages is not clear, they may be connected with seed dispersal or with water uptake during germination (Young and Evans, 1973; Fahn, 1979; Garwood, 1985). Pectins are normal constituents of plant cells and consequently also of seeds.

The carbohydrate composition of barley seeds is shown in Table 2.7, as is the division of these constituents among the various parts of the seed. The data do not include starch which makes up 58-65% of the seed and is located almost entirely in the endosperm.

TABLE 2.7 *Carbohydrate composition of barley (McLeod, 1960)*

	% Dry-weight of tissue		
	Husk	Embryo	Endosperm
Sucrose	0	14.0	0.2
Raffinose	0	10.0	0.08
Hexoses	0	0.2	0.26
Total glucan	<0.02	0	1.7
Total pentosan	3.6	0.4	1.0
Galactan	0	0.3	0
Uronic acid	+	+	0
Crude cellulose	30	7.0	0.4

Many secondary plant products are present in seeds in the form of glycosides. Thus bitter almonds contain amygdalin-mandelonitrile gentiobioside. Black mustard seeds contain sinigrin, the glycoside of a mustard oil and *Nigella* contains damesenine. Various alkaloids, tannins and leucoanthocyanins also occur as glycosides in seeds. Many toxic compounds have been reported in seeds (Bell, 1978). Among these are the so-called lectins (Pusztai *et al.*, 1983), capable of causing agglutination of blood and hence sometimes called phytohaemagglutinins, e.g. concanavalin in *Canavalia ensiformis*. Many lectins contain carbohydrates. A toxic glycoprotein, ricin, is present in *Ricinus communis*. The Sapindaceae have been shown to contain cyanolipids capable of liberating cyanide. Much speculation exists about the role of such compounds and there is at least some evidence that they protect seeds against their predators. The use of seeds by man has been accompanied by his ability to destroy some of the toxic compounds. Cooking is understandably one of the ways in which such compounds may be neutralized. This would also be true for the proteinase and α-amylase inhibitors which are present in many seeds, which are low molecular weight proteins.

II. Lipids

Seed lipids are the source of oils for human consumption as well as for industry. Lipids are generally present in the form of the glycerides of fatty acids: CH_2OR_1. $CHOR_2.CH_2OR_3$ where R_1,R_2 and R_3 may be the same or different fatty acids.

Most seed fatty acids are unsaturated and the most commonly-occurring ones are oleic, linoleic and linolenic acids. In addition, however, other organic acids both saturated and unsaturated, such as acetic, butyric, palmitic, stearic, lauric and myristic acids and many others, occur as glycerides. Ground-nuts contain the glyceride of arachidic acid.

In recent years improved analytical methods have shown the presence of very diverse seed triglycerides in seed oils. In addition to the well-known mono- and di-unsaturated acids, polyolefinic acids and acetylenic acids are now known to occur in seeds, sometimes in large amounts, e.g. α-parinaric acid (C 18:4) comprises about 40 – 50% in the lipids of *Impatiens* (Nozolillo *et al.*, 1986). Oxygenated fatty acids seem to be present in seeds quite frequently (Wolff, 1966).

The lipids are found in the seeds both as fats and as oils, depending on the relative amounts of saturated and unsaturated fatty acids occurring in the glycerides. Also the saturated and unsaturated fatty acids are not esterified at random: R_2 is very rarely a saturated acid while R_1 and R_3 may be similar although not necessarily identical as to whether they are saturated or unsaturated acids (Broekeraff and Yurkowski, 1966). An example of the composition of seed lipids from soybeans is given in Table 2.8.

TABLE 2.8. *Composition of oils extracted from different seeds (compiled from various sources)*

Fatty acid		% Triglyceride fatty acid			
		Cotton seed oil	Corn oil	Rape seed *B. napus* cv. Regent	Soybean
Myristic acid	14:0	0·8	—	—	—
Palmitic acid	16:0	23·7	11·5	3·4	10·6
Palmitoleic acid	16:1	0·65	1·0	—	—
Stearic acid	18:0	2·55	2·0	1·4	4·1
Oleic acid	18:1	17·4	24·1	61·7	24·2
Linoleic acid	18:2	54·5	61·9	20·0	52·6
Linolenic acid	18:3	—	0·7	10·2	7·6
Arachidic acid	20:0	—	0·2	2·5	—
Behenic acid	22:0	—	—	0·7	—

Other lipid materials found in seeds are esters of higher alcohols, sterols, phospholipids and glycolipids, tocopherols and squalene. An exceptional case of lipids contained in seed is that of the jojoba (*Simmondia chinensis*). Its storage lipid is a wax in which long chain fatty acids (20 – 22C) are esterified with the corresponding long chain alcohols. These wax esters are hydrophobic and are located in the cells of the cotyledons as oil bodies which are apparently membrane bound (Muller *et al.*, 1975). The composition of this wax is such that, provided the plant can grow economically and in quantity, it may replace many other wax sources (Moreau and Huang, 1977).

A highly toxic oil has been isolated from seeds of *Dichapetalum toxicarium*. This lipid has been shown to contain fluoro-oleic acid (Peters *et al.*, 1960). The compound is so toxic that the seeds of this plant are often referred to as "ratsbane". It is interesting that the leaves of this plant also contain fluoro-acetic acid. A large number of other toxic lipids are known, between them cyano-lipids and phenolic-lipids (Seigler, 1979).

Because of the recent interest in the properties of cellular membranes the detailed composition of the phospholipids in some seeds is under intensive investigation (Morrison, 1979; Di Nola and Mayer, 1986).

III. Proteins

For many years the classification of proteins was based on their solubility in water and salt solutions (Osborne, 1919). Thus proteins soluble in water and salt solution were defined as albumins; those soluble in salt solution but very slightly soluble in water were defined globulins. Prolamins are soluble in 70 – 80% (aqueous) ethanol but not soluble in water or absolute alcohol, while glutelins are defined as the proteins that are soluble only in acids and alkalis. Each of these groups contains numerous proteins differing between them in numerous features, such as the number of sub units, charge, molecular weight etc. There is no sharp division between globulins and albumins as the solubility of proteins of both these groups increases with the increasing salt concentration. Many of the proteins, especially of albumins, are enzymes and are probably metabolically active. Proteins which are metabolically inactive are regarded as storage proteins and are usually located in special organelles — the protein bodies, or microbodies — (see Chapter 1). The microbodies often contain globoids which may be quite large (Pernollet and Mosse, 1983).

The terms "storage proteins" and "globulins" are often interchanged, especially for legume proteins (Casey *et al.*, 1986).

More recently, a different criterion has been used to classify proteins — their coefficient of sedimentation (Casey *et al.*, 1986). However, separations based on this criterion also did not yield proteins which were electrophoretically pure and each fraction contained numerous proteins. A more reliable classification should be based on DNA and amino acid sequencing and the degree of homology between certain regions in the molecules of different proteins. This may also prevent the naming of very similar proteins by different names.

The storage proteins of seeds vary according to species. Thus, in wheat, at least four different proteins occur, glutelins, prolamins, globulins and albumins. The glutelins and prolamins form the major components of the protein, while the globulins contribute only 6 – 10% of the total and the albumins 3 – 5%. The total active proteins, globulins and albumins, account for no more than 15% of the total proteins in wheat. The distribution between metabolically active and inactive proteins is similar in most cereals. Albumins can also act as storage proteins.

From the foregoing it is clear that it is preferable to refer to definite proteins whose properties have been investigated. The best-known prolamins are those found in wheat – gliadin, in barley – hordein and in maize – zein. Amino acid sequences and secondary and tertiary structures of a number of prolamins are now known. The quaternary structures of zein from *Zea* and hordein from barley have been proposed (Ingversen, 1983). Detailed studies of 7S and 11S "globulins" (vicilins and legumins) in legume storage proteins have been carried out (Casey *et al.* 1986) and compared with similar proteins from cereal seeds and from seeds of various dicotyledons. Thus 11 – 12S proteins are in the main storage proteins in *Oriza sativa* and *Avena sativa*, their sub-unit composition, biosynthesis and processing being very similar to those of legumes. However, rice and oats both differ from other cereals. In both, in addition to the fact that 11 – 12S proteins are the main storage ones, the prolamine content is very low.

In dicotyledonous plants prolamins seem to be almost absent. Glutelins are absent in some species and in others constitute up to 50% of the total proteins of the seeds. Albumins and globulins in these seeds are usually well-defined and a number of the globulins have been obtained in crystalline form after extraction with hot salt solutions. Among those which have been investigated in detail are legumin and vicilin from peas which have been obtained electrophoretically homogenous, arachin and conarachin from peanuts, glycinin from soybeans and edestin from hemp seeds. Vicilin from *Phaseolus aureus* is a glycoprotein, containing neutral sugars and a small amount of glucosamine. Legumin, previously also regarded as a glycoprotein, appears merely to co-purify with a low molecular weight glycoprotein, but can be separated from it (Hurkman and Beevers, 1980). Both proteins are made up of non-identical subunits, three for legumin and four for vicilin (Ericson and Chrispeels, 1973). The structure of some of these proteins is now well established and proposals for their quaternary assembly have been made (Pernollet and Mosse, 1983).

The use of electrophoretic techniques has revealed that the individual storage proteins show a great deal of heterogeneity, both in the charge of their subunits and in their molecular weight. This was shown, for example, for gliadin, glutenin, cucurbitin, legumin and appears to be a general phenomenon (Derbyshire *et al.*, 1976; Wall, 1979; Blagrove and Lilley, 1980). This heterogeneity is of considerable practical interest. In seed quality control it is important to detect the occurrence of mixtures of seeds of different varieties of the same species. The work of Autran (1975) has shown that the storage proteins of each seed variety has a highly characteristic electrophoretic pattern, which permits its identification. Thus fraudulent additions can be easily detected. Identification of cultivars using electrophoretic techniques is becoming now a standard tool in the seed industry (Vaughan, 1983).

The heterogeneity of seed storage protein is much greater than that observed for example in enzyme proteins. It seems probable that in these proteins minor alterations in structure do not impair their functionality and hence greater variability is observed.

Seed storage proteins generally have a high nitrogen content, high proline content and are often low in their content of lysine, tryptophan and methionine. Table 2.9 shows the protein content of some mono- and dicotyledonous seeds, and Table 2.10 shows the amino acid composition of some seed proteins.

TABLE 2.9. *The protein content of various seeds (from Brohult and Sandegren, 1954)*

	Total protein	Various fractions as % of total protein			
	% Dry seeds	Albumin	Globulin	Prolamin	Glutelin
Triticum vulgare	10 – 15	3 – 5	6 – 10	40 – 50	30 – 40
Hordeum vulgare	10 – 16	3 – 4	10 – 20	35 – 45	35 – 45
Avena sativa	8 – 14	5 – 10	80	10 – 15	5
Secale cereale	9 – 14	5 – 10	5 – 10	30 – 50	30 – 50
Cucurbita pepo	12	very little	92	very little	small amounts
Nicotiana sp.	33	24	26	very little	50
Gossypium herbaceum	20	very little	90	very little	10
Glycine hispida	30 – 50	small amounts	85 – 45	very little	very little
Lupinus luteus	40	1	78	very little	16

TABLE 2.10. *Amino acid composition of some proteins from seeds (amino acids as % of total residues recovered. Modified from Pernollet and Mosse, 1983)*

Amino acid	Legumin *(Pisum)*	Vicilin *(Pisum)*	Zein *(Zea)*	w-Gliadin *(Triticum)*
Glycine	7·8	5·2	1·8	1·6
Alanine	5·9	4·5	13·2	0·6
Valine	5·8	5·7	4·1	0·5
Leucine	8·3	10·3	18·9	3·4
Isoleucine	4·5	5·5	3·9	3·8
Serine	6·3	7·8	6·4	3·4
Threonine	3·7	3·0	3·0	1·2
Tyrosine	2·2	2·3	2·4	1·1
Phenylalanine	3·8	5·0	5·2	9·1
Tryptophan	0·7	0·05	0	—
Proline	5·1	4·4	10·1	22·9
Methionine	0·8	0·13	0·8	0·1
Cysteine	0·7	0·05	1·0	0·4
Lysine	5·5	7·2	0·1	0·4
Histidine	2·1	1·6	0·9	1·0
Arginine	7·7	5·7	1·0	0·7
Asparagine	11·7	13·4	4·9	0·5
Glutamine	17·6	18·2	21·3	50·1

Proteinase inhibitors, such as the soybean trypsin inhibitors, are of widespread occurrence in seeds and are often present in large amounts. These inhibitors are polypeptides or low molecular weight proteins (Vogel *et al.*, 1968; Ryan, 1973; Mikola, 1983). Their biological function in the seeds is still unclear.

IV. Other Components

In addition to the major compounds mentioned, seeds contain a large number of other substances. All seeds contain a certain amount of minerals (see Table 2.4). The mineral composition of seeds is essentially similar to that of the rest of the plant and usually comprises all the essential elements. However, K^+/Ca^{2+} ratio varies considerably, depending on the species and on the nature of the environment. In the Protaceous group the K^+/Ca^{2+} ratio is especially high (Kuo *et al.*, 1982).

A special feature of many seeds is that a very large part of the phosphate occurs as phytin, the calcium and magnesium salt of inositol hexaphosphate. The phytin is apparently synthesized in the cytoplasm, in association with the endoplasmic reticulum and then moved to the globoids in the developing protein bodies (Greenwood, 1983 as cited by Lott, 1984). In lettuce seeds 50% of the total phosphorus occurs as phytin, 6 – 10% as free phosphate and the remainder in other phosphorus-containing compounds such as nucleotides, sugar-phosphates (20 – 25%), phospholipids, nucleoproteins and other compounds (20 – 25%). In addition, the nucleic acids constitute an extremely important part of the phosphorus-containing compounds. The nucleic acids occur partly in their free form and partly in the form of nucleoproteins. The phosphorus composition of *Amaranthus* is shown in Table 2.11.

TABLE 2.11. *Phosphorus compounds in Amaranthus retroflexus (Woo, 1919)*

	Phosphorus as % of total ash
Total P	4·6
Inorganic P	0·13
Lipid P	0·18
Soluble organic P	0·33
Phosphoprotein P	1·8
Nucleoprotein P	2·5

Seeds contain, besides proteins, numerous other nitrogenous compounds such as alkaloids, free amino acids and amides. The amides found are glutamine and asparagine, as well as γ-methylene glutamine, for example in ground nuts. The free amino acids found in seeds are usually the same as those forming part of the protein structure. Many of the non-protein amino acids, present in various tissues of plants, also occur in the seed of the plants, while others have been shown to occur especially in the seeds. Among these compounds are γ-methylene glutamic acid, γ-aminobutyric acid, β-pyrazol-1-ylalanine, lathyrine, pipecolic acid and many others. Some of these acids are restricted to a single species, while others are of very widespread distribution among many species (Fowden, 1970; Bell, 1978; Fowden *et al.*, 1979). Some of the non-protein amino acids are highly toxic to animals or man. There are some reports that the presence of at least some of these substances makes the seeds less edible to predators (Rehr *et al.*, 1973; Janzen, 1978).

Some examples for alkaloids present in the seeds are piperine in *Piper nigrum* seeds, ricinine in castor oil beans, hyoscine in seeds of *Datura* and lupinidine (sparteine) in lupin seeds. The occurrence of an alkaloid in the plant does not necessarily indicate its presence in the seed in comparable amounts. An example of this is *Coffea*, where the caffeine content of the bean is much higher than that of the vegetative plant. Cacao, *Theobroma cacao*, contains a very large amount of theobromine in the seeds, as well as a smaller amount of caffeine. Strychnine and brucine are found in the seeds of *Strychnos nux-vomica* where they amount to about 2–3% of the weight of the seed.

Various organic acids such as tricarboxylic acid cycle intermediates, as well as malonic acid, have been detected in seeds of many species.

Phytosterols occur in a number of seeds. The best-known ones are the sitosterols and stigmasterols from soybeans. The latter is pharmaceutically important as it is used as a precursor of progesterone.

Although seeds contain a number of pigments, chlorophyll is usually absent from ripe seeds, although it does occur in the seeds of Gymnosperms. Protochlorophyll, however, occurs in the Cucurbitaceae. Young, immature seeds are green and it is therefore not surprising that breakdown products of chlorophyll may be present in some seeds. Other pigments found are β-carotene and various other carotenoids. The seed coats of many seeds contain anthocyanins or leucoanthocyanins. Flavonoid pigments are also known to be present in various seeds. Many of these compounds occur as glycosides. Thus the seed coat of *Phaseolus vulgaris* may contain leuco-delphinidin, leuco-pelargonidin as well as delphinidin, petunidin and malvidin as the corresponding glycosides, and also at least two flavonol glycosides. Cotton seeds contain a yellow pigment, gossypol, which is located in special pigment glands.

Various phenolic compounds such as coumarin derivatives, chlorogenic acid and simple phenols such as ferulic, caffeic and sinapic acids, occur in many seeds. These compounds may give rise, by oxidation and condensation, to pigments of the melanin type. Another phenolic type constituent is tannin, e.g. in *Arachis* and sorghum seeds. The phenolic compounds may act as germination inhibitors and they probably have a regulatory role in germination, as well as a possible role in the defence mechanisms against predators.

All seeds contain vitamins. Detailed information concerning the vitamin content is available especially for seeds which are used for food or fodder. Data on the vitamin B content of many seeds are available and attempts are continuously made to raise the vitamin content of such seeds by selective breeding. Some figures are given in Table 2.12. Tocopherols are present in the oil of many seeds. Most of the known vitamins have in fact been shown to occur in some kinds of seeds.

As a result of the development of modern analytical methods, such as gas chromatography, HPLC and immunochemical assays, in addition to paper chromatography and biological assays, it has become possible to study the growth substance content of various seeds. The occurrence of indolylacetic acid or its derivatives has been demonstrated in a number of seeds (Elkinowy and Raa, 1973; Cohen and Bandurski, 1982), while in other cases various indole derivatives have been found. Gibberellic acid and gibberellin-like substances have been found in runner beans, in lettuce seeds, barley and in the seeds of many other plants (Mounla and Michael, 1973). Cytokinins, both free and bound, occur in seeds. The first natural cytokinin, zeatin, was isolated from immature maize kernels. The plant growth inhibitor

TABLE 2.12. *The vitamin content of some seeds. The figures are for the air-dry seeds (from food composition tables, FAO, 1954)*

	mg/100 g				I.U.
	Thiamin	Riboflavin	Nicotinic Acid	Ascorbic Acid	Vitamin A
Wheat (Durrum)	0·45	0·13	5·4	0	0
Rice (hulls removed)	0·33	0·05	4·6	0	0
Barley	0·46	0·12	5·5	0	0
Maize	0·45	0·11	2·0	0	450
Ground-nuts (shelled)	0·84	0·12	16·0	0	30
Soybean	1·03	0·30	2·1	0	140
Broad bean	0·54	0·29	2·3	4	100
Peas	0·72	0·15	2·4	4	100
Sunflower	0·12	0·10	1·4	0	30
Chestnuts (fresh)	0·21	0·17	0·4	24	0

abscisic acid, ABA, has been found in seeds of *Fraxinus* and the achenes of *Rosa*. Probably it occurs in most species, both in free and bound form. Ethylene is formed in seeds very rapidly after the beginning of imbibition and must be added to the list of growth substances occurring in seeds. In addition other growth-promoting and growth-inhibiting substances have been shown to be present in various seed extracts, but their nature has not yet been elucidated. The possible importance of these substances in the regulation of dormancy will be discussed in a later chapter.

Bibliography

Anderson R. J. and Kulp, W. L. (1921) *N. Y. Agr. Exp. Sta. Bull.* **81**.

Aspinall G. O., Rashbrook R. B. and Kessler G. (1958) *J. Chem. Soc.* Part 1, 215.

Autran J. C. (1975) *Inds. Agricol. Aliment.* **9 – 10**, 1075.

Becker R., Wheeler E. L., Lorenz K., Stafford A. E., Grosjean O. K., Betschart A. A. and Saunders R. M. (1981) *J. Food Science* **46**, 1175.

Bell E. A. (1978) in *Biochemical Aspects of Plant and Animal Coevolution*, p. 143 (Ed. J. B. Harborne), Academic Press, London.

Blagrove R. J. and Lilley G. G. (1980) *Eur. J. Biochem.* **103**, 577.

Broekeraff H. and Yurkowski M. (1966) *J. Lipid Res.* **7**, 62.

Brohult S. and Sandegren E. (1954) in *The Proteins*, vol. 2A, p. 487, Academic Press, New York.

Casey, R., Domoney C. and Ellis N. (1986) in *Oxford Survey of Plant Molecular and Cell Biology*, Vol. 3, 1. Oxford University Press, Oxford.

Circle S. J. (1950) in *Soybeans and Soybean Products*, vol. 1, p. 275. Interscience, New York.

Cohen J. D. and Bandurski R. B. (1982) *Ann. Rev. Plant Physiol.* **33**, 403.

Coutinho L. M. and Struffaldi Y. (1972) *Phyton* **29**, 25.

Crouch P. A. and Vander Kloet S. P. (1980) *Can. J. Bot.* **58**, 84.

Czapek F. (1905) *Die Biochemie der Pflanzen*, Gustav Fisher, Jena.

Derbyshire E., Wright D. J. and Boulter D. (1986) *Phytochemistry* **15**, 3.

Di Nola L. and Mayer A. M. (1986) *Phytochemistry* **25**, 2725.

Elkinowy M. and Raa J. (1973) *Physiol. Plant.* **29**, 250.

Ericson M. C. and Chrispeels M. J. (1973) *Plant Physiol.* **52**, 98.

Fahn A. (1979) *Secretory Tissues in Plants*, Academic Press. London.
FAO: Food Composition Tables, Minerals and Vitamins, Rome, Italy (1954).
Fowden L. (1970) in *Progress in Phytochemistry*, vol. 2, p. 203. Interscience, London.
Fowden L., Lea P. J. and Bell E. A. (1979) *Adv. Enzymology* **50**, 117.
Garwood N. C. (1985) *Am. J. Bot.* **72**, 1095.
Gayler K. R. and Sykes G. E. (1985) *Plant Physiol.* **78**, 582.
Greenwood J. S. (1983) Ph. D. Thesis, cited by Lott, J. N. A. (1984) in *Seed Physiology* vol. 1 (Ed. D. R. Murray), Academic Press, Sydney.
Gutlin Schmitz P. H. (1957). Dissertation of the University of Basel; Uber die Anderung des Aneuringehaltes wahrend der Keimung un Samen verschiedener Reserve Stoffe.
Hurkman W. J. and Beevers L. (1980) *Planta* **150**, 82.
Ingversen J. (1983) in *Seed Proteins*, p. 193 (Ed. J. Daussant, J. Mosse and J. Vaughan), Academic Press.
Iwanoff N. N. (1927a) *Biochem. Z.* **182**, 188.
Iwanoff N. N. (1927b) *Bull. Bot. and Plant Breeding* **17**, 225.
Janzen D. H. (1978) in *Biochemical Aspects of Plant and Animal Coevolution* , p. 163 (Ed. J. B. Harborne), Academic Press, London.
Keusch L. (1968) *Planta* **78**, 321.
Kuo J., Hocking P. J. and Pate J. S. (1982) *Aust. J. Bot.* **30**, 231.
Matheson M. K. (1984) in *Seed Physiology,* Vol. 1 (Ed. D. R. Murray), Academic Press, Sydney.
Mikola J. (1983) in *Seed Proteins*, p. 35 (Ed. J. Daussant, J. Mosse and J. Vaughan), Academic Press, London.
McLeod A. M. (1960) *Wallerstein Lab. Comm.* **23**, 87.
Mertz E. T., Nelson D. E., Bates L. S. and Vernon O. A. (1966) *Adv. Chem.* no. 57, p. 228, American Chemical Society.
Moreau R. A. and Huang A. H. C. (1977) *Plant Physiol.* **60**, 329.
Morrison W. R. (1979) in *Recent Advances in the Biochemistry of Cereals*, p. 13 (Eds D. L. Laidman and R. G. Wyn Jones), Academic Press, London.
Morse W. J. (1950) in *Soybeans and Soybean Products*, vol. 1, p. 135. Interscience, New York.
Mounla M. A. Kh. and Michael G. (1973) *Physiol. Plant.* **29**, 274.
Muller L. L., Hensarling F. P. and Jacks T. J. (1975) *J. Am. Oil Chem. Soc.* **52**, 164.
Munck L., Karlsson K. E., Hagberg A. and Eggum B. O. (1970) *Science* **168**, 985.
Nozolillo C., Rahal H. and Liljenberg C. (1986) *Am. J. Bot.* **73**, 96.
Osborne T. B. (1919) *The Vegetable Proteins*, Longmans, Green and Co., London.
Payne P. I. (1983) in *Seed Proteins*, p. 223 (Ed. J. Daussant, J. Mosse and J. Vaughan), Academic Press, London.
Pernollet J. C. and Mosse J. (1983) in *Seed Proteins*, p. 155 (Ed. J. Daussant, J. Mosse and J. Vaughan), Academic Press, London.
Peters R. A., Hall R. J., Ward P. F. and Sheppard N. (1960) *Biochem. J.* **77**, 17.
Pusztai A., Croy R. R. D., Grant G. and Steward J. C. S. (1983) in *Seed Proteins*, p. 53 (Ed. J. Daussant, J. Mosse and J. Vaughan), Academic Press, London.
Raboy V. and Dickinson D. B. (1984) *Plant Physiol.* **75**, 1094.
Raboy V., Hudson S. J. and Dickinson D. B. (1985) *Plant Physiol.* **79**, 123.
Randall P. J., Thomson J. A. and Schroeder H. E. (1979) *Aust. J. Plant Physiol.* **6**, 11.
Rehr S. S., Bell E. A., Janzen D. H. and Feeny P. P. (1973) *Biochem. Syst.* **1**, 63.
Reid J. S. G. (1984) *Adv. Bot. Res.* **11**, 125, Academic Press, London.
Reid J. S. G. (1985) in *Biochemistry of Plant Cell Walls*, p. 259 (Ed. C. T. Brett and J. R. Hillman), Cambridge University Press, Cambridge.
Reid J. S. G. and Meier H. (1970) *Phytochemistry* **9**, 513.
Ryan C. A. (1973) *Ann. Rev. Plant. Physiol.* **24**, 173.
Seigler D. S. (1979) in *Herbivores*, p. 449 (Ed. G. A. Rosenthal and D. M. Janzen), Academic Press, New York.
Smith A. K. and Circle S. (1978) *Soybean: Chemistry and Technology*, vol. 1, Avi Publishing Co.
Tookey H. L., Lohmar R. L., Wolff I. A. and Jones Q. (1962) *Agr. Food Chem.* **10**, 131.
Vaughan J. G. (1983) in *Seed Proteins*, p. 35 (Ed. J. Daussant, J. Mosse and J. Vaughan), Academic Press, London.
Vogel R., Trautschold J. and Werle E. (1968) *Proteinase Inhibitors*, Academic Press, New York.
Wall J. S. (1979) in *Recent Advances in the Biochemistry of Cereals* (Ed. Laidman and R. W. Wyn Jones), Academic Press, London.
Wehmer C. (1929) *Die Pflanzenstoffe*, 2nd Ed., Jena.

White H. L. (1966) *J. Exp. Bot.* **17**, 105.
Wolff I. A. (1966) *Science* **154**, 1140.
Woo M. L. (1919) *Bot. Gaz.* **154**, 1140.
Woodworth C. M., Leng E. R. and Jugenheimer R. W. (1952) *Agron. J.* **44**, 60.
Young J. A. and Evans R. A. (1973) *Weed Science* **21**, 52.

3

Factors Affecting Germination

I. Viability and Life Span of Seeds

Most seeds are fairly resistant to extreme external conditions, provided they are in a state of desiccation. In the dry state, seeds can retain their ability to germinate, or viability, for considerable periods. The length of time for which seeds can remain viable is determined genetically.

However, environmental factors and storage conditions have a decisive effect on the life span of any given seed, i.e. whether the seed will remain viable for the full period determined by its genome or whether it will lose its viability at some earlier stage. In general, viability is retained best under conditions in which the metabolic activity of seeds is greatly reduced, i.e. low temperature and high carbon dioxide concentration. In addition, other factors are of great importance in determining viability, particularly those which determine seed dormancy. Numerous attempts have been made to determine the "maximal" life span of the seeds (Roos, 1986). The best documented claims for longevity are those of seeds with known collection dates, taken from Herbaria and those from controlled experiments.

Becquerel (1932, 1934) tried to germinate seeds which were taken from the Herbarium of the National Museum in Paris, in order to estimate their life span. He came to the conclusion that *Mimosa glomerata* seeds remained viable for some 221 yr, while various other leguminous seeds remained viable for periods of 100 – 150 yr, e.g. *Astragalus massiliensis, Dioclea paucifera* and *Cassia bicapsularis.* Turner (1933) tested seeds from collections kept at Kew. He also cites many examples in which life span was estimated from data on length of burial period and similar circumstantial information. Some of the data of Turner and Becquerel are given in Table 3.1. Many seeds (e.g. charlock) kept better buried in soil, than in jars on the laboratory shelves.

TABLE 3.1. *Life span of seeds in years (from data of Becquerel, 1932 and 1934; and Turner, 1933)*

Anagallis foemina	60	*Lotus uliginosus*	81
Anthyllis vulneraria	90	*Medicago orbicularis*	78
Cassia bicapsularis	87	*Nelumbium luteum*	56
Cytisus biflorus	84	*Stachys nepetifolia*	77
Ipomoea sp.	43	*Trifolium arvense*	68
Lavatera olbia	64	*Trifolium pratense*	81
Lens esculenta	65	*Trifolium striatum*	90

In contrast to the seeds listed in Table 3.1, which have a long life span, other seeds are characterized by a very short life span. For example *Acer saccharinum, Zizana aquatica, Salix japonica* and *S. pierotti* lose their viability within a week if kept in air. *Ulmus campestris* and *U. americana* remain viable for about 6 months. *Hevea*, sugar cane and *Boehea, Thea*, cocos and other tropical crop seeds remain viable for less than a year (Crocker, 1938).

Several attempts have been made to establish the viability of seeds by controlled experiments. The seeds were usually buried in the soil in some suitable containers and samples removed at different periods.

Thus Beal began experiments in 1879 and these were continued for a period of 30 yr, seeds being removed and tested every 5 yr. After 30 yr a considerable number of species were still viable, namely *Amaranthus retroflexus, Brassica nigra, Capsella bursa-pastoris, Lepidium virginicum, Oenothera biennis, Rumex crispus* and *Setaria media*. Subsequently, seeds were tested every 10 yr. The last sample was tested in 1980, there are still six samples buried (Roos, 1986). The 100 yr of the experiment were summarized by Kivilaan and Bandurski (1973, 1981). Two species of *Verbascum, V.blattaria* and *V.thapsus*, were found to be viable; the seedlings survived and produced normal seed. In addition seeds of *Malva rotundifolia* germinated in 1980, although this species did not germinate in the previous tests (Kivilaan and Bandurski, 1981). A similar experiment was conducted by Duvel (1905), in which seeds of 107 different species were tested. This experiment was terminated in 1941 and its final results summarized by Toole and Brown (1946). After 39 yr in the soil seeds of 36 species were still able to germinate. A long-term experiment planned to last 300 yr was initiated by Went and Munz (1949).

In the experiments of Beal and in those of Becquerel the seeds were maintained under relatively dry conditions during storage. Generally seeds remain viable for longer periods if they are dry. For example, lettuce seeds kept much longer if their moisture was reduced from air-dry to half this value (in Ithaca, USA) (Griffiths, 1942). Moisture content was more critical than temperature during storage. Raising moisture content from 5 to 10% caused a more rapid loss of viability than a temperature rise from 20° to 40°C.

Similar data have been obtained for clover. The seeds remained viable for 3 yr at all temperatures up to 38°C when their moisture content was 6%. At 8% moisture content they remained viable except at 38°C. When the moisture content of the seeds was raised to 12 – 16% the seeds remained viable only up to 30°C. At 16% moisture content viability at 22%C fell within 3 months (Ching *et al.*, 1959).

The moisture content of seeds is determined, at least in part, by the relative humidity of the air and by the temperature of storage. Generally, as the temperature and relative humidity increase, the moisture content of the seeds increases up to a maximal value and then falls again as the temperature increases further. Loss of germinability usually occurs before the maximal water content is reached (Barton, 1961). Other attempts at long-term experimentation for testing the longevity of seeds are summarized by Roos (1986).

Viability is retained for very long periods of time, especially in seeds having a hard seed coat, as in the Leguminosae where viability is often retained for several decades. The most extreme cases of retention of viability is the case of *Nelumbo nucifera*, the Indian lotus. *Nelumbo* seeds were found on several occasions in several places. Carbon dating of the same seeds, on different occasions gave different results. Thus *Nelumbo* seeds found in the mud of a lake bed in Manchuria and germinated after their seed coat was broken, were carbon dated and were estimated to be 250 – 400 yr old. However, carbon dating of the same seeds on a different occasion estimated them to be up to 1000 yr old (Priestley and Posthumus, 1982; Roos, 1986). Carbon dating of seeds is open to many serious errors, and must be treated with great reserve (Godwin, 1968). Often not the seeds themselves but the objects found in their vicinity are carbon dated. *Canna* seeds also appear to be very long lived. The age of the seeds found in Argentina in an excavated tomb was estimated by carbon dating to be approximately 600 yr (Lerman and Cigliano, 1971). In contrast to these cases, where old seeds did germinate, the claims of viability of grains found in the Egyptian pyramids have been shown to be spurious.

Attempts have been made to predict the viability of seeds and the germination percentage following storage. Equations relating viability, moisture content and storage temperatures for various cereal crops were formulated and nomographs constructed, which permitted prediction of the longevity of seeds of species which desiccate naturally (Roberts, 1972, 1986). The original equations have been revised in order to extend the range of moisture content and storage temperatures used for storage of seeds, and to take into account the initial quality of seeds. The equation arrived at is:

$$v = K_i - p/10^{K_E - C_W \log m - C_H t - C_Q t^2}$$

where v = probit germination after a storage period p, m = moisture content, t = storage temperature, k_1 is a measure of initial seed quality and K_E, C_w, C_Q and C_H are species specific constants (Ellis and Roberts, 1981; Roberts, 1986).

Although this equation seems to fit the behaviour of many seeds over wide ranges of storage conditions, it is complicated and contains empirical constants specific to each species. Other attempts to predict viability are made periodically. Priestley *et al.* (1985) predicted the time needed for viability to drop by 50% in various species by analysing known data on deterioriation, using a model and probit analysis. Good agreement between estimates and known data was observed.

Although from the above data it appears that dry conditions are essential for retention of viability, many seeds remain viable when submerged in water. Shull (1914) was able to show that 11 out of 58 species tested were still viable after 4½ yr submergence, for example *Asclepias syriaca, Juncus tenuis, Plantago rugelii* and *Solidago rugosa*. *Juncus* species remained viable even after seven years of submergence (Shull, 1914).

Even seeds such as lettuce which store well under very dry conditions retain their viability when fully imbibed. Lettuce seeds stored fully imbibed at 30°C lost their viability less rapidly than seeds with a moisture content of 5.1% stored dry (Villiers, 1974). When the moisture content of such seeds rose to 13.5% viability was lost after 2 months. Similar results were obtained for *Fraxinus americana*. These results were ascribed to the operation of repair mechanisms in the fully imbibed seeds which cannot operate in partially hydrated ones.

Loss of viability in seeds may be due to several factors. There is little doubt that during the prolonged storage of seeds lesions in their DNA and its fragmentation can occur. It has also been shown that at least in *Secale* embryos some of these lesions are repaired during the imbibition phase of germination (Osborne *et al.*, 1980/81). DNA ligase activity is necessary for repair of such lesions. Evidence for the ability of seeds to repair lesions in their DNA exists only for monocotyledonous species. Such lesions appear to occur also during seed dormancy. The rehabilitation of the genome may be an essential step for effective germination (Osborne *et al.*, 1984). The appearance of lesions in RNA, during ageing, has also been reported.

Loss of viability during storage is accompanied by, and often ascribed to, various chemical changes which may be enzymatic in their nature. Among the most striking ones are changes in lipids induced by peroxidative processes. Lipoxygenases apparently can be active in seeds with very low water content (Fig. 3.1). Lipoxygenation of linoleic (18:1) and linolenic (18:3) acids apparently generates free radicals which could attack the intrinsic membrane proteins and cause damage to the membranes. As some of these intrinsic proteins may be the ion transport channels, lipoxygenation could be one of the causes of the increased leakiness typical of deterioriating seeds (Bewley, 1986).

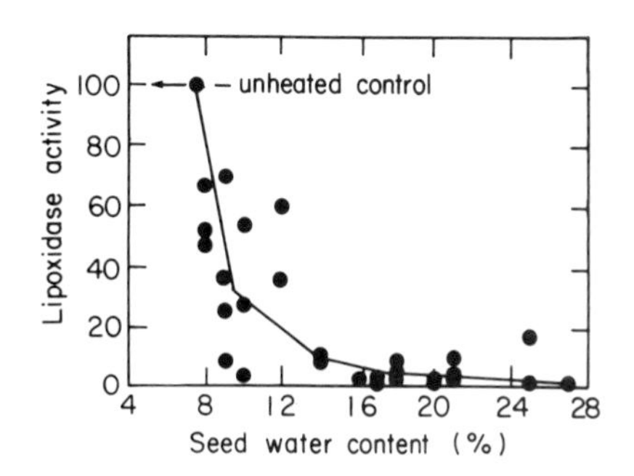

FIG. 3.1. Lipoxygenase activity of soybean seeds having different water contents. The seeds were equilibrated to different water contents and then heated for 10 min at 90°C. Enzyme activity was then determined. At low water contents, below 10% the enzyme was resistant to heating but above 10% enzyme activity was lost (from data of Leopold and Vertucci, 1986).

A detailed and informative discussion of these problems may be found in the book by Priestley (1986) and in that edited by McDonald and Nelson (1986).

Loss of viability is also accompanied by morphological changes such as discolouration or the loss of hair from the seed coat. Ultrastructural changes have also been observed (Priestley, 1986).

Although in many plant species seeds are desiccated during maturation, and the desiccation process is often considered as a necessary step to condition the seeds for subsequent germination, other seed species as mentioned above lose viability if desiccated. Such seeds, which cannot be dried down without loss of viability, are often referred to as "recalcitrant". Examples of such seeds include many of the tropical species such as *Cocoa, Hevea, Duria, Mango* and *Symphonia.* The description of cassava and coffee as recalcitrant is probably erroneous. The germination behaviour of some recalcitrant species is described by Corbineau and Come (1988a, b, c).

As we have seen, the storage conditions required to maintain viability for different seeds are different. Sometimes drying causes very rapid loss in viability, while in other cases the seeds will remain viable only on drying. Similar differences in response to storage conditions are known for oxygen and carbon dioxide concentration. Therefore it is very difficult to formulate any general rule. This problem is also discussed by Owen (1956) and by Roberts (1972).

Loss of viability is not a sudden abrupt failure to germinate of all the seeds in a certain population. Rather the percentage of seeds which will germinate in any given population will slowly decrease. Moreover, even if a seed loses its viability this does not imply that all metabolic processes stop simultaneously or that all enzymes are inactivated. Only the sum total of processes which lead to germination no longer operates properly. This was illustrated for cocoa beans by Holden (Table 3.2). For this reason all chemical or histochemical methods devised to test viability are only partially satisfactory. Such tests can only check for one definite reaction which may to some extent be correlated with the eventual ability of the seed to germinate. Most of these tests are based on the activity of certain oxidizing enzymes. The best correlation has been found with the activity of enzymes reacting with redox dyes, such as tetrazolium, but even here positive results do not always indicate 100% germination of the seed population. An X-ray contrast technique has also been used quite successfully to predict seed viability and seed quality (Kamra, 1964, 1976). General chemical changes in seed composition, as viability is lost, are discussed by Owen (1956) and Abdul-Baki and Anderson (1972).

TABLE 3.2. *Enzyme activity and viability in cocoa beans. The beans were allowed to ferment which raises their temperature and eventually kills them (after Holden, 1959)*

	Enzyme activity as % of unfermented beans			
Time of fermentation (hr)	20	44	68	92
Temp. °C	33	42	44	44
% gemination	100	0	0	—
Amylase	111	111	32	<5
β-Glucosidase	103	46	0	0
Catalase	160	20	0	—
Peroxidase	75	9	<5	<5
Polyphenol oxidase	73	17	15	11

In the description of seed behaviour the term "vigour" is frequently used. Seed vigour is the potential growth rate of the seedling and relates to seedling emergence. A seed which shows loss of vigour is one in which there is a reduction in the potential maximal growth rate. Loss of vigour does not necessarily mean an inability to germinate or to develop into a normal plant, although it is frequently accompanied by repression of early biochemical events during germination.

The concept of seed "vigour" although clearly not directly related to either viability or germination is of great practical importance. In agriculture seed vigour, which is a way to estimate seedling emergence in the field, may be more important than actual germination percentage.

The very long life span of many species of wild plants must be contrasted with the relatively short one of many cultivated species. In the latter a high germination percentage is important from a practical point of view, and is usually retained for relatively short periods. Some data are given in Table 3.3.

TABLE 3.3. *Changes in percentage germination of seeds during storage (compiled from various sources)*

	Years of storage in which seeds germinate		
	70 – 100%	30 – 60%	less than 30%
Wheat	9	—	13
Rye	7	—	12
Barley	8	—	12
Oats	11	12	—
Melon	11		
Cucumber		9	
Spinach			5
Tobacco			11
Sunflower			9
Buckwheat			8
Alfalfa		11	
Clover (red)		4	
Clover (white)		2	
Peas	3		
Timothy grass		5	
Carrots	1	7	15
Eggplant	5	7	10
Lettuce	3	—	5
Onion	—	1	3
Pepper		1	5
Tomato	7	—	10
Flax		18	
Radish		10	

It must be stressed that seed viability is not only a function of the conditions during seed storage. A variety of factors to which the parent plant is exposed during seed formation and ripening can also profoundly affect subsequent viability of seeds, after dispersal or harvest. Such factors include water supply, temperature, mineral nutrition and light. However, these environmental factors are secondary in importance, compared to the genetic control of seed viability. Storage of seeds under conditions which will preserve their viability is of great importance in agricultural practice. Until recently, this problem was solved empirically, using agrotechnical practices which were worked out for every type of crop seeds. The development of biotechnology, or cryopreservation and of vegetative propagation with the aid of tissue and somatic embryo culture, together with the establishment of germplasm banks, may help to overcome some of the practical and genetic limitations of seed storage and preservation of viability.

II. External Factors Affecting Germination

In order that a seed can germinate, it must be placed in environmental conditions favourable to this process. Among the conditions required are an adequate supply of water, a suitable temperature and composition of the gases in the atmosphere, as well as light for certain seeds. The requirement for these conditions varies according to the species and variety and is determined both by the conditions which prevailed during seed formation and even more so by hereditary factors. Frequently it appears that there is some correlation between the environmental requirement for germination and the ecological conditions occurring in the habitat of the plant and the seeds. In the following these various factors will be considered in detail.

1. Water

The first process which occurs during germination is the uptake of water by the seed. This uptake is due to imbibition. The extent to which imbibition occurs is determined by three factors, the composition of the seed, the permeability of the seed coat or fruit to water and the availability of water, in the liquid or gaseous form, in the environment.

Imbibition is a physical process which is related to the properties of colloids. It is in no way related to the viability of the seeds and occurs equally in live seeds and in seeds which have been killed by heat or by some other means. During imbibition molecules of solvent enter the substance which is swelling, causing solvation of the colloid particles and, in addition, occupying the free capillary spaces and the intermicellar spaces of the colloid. The swelling of the colloid results in the production of considerable pressures, called *imbibition pressure.* The imbibition pressure developed by seeds may reach hundreds of atmospheres and is of great importance in the process of germination as it may lead to the breaking of the seed coat, thus enabling the embryo to emerge, and also makes room in the soil for the developing seedling. The magnitude of the imbibition pressure is also an indication of the water retaining power of the seed and therefore determines the amount of water available for rehydrating the seed tissues during germination. In seeds we are dealing with the imbibition of water by hydrophilic colloids. Colloids are characterized by the size of the particles in the dispersion phase, and imbibition is a property of colloids which are in the form of a gel, i.e. where the colloidal particles constitute a more or less continuous micellar network, showing a certain amount of rigidity.

The gels occurring in nature, and in seeds in particular, are usually polyelectrolytes and contain a large number of ionic groups. It is therefore possible to treat the imbibition of water into such polyelectrolytes according to the theories of a Donnan equilibrium. This treatment has led many authors to the view that imbibition as a whole should be treated as a special case of osmosis and that the driving forces involved are in fact the same as those concerned in osmosis. These studies consider that part of the swelling colloid, due to its immobility, acts as a semi-permeable membrane and the bulk of it as the osmotic system (Glasstone, 1946; Haurowitz, 1950; Katchalski, 1954; Jirgensons, 1958).

In seeds the chief component which imbibes water is the protein. However, other components also swell. The mucilages of various kinds will contribute to swelling, as will part of the cellulose and the pectic substances. Starch on the other hand does not add to the total swelling of the seeds, even when present in large amounts. Starch only swells at very acid pH or after treatment with high temperatures, conditions which do not occur in nature. The swelling of seeds (Table 3.4) therefore reflects the storage materials present in the seeds. Seed swelling depends on the pH of the solution but does not strictly follow the behaviour expected if only ampholytes were swelling. Proteins being "Zwitter-ions" show a minimum of imbibition at their isoelectric point, the imbibition rising with pH on either side of this point. Other colloids show a dependence of imbibition on pH which can be related to their dissociation constant. For agar, for example, maximal dissociation is near neutrality and maximal swelling also occurs around pH 7.0. Imbibition is dependent on temperature and proceeds more rapidly at higher temperatures. However, the effect of temperature on imbibition is probably complex. The viscosity of water decreases with increased temperature and its kinetic energy increases. The kinetic energy is directly proportional to the absolute temperature, while the molecular velocity varies as the square root of the absolute temperature. As mentioned above, imbibition may be considered as a special case of osmosis. Modern views on osmotically-induced flow assume that the bulk of the flow is hydrodynamic mass flow (Pappenheimer, 1953) through the pores of the membrane, and not diffusion. Therefore any effect of temperature on the structure of the colloid and the dimensions of its intermicellar spaces might affect the rate of imbibition. The final volume obtained by seeds imbibing at low temperature is greater than that resulting from the rapid imbibition at higher temperature (Fig. 3.2). However, experimental evidence shows that these differences are very slight.

TABLE 3.4. *Imbibition by various seeds*
Imbibition at 28°C expressed as percentage of the original weight of the seeds (after Levari, 1960)

Time in hours	Lettuce	Wheat	Sunflower	*Vicia sativa* (vetch)	*Zea mays*
1	170	114	124	112	111
2	185	120	137	143	116
4	197	127	147	168	—
6	201	133	153	181	124
10	213	140	154	182	—
16	225	—	—	—	136
24	237	151	—	—	137
32	252	155	—	—	—
40	254	—	—	—	—
48	270	161	—	—	—

Shull (1920) studied the imbibition of *Xanthium* seeds and split peas. He noted that swelling of the seeds was essentially similar to that of colloids. The Q_{10} value of imbibition, in both cases, was between 1.5 and 1.8. Shull concluded that no chemical change was involved in the effect of temperature on imbibition and that it was not markedly affected by the presence of a semi-permeable membrane.

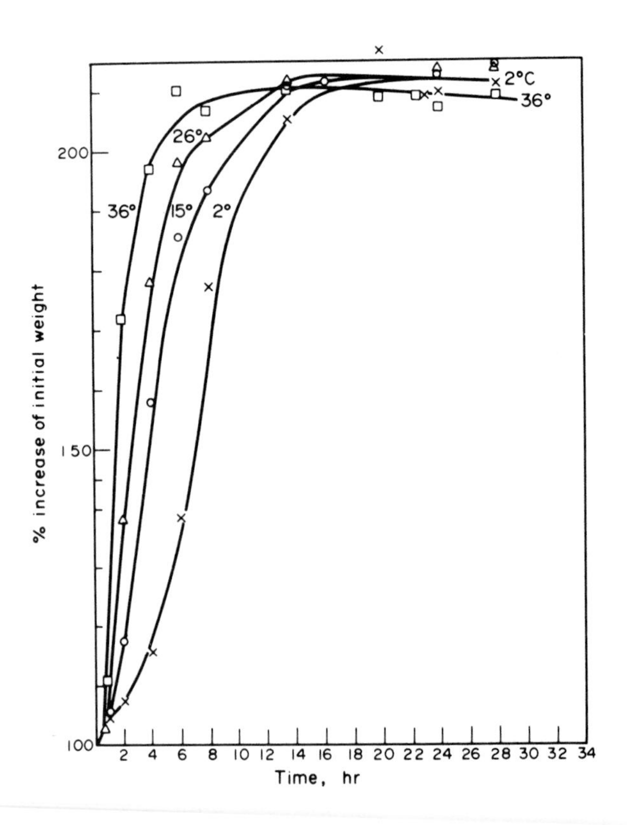

FIG. 3.2. Imbibition of heat-killed peas at different temperatures. ×——× 2°C, △——△ 26°C, □——□ 36°C, ○——○ 15°C.

Rupley and Siemankowski (1986), based on measurements of sorption isotherms and enthalpy calculations of a Lysozyme-water system, describe the hydration of the protein as being in three stages. In the first stage (the lowest hydration), the water added to the dry protein macromolecule dissolves in the protein; this situation persists in the range of 0– 0.07g water.g – 1 protein. The second stage is in the range of up to 0.25 g.g – 1 when the water molecules actually cover the surface of the macromolecule. In the third stage there is separation of phases and the macromolecule is actually dissolved in the bulk of the water. The structure and the nature of the surface of the macromolecule are very important in this interaction. Vertucci and Leopold (1984, 1986) consider the process of imbibition of seeds and the amount and strength of water binding as also having three stages, which are very similar to those described by Rupley and Siemankowski (1986). These stages may differ in different seed species according to the nature of their proteins. At each level of hydration, only certain physiological functions can take place. Thus at the lowest hydration level almost no enzymic activity can be detected. The only enzyme which was active at such a low hydration level was lipoxygenase (Leopold and Vertucci, 1986). The enzyme activity which eventually leads to germination occurs at the third, highest, state of hydration which was found to be >26% in peas, >20% in corn, but only >14% in apple seeds (Vertucci and Leopold, 1986).

The availability of water for imbibition depends to a large extent on the composition of the medium in which germination takes place. This is of significance under natural conditions where the solution in which the seeds are found is not usually pure water. As the concentration of solutes in the solution increases, imbibition decreases, largely due to osmotic effects. In addition, however, a direct effect of the ions on the seed is also frequently observed. Toxic effects may be present, for example under very saline conditions. However, the ability of seeds to absorb water from the soil as compared to water uptake from solution is determined not only by the osmotic potential of the soil solution, but also by the matric potential of the soil. Contact of the seed surface with soil particles is very important in this respect. Although the water potential of seeds is usually very low, compared to that of the soil, seeds can also lose water to the air, which also has a low water potential. The final uptake of water will then be determined by the relative values of these water potentials. Normally only the water potential in the immediate vicinity of the seeds will determine their imbibition.

The entry of water into the seed is not determined only by its availability in the germination medium. The entry of water into seeds is determined in the first instance by the permeability of the seed coat or the fruit coat. Seeds which are surrounded by an impermeable seed coat will not swell even under otherwise favourable conditions. Impermeable seed coats are frequently found in Leguminosae as well as in other groups. The seed coat is usually a multi-layered membrane containing a number of cell layers (see Chapter 1). Frequently it shows selective permeability toward certain substances. The impermeability of the seed coat, or its selective permeability, is frequently the cause of dormancy (see Chapter 4). Water uptake into seeds must be regulated. If the water uptake is too rapid damage to the embryo may result. This appears to be the case for pea seed whose seed coat has been removed (Powell and Matthews, 1978). Various external factors can cause changes in the permeability of the seed coat. For example, heat-killed seeds often imbibe water more rapidly than the corresponding viable seeds, probably because the permeability of the seed coat is increased by the heat treatment. Denny (1917,a,b) showed that the seed coats of various species have different permeabilities to water. He was able to relate these differences to the composition of the seed coat and especially to their lipoid components (Table 3.5). The conditions under which seeds are drying has been shown to affect the permeability of the seed coats (Marbach and Mayer, 1974, 1975). When seeds of hardseeded legumes were dried in the absence of oxygen, permeable coats were obtained (Table 3.6). In the case of seeds of *Lathyrus* and *Robinia* the precise state of harvest of the seed prior to drying was of considerable importance. If harvested too late conditions of drying had no further effect. The effect of drying in the case of *P. elatius* could be related to the presence in the seed coat of a catechol oxidase, which in the presence of air or oxygen led to the oxidation of phenolic compounds, which rendered the seed coats impermeable to water (Marbach and Mayer, 1974, 1975). In *Sida spinosa*, the formation of substances inducing seed coat impermeability involve peroxidase activity (Egley *et al.*, 1983). Such effects may be the basis of part of the parental effects on seed behaviour. Availability of oxygen to the mature seeds might differ in different parts of a given fruit.

TABLE 3.5. *Effect of extracting seed coats with hot water or hot alcohol on the water permeability of the isolated seed coat (compiled from data of Denny, 1917a, b)*

Seed	Solvent	% Increase in permeability	Probable seed coat constituent restricting water permeability
Arachis hypogea	Hot water	170	Tannins
Arachis hypogea	Hot alcohol	80	Lipids (2% of seed coat)
Prunus amygdalus	Hot water	500	Pectic substances
Prunus amygdalus	Hot alcohol	350	Lipids
Citrus grandis	Hot water	0	Fatty substances surrounded by thick pectinized walls
Citrus grandis	Hot alcohol	0	
Cucurbita maxima	Hot water	0	Lipids
Cucurbita maxima	Hot alcohol	700	

TABLE 3.6. *Effect of conditions during seed drying, on seed coat permeability (compiled from data of Marbach, 1976)*

Species	Condition of drying	Air	Oxygen	Nitrogen	Vacuum
Pisum elatius	Permeability	0	0	100	100
	Seed coat colour	Brown	Brown	Green	Green
Cercis siliquastra	Permeability	40	50	82	87
	Seed coat colour	Brown	Brown	Green-Brown	
Robinia pseudoacacia	Permeability	20	16	50	53
Lathyrus blepharicarpus	Permeability	22	21	68	60
	Seed coat colour	Dark Brown	Dark Brown	Green	Green

Permeability = % of seeds imbibing water.

Permeability of the seed coat is generally greatest near the micropylar end of the seed, where it is almost invariably thinner than the rest of the seed coat. However, in some cases the micropyle does not contribute to permeability. In *Quercus* species the pericarp is normally impermeable and unless it is damaged water penetrates through the capscar (Bonner, 1968).

In *Sida spinosa*, the site of water entry is the chalazal area, which becomes permeable after the seed coat, in this area, separates from the underlying layer of thin-walled subpalisade cells (Egley *et al.*, 1986). In sugar maple seeds, *Acer saccharum*, two portals for water entry were demonstrated, one at the chalazal and the other at the micropylar areas (Janerette, 1979).

The precise path of water entry into pea seeds has been examined by Spurny (1973), who related water impermeability to the suberin incrustation of macrosclereids.

The presence of mucilages in the seed coat improves the ability to imbibe water and the artificial addition of mucilages has essentially the same effect. Mucilage reduces the sensitivity of the seed to soil water tension (Harper and Benton, 1966). Various attempts have been made to relate the imbibition of seeds to temperature and seed quality. Blacklow (1972) developed an equation for the imbibition of *Zea*:

$$\frac{dW}{dt} = Kf(t) - W + b$$

where K is a measure of the permeability of the seeds to water during the exponential phase of water uptake, W the water content of the seeds, $f(t)$ the water capacity of the seeds and b a measure of the linear phase of water uptake. With the aid of this mathematical model, Blacklow was able to predict the imbibition of seeds under various conditions.

It has always been difficult to interpret the results of laboratory experiments on water uptake in relation to behaviour of seeds in the field. Under field conditions factors other than water potential might affect water uptake and therefore germination. Hadas (1977a) developed a method for testing and predicting field germination of seeds. This method is based on germinating seeds in solutions of various water potentials. The time needed for the onset of germination and for its completion was determined for each water potential, as well as the time needed to reach a minimal hydration level. Germination behaviour was predicted from curves giving the presumed time needed to reach a minimal hydration value, based on the water potential, assumed to prevail in the seed. Predicted and observed values were always in very good agreement (Fig. 3.3).

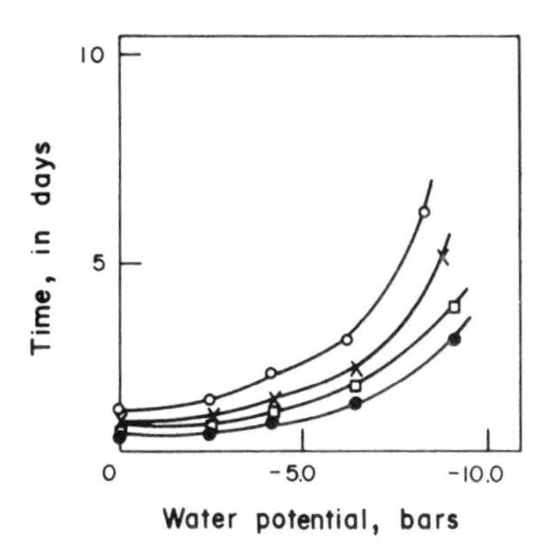

FIG. 3.3. Relation between water potential of germination medium and time to achieve given germination percentage. Seeds of *Cicer arientinum* germinated in aerated solution of polyethylene glycol of various concentrations. ●——● 100%, □——□ 75%, ×——× 50%, ○——○ 25%, germination (after Hadas, 1977c).

Since in the soil not only the water potential but also aggregation of soil particles change, this question was also studied (Hadas, 1977b). If such aggregates were small compared to seed size, contact between seed and soil was not an important factor in water uptake by the seeds, provided water diffusivity in the soil is high. However, when the soil is dense or friable, other factors may affect germination behaviour. When soil aggregates are large compared to seed size and the water potential low, the contact between the seed and the soil becomes an important factor. Hadas suggests that the pressure balance for a germinating seed is given by the following equation:

$$\text{Soil water potential components} + \text{seed's osmotic potential} \\ + \text{seed cell wall pressure} + \text{soil's normal stress} = 0$$

The differential uptake of water by different parts of the seed may be greatly accentuated by the degree of contact between seeds and soil particles (Hadas, 1982). The degree of hydration of seeds may function to regulate germination in various ways, e.g. gauge of soil depth or as a sensor of rainfall or for integrating total soil moisture (Koller and Hadas, 1982).

It is generally true that seed germination is more sensitive to water stress than subsequent growth (Hegarthy and Ross, 1980/81). However, this is not so in all cases. For example the stress imposed by salinity on pea plants depressed epicotyl emergence much more than root protrusion, which is usually taken as a measure of germination (Hasson and Poljakoff-Mayber, 1980/81).

Water taken up by the seed is not necessarily evenly distributed within the seed, so that different parts of the seed, such as cotyledons and embryonic axis for instance, may be hydrated at different rates (Mayer, 1986). This has been demonstrated directly using high resolution imaging by NMR which permits the direct observation of water in the intact imbibing seed (Connelly *et al.*, 1987). Possibly even different organelles of a given cell may hydrate at different rates. Since during seed dehydration, membranes are distorted or folded, hydration must be an important part in the restoration of normal compartmentation of the cells of the seed.

The difference in rate of hydration of the various proteins in the different parts of seed may also lead to differences in the sequence of the processes triggered by the various levels of hydration (Vertucci and Leopold, 1986). Some of these processes may be growth substances production and/or transport which may act first as sensors and then as regulators and coordinators of the processes leading to germination (Mayer, 1986). It is well established by now that there are critical minimal hydration levels which enable the seed to sense changes in external conditions, such as temperature or light (Vertucci and Leopold, 1986; Vertucci *et al.*, 1987).

2. Gases

Germination is a process related to living cells and requires an expenditure of energy. The energy requirement of living cells is usually met by the utilization of ATP, which is formed due to aerobic oxidation processes, although some ATP is formed as a result of glycolysis (Hourmant and Pradet, 1981; Raymond *et al.*, 1983). Consequently seed germination is greatly affected by the composition of the ambient atmosphere. Most seeds germinate in air, i.e. in an atmosphere containing 20% oxygen and a low percentage, 0.03% of carbon dioxide. However, many authors have shown that certain seeds respond to an increase in the oxygen tension above 20% by increased germination, e.g. *Xanthium* and certain cereals. On the other hand, a number of seeds show increased germination as the oxygen content of the air is decreased below 20%. Such seeds, for example, are *Typha latifolia* and *Cynodon dactylon* which germinate better in the presence of about 8% oxygen (Morinaga, 1926a,b).

Rice and *Echinocloa crus-galli* are among the few species which can germinate and grow under anaerobic conditions (Kennedy *et al.*, 1980). However, although rice germinates in flooded paddies, the anaerobic conditions there may lead to formation of abnormal seedlings. It has been shown that such abnormalities may be prevented by presence of oxygen (Tang *et al.*, 1959; Chu and Tang, 1959).

Cultivated species generally require oxygen for germination. In some species, such as lettuce, sunflower and radish, germination stopped at oxygen pressure below 2 KPa (Kilopascal), and at this point their energy charge, which may be considered as a measure for ATP supply, dropped below 0.6. In maize and rice, germination was slowed down but still occurred at 0.1 KPa, and at this value their energy charge was still above 0.6 (Al-Ani *et al.*, 1985). In rice the energy charge changes very rapidly, within one minute, when they are transferred from anaerobic conditions to an atmosphere containing oxygen.

Thornton (1943) showed that the germination of many cultivated seeds, if maintained at 20% oxygen, was unaffected by increased concentration of carbon dioxide. Seeds of *Daucus carota* and *Rumex crispus* respond to increased oxygen concentrations in the dark by increased germination; no germination at 20% 0_2, 3% at 40% 0_2, and 24% at 80% 0_2 (Gardner, 1921).

Many seeds will show lower germination if the oxygen tension is decreased appreciably below that normally present in the atmosphere. Some species germinate as well in 2% 0_2 as in air, e.g. *Celosia, Portulaca* and cucumber (Siegel and Rosen, 1962). On the other hand germination of lettuce seeds was only 58% in a mixture of 5% 0_2/95% N_2 but increased to 79% and 96% when the oxygen content of the mixture was raised to 15% and 20% respectively, N_2 being 85% and 80% respectively. Seedling growth was, however, depressed even at 15% 0_2 (Harel and Mayer, 1963). It is important to note that quite different results are obtained if the dilution of the air is done with nitrogen or with hydrogen. For example, Morinaga found that while white clover, *Trifolium repens*, seeds germinate 52% in air, they germinate only 47% in air diluted with 60% nitrogen, but 70% in the presence of air equally diluted with hydrogen. In none of these cases is anything known about the oxygen tension within the seeds. At least in one case — *Trifolium subterraneum* — pre-incubation of seeds in the absence of oxygen led to higher subsequent germination in air (Ballard and Grant-Lipp, 1969). This case shows not a reduced requirement for 0_2, but rather a case of dormancy breaking. A similar requirement for anaerobiosis in the dark part of the inductive cycle has been demonstrated by *Eragrostis ferruginea* (Fujii, 1963).

Germination may be inhibited by high 0_2 concentration. Dormant seeds of *Oldenlandia corymbosa* germinate at 40°C in the light only between 5 – 10% 0_2. Above and below these concentrations germination is strongly inhibited. Non-dormant seeds show no such sensitivity to 0_2 concentration (Corbineau and Come, 1980/81). Here we are probably dealing with a case of dormancy induced by high 0_2 concentration. Dormancy is often ascribed to the impermeability of the seed coat to certain gases. Such differential permeability of seed coats was demonstrated by

Brown (1940) and is discussed in Chapter 4. The determining factor which may be involved in the effect of oxygen on germination will be considered in Chapter 4 on dormancy, as well as in Chapter 7.

The effect of carbon dioxide is usually the reverse of that of oxygen. Most seeds fail to germinate if the carbon dioxide tension is greatly increased, as shown, for example, by Kidd (1914), for *Hordeum vulgare* and *Brassica alba*. However, in some cases at least, there appears to be a minimal requirement for carbon dioxide in order that germination can occur. This seems to be so for *Atriplex halimus* and *Salsola*, as well as for lettuce. Other *Atriplex* species are resistant to high carbon dioxide concentrations, as shown by Beadle (1952) and as found for cultivated seeds provided that the 0_2 concentration is kept constant. Very high carbon dioxide concentrations which prevent germination seem to have a favourable effect on the keeping of seeds. Kidd (1914) showed that the life span of *Hevea brasiliensis* seeds was prolonged by sealing the seeds in an atmosphere containing 40 – 45% CO_2. However, an atmosphere of nitrogen was even more effective. For lettuce and onion seeds storage under carbon dioxide decreased the number of chromosomal aberrations which occur in the mitosis during germination subsequent to storage (Harrison and McLeish, 1954). Cases are known where increases in CO_2 concentration increase germination, e.g. *Phleum pratense* (Maier, 1933). Dormancy breaking by CO_2 has been demonstrated for *Trifolium subterraneum* seeds. Generally 2.5% was optimal and inhibition occurred only when the CO_2 concentration reached 10 – 15%. The CO_2 effect was, as might be expected, temperature dependent (Ballard, 1967). For carbon dioxide, as for oxygen, the determining factor probably is the internal concentration. This is determined by the permeability of the seed coat to carbon dioxide accumulating within the seed (see also dormancy). Water content of the seeds is probably an important factor in the response to all gases.

Another gas which is produced by seeds during germination is ethylene. Small amounts of ethylene have long been known to promote the germination of some seeds. This effect has been ascribed to dormancy breaking or to effects on the rate of growth immediately after the onset of germination (Esashi and Leopold, 1968). Ethylene production begins very soon after the onset of imbibition (Abeles and Lonski, 1969). It has been shown that when ethylene formation by the seeds, from imbibition onwards, is inhibited by 2,4-norbornadiene, germination is also inhibited. This inhibition may be completely reversed by addition of ethylene. Thus it does appear that endogenous ethylene has some definite function, at least in the germination of some seeds (Kepeczynski and Karssen, 1985), for example *Amaranthus caudatus*. Ethylene interacts with growth hormones and its effect is regarded as that of a growth regulator. In lettuce it promotes germination in seeds treated with either red or far-red light, but not in the dark.

It might be supposed that because of the sensitivity of seeds to gases, pressure would have a marked effect on germination. However, pressures up to 200 atmospheres seem to have little effect on germination of seeds (Vidaver, 1972).

3. Temperature

Different seeds have different temperature ranges within which they germinate. At very low temperatures and very high temperatures the germination of all seeds is prevented. The precise sensitivity is very different according to the species. A rise in temperature does not necessarily cause an increase in either the rate of germination or in its percentage. Germination as a whole is therefore not characterized by a simple temperature coefficient. This can be understood if it is appreciated that germination is a complex process and a change in temperature will affect each constituent step individually, so that the effect of temperature which is observed will merely reflect the overall resultant effect.

In studying the effect of temperature on germination one must distinguish between the resistance of dry seeds to various temperatures and the effect of temperature on the actual germination. Many dry seeds are fairly resistant to extreme temperatures. It has been shown that the viability of seeds is not affected by placing them at the temperature of liquid air. This property is utilized for the long-term preservation of seeds in "germ banks". Even in the hydrated state seeds can survive temperatures from −12° to −18°C if they undergo supercooling, i.e. lowering of the temperature while ice crystal formation is prevented. This avoidance of injury in lettuce by supercooling depends on the integrity of the endosperm (Juntilla and Stushnoff, 1977). High temperatures up to about 90°C for prolonged periods of time are tolerated by many seeds such as radish, turnips and poppy, as shown by David (1936) who studied oil-containing seeds. In other cases the temperature tolerance is lower. Above 90°C the viability of the seeds is greatly lowered. There probably is some correlation between the storage materials in the seed and their heat resistance. Although seeds may still be viable after treatment at high temperature, the subsequent development of the seedling is often adversely affected (Levitt, 1956). In the imbibed state seeds seem to be more sensitive to temperature. *Solanum nigrum* progressively lost its ability to germinate after 48 hr at 50°C or within 6 hr at 55°C (Givelberg *et al.*, 1984).

In the range of temperature within which a certain seed germinates there is usually an optimal temperature, below and above which germination is delayed but not prevented. The optimal temperature may be taken to be that at which the highest percentage of germination is attained in the shortest time. The minimal and maximal temperatures for germination are the highest and lowest temperatures at which germination will still occur. The maximal temperature at which germination occurs may be as high as 48°C, for example, for *Cucumis sativa* (Knapp, 1967).

The minimal temperature is frequently ill-defined because visible germination is so slow that the experiments are often terminated before germination could in fact have occurred. In many cases it is stated that the optimal temperature for germination shifts with length of the germination period. In these cases the term "optimal" is used ambiguously and in fact refers to the optimal temperature when germination is observed after some definite, arbitrary, time interval. If a different time interval is chosen, then a different temperature may be optimal according to this usage of the term (Fig. 3.4).

The reason why seeds fail to germinate at low temperatures is still obscure. Various explanations have been suggested, such as changes in membranes or denaturation of protein. There has been no real attempt to explain the way in which seeds sense the ambient temperature. The possibility that membranes are themselves the temperature sensors of seeds has been suggested by Mayer and Marbach (1981) and Mayer (1986). This idea is supported by the findings that the properties of seed

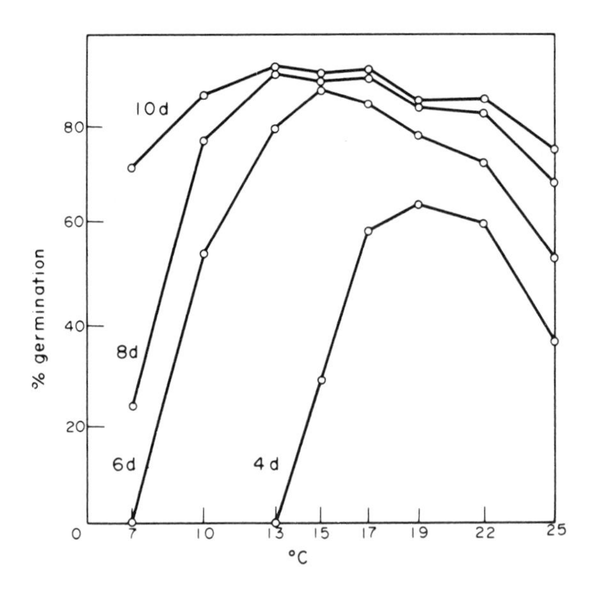

FIG. 3.4. Effect of temperature on germination of oats (after Edwards, 1932 and Attenberg, 1928). The germination percentage as determined at the different temperatures after various periods of time: 4, 6, 8, and 10 days.

membranes changed discontinuously with changes in temperature, and that leakiness of seed membranes at high temperature has been demonstrated (Hendricks and Taylorson, 1979). Studies on the ultrastructure and phospholipid metabolism of pea seeds briefly exposed to 5°C during imbibition indicated that the phospholipid turnover and synthesis showed a kind of "memory" of the cold which persisted after the treatment had stopped (Hodson *et al.*, 1987). A "memory" effect could also be discerned in changes in structure, particularly of the plasma membrane and in the mitochondria.

The temperatures at which germination can still occur can be modified. Thus in lettuce, abscisic acid, although slowing down the rate of germination, reduced the temperature at which the seeds failed to germinate by some 3 – 4°C while kinetin raised the temperatures by as much as 10°C (Reynolds and Thompson, 1971). Thus even the range of temperatures at which germination can occur is not absolute and can be modified by exogenous compounds. It is also reasonable to assume that the endogenous growth substance levels also play a role in determining the temperature range of germination.

The temperature range within which different seeds germinate is determined by the source of the seeds, by genetic differences within a given species, e.g. varietal differences, as well as by the age of the seeds. A few examples of ranges of temperature in which germination occurs are given in Table 3.7, but in view of the preceding discussion such data must be treated with great reserve.

In contrast to those seeds which germinate readily if held at one specific temperature, instances are known where a periodic alternation of temperature is required

TABLE 3.7. *Temperature ranges in which germination occurs for different seeds (compiled from various sources)*

Seeds	Temperature		
	Minimum	Optimum	Maximum
Zea mays	8 – 10	32 – 35	40 – 44
Oryza sativa	10 – 12	30 – 37	40 – 42
Triticum sativum	3 – 5	15 – 31	30 – 43
Hordeum sativum	3 – 5	19 – 27	30 – 40
Secale cereale	3 – 5	25 – 31	30 – 40
Avena sativa	3 – 5	25 – 31	30 – 40
Fagopyrum esculentum	3 – 5	25 – 31	35 – 45
Cucumis melo	16 – 19	30 40	45 – 50
Convolvulus arvensis	0·5 – 3	20 – 35	35 – 40
Lepidium draba	0·5 – 3	20 – 35	35 – 40
Solanum carolinense	20	20 – 35	35 – 40
Nicotiana tabacum (Florida cigar wrapper)	10	24	30
Delphinium (annual)	—	15	20 – 25

for germination to occur, as in *Oenothera biennis, Rumex crispus, Cynodon dactylon, Nicotiana tabacum, Holcus lanatus, Agrostis alba, Poa trivialis* and many others. The most usual cases of such alternations are diurnal ones, between a low and a high temperature. For example, *Agrostis alba* seeds germinated to 69% if alternated between 12° and 21°C, and 95% for an alternation between 21° and 28°C or 21° and 35°C. At constant temperature the germination was 49% at 12°C, 53% at 21°C, 72% at 28°C and 79% at 35°C (Lehman and Aichele, 1931).

Fluctuating temperatures can now be investigated using thermogradient bars on which the seeds may be exposed to different temperatures and on which not only the actual temperature can be varied with great precision, but also the length of exposure to each of the temperatures, between which the seeds are alternating. From experiments using this procedure, with seeds of *Dactylis glomerata*, it was shown that the proportion of light germination in the population increased with increasing amplitude of the temperature alternations. Also the length of exposure to each temperature and the actual amplitude of the cycle all had significant effects on germination (Probert *et al.*, 1986). Some of these interactions of fluctuating temperatures with light may be relevant to the behaviour of seeds in their natural environment. This has been studied using different populations of the same species and different species of the same genus (Okagami and Kawai, 1982; Probert *et al.*, 1986).

The mechanisms by which alternating temperatures act is still poorly understood. Stimulation by alternating temperatures has been variously ascribed to an effect of temperature on sequential reactions during germination or to mechanical changes occurring in the seed. It is possible to assume that alteration of temperature causes a change in the structure of some macromolecules in the seed, which in its original form prevents germination in some way. *Lycopus europaeus* seeds have an absolute requirement for fluctuating temperature, but there was no evidence for a critical temperature requirement. Successive temperature cycles were cumulative in their effect on germination (Thompson, 1969). This would be consistent with the view that structural changes are involved.

Interdependence between temperature and light are known for celery, *Amaranthus* and other seeds, where light promotes germination at unfavourable high temperatures, but not at low ones. Often sensitivity to light is enhanced by prolonged imbibition at high temperatures, for example, in *Amaranthus* (Chanoeuf-Hannel and Taylorson, 1985).

Physalis franchetti seeds germinate well in the dark between 5° and 15°C. Between 15°C and 35°C they require light for their germination and the light requirement increases with the temperature (Baar, 1912). A very similar situation exists for lettuce seeds.

The interaction of temperature and light is also demonstrated by work on *Rumex crispus* (Taylorson and Hendricks, 1972). These seeds will germinate when exposed to a temperature shift from 20°C to 35°C for short periods. High temperature, 30°C, reduces germination. Exposure to 30°C during imbibition reduces the subsequent response to temperature shifts. This effect of high temperature is reversed by exposure to low energies of red light (R). It appears that germination in the dark, caused by the temperature shift, is due to an effect of the temperature shift on pre-existing phytochrome in the seeds in the P_{FR} form. Light and temperature interactions have been reported in many other seeds.

Temperature effects are also known in relation to after-ripening of seeds, as well as to secondary dormancy. This point will be considered in Chapter 4.

4. Light

Among cultivated plants there is very little evidence for light as a factor influencing germination. The seeds of most cultivated plants usually germinate equally well in the dark and in the light. In contrast, among wild plants much variability in the behaviour toward light is observed. Seeds may be divided into those which germinate only in the dark, those which germinate only in continuous light and those which are indifferent to the presence or absence of light during germination. Daily illuminations have also been shown to affect germination, the effects being similar to those of photoperiodism in flowering.

The importance of light as a factor in the germination of seeds has long been recognized. Already at the end of the 19th and the beginning of the 20th century, a number of papers by Cieslar (1883), Gassner (1915) and by Lehman (1913) analysed some of these phenomena. Light sensitivity of seeds may be related to their germination in their natural habitat. Under natural conditions seeds may be shed so as to fall on the soil, or enter the soil, or be covered by leaf litter, thus exposed to different conditions of light during germination. Among the species which have been investigated for their light response during germination at least half showed a light requirement. For example, Kinzel (1926) lists hundreds of plant species which he divided into several categories. His first group include those germinating at or above 28°C in the light (about 270 species) and those germinating at the same temperature in the dark (114 species). Two further categories germinate in the light (190 species) or in the dark (81 species) after severe frost, and others showed similar behaviour after mild frost (52 species in the light and 32 species in the dark). A group of seeds which are indifferent to light or dark includes 33 species. A selection from this data on light sensitivity is given in Table 3.8.

TABLE 3.8. *Response of seeds of different species to light (from data of Kinzel, 1926)*

A—seeds whose germination is favoured by light
B—seeds whose germination is favoured by dark
C—seeds indifferent to light or dark

A	B	C
Adonis vernalis	*Ailanthus glandulosa*	*Anemone nemorosa*
Alisma plantago	*Aloë variegata*	
Bellis perennis	*Cistus radiatus*	*Bryonia alba*
Capparis spinosa	*Delphinium elatum*	*Cytisus nigricans*
Colchicum autumnale	*Ephedra helvetica*	
Erodium cicutarium	*Evonymus japonica*	*Datura stramonium*
Fagus silvatica	*Forsythia suspensa*	*Hyacinthus candicans*
Genista tinctoria	*Gladiolus communis*	
Helianthemum chamaecistus	*Hedera helix*	*Juncus tenagea*
Iris pseudacorus	*Linnaea borealis*	*Linaria cymbalaria*
Juncus tenuis	*Mirabilis jalapa*	
Lactuca scariola	*Nigella damascena*	*Origanum majorana*
Magnolia grandiflora	*Phacelia tanacetifolia*	*Pelargonium zonale*
Nasturtium officinale	*Ranunculus crenatus*	*Sorghum halepense*
Oenothera biennis	*Silene conica*	*Theobroma cacao*
Panicum capillare	*Tamus communis*	*Tragopogon pratensis*
Resedea lutea	*Tulipa gesneriana*	*Vesicaria viscosa*
Salvia pratense	*Yucca aloipholia*	
Suaeda maritima		
Tamarix germanica		
Taraxacum officinale		
Veronica arvensis		

Such a classification of light requirement is probably an oversimplification, as light requirement varies during storage. In some species a light requirement only exists immediately after harvesting (e.g. in *Salvia pratensis, Saxifraga caespitosa* and *Epilobium angustifolia*), while in other species this effect persists at least for a year (e.g. *Epilobium parviflorum, Salvia verticillata* and *Apium graveolens*), while in yet other species it only develops during storage.

In the studies of the effect of light on germination, there is an inconsistency in use of terminology. The energy per Avogadro's number of photons (i.e. 6.022×10^{23} photons) used to be called the "Einstein", but it is now referred to as "mole". Lately in analysing the effect of light on germination, it is customary to use the terms "fluence" and "fluence rate". The fluence rate is the number of photons reaching a unit area per unit time and it is expressed as mol. $m^{-2} s^{-1}$. This term is equivalent to the "photon flux density". Fluence = fluence rate multiplied by the time and is expressed as Jm^2 and is actually the amount of energy for a given area (10^{-7} J = 1 erg). The amount of energy obviously depends on the wavelength of light. In the following discussion, we shall not make an effort to unify the terminology. We will report the units of light as they were reported in the original papers.

Light requirement is extensively studied under laboratory conditions. As is to be expected, different spectral zones affect germination quite differently. The early work on the effect of light distinguished only between fairly wide spectral bands. These showed that light below 290 nm inhibited germination in all seeds tested. Between 290 nm and 400 nm no clear-cut effects on germination were detected. In the visible range, 400 – 700 nm, it was shown that light in the range 560 – 700 nm and especially red light, usually promoted germination, while blue light was said to inhibit. Flint and McAllister (1935, 1937) determined these spectral ranges more accurately for lettuce seeds and showed that the most effective light in promoting germination is that having a wavelength of 670 nm and that a germination inhibiting zone of the spectrum has its maximal activity at 760 nm. This latter was far more effective than the inhibiting zone in the blue region of the spectrum.

Kincaid (1935) studied the effect of light on the germination of tobacco seeds. He found that as little as 0.01 sec of sunlight was effective in stimulating germination and even moonlight could stimulate. Although the action spectrum was not studied in detail, a filter transmitting between 435-580 nm with a maximum at 544 nm was very effective. The seed coat had an absorption maximum at 510 nm. Resuehr (1939) studied the germination of *Amaranthus caudatus.* The action spectrum showed three inhibitory zones, at about 450 nm, between 475 and 490 nm and between 700 and 750 nm. Two promoting zones were observed, at 640 nm where strong stimulation was obtained and at 675-680 nm, where slight stimulation was noted. For the usually light-inhibited seeds of *Phacelia*, Resuehr observed light stimulation of germination at 640 nm, and five inhibitory zones, at 350 nm, 450 nm, 475 – 490 nm, and above 1000nm. Orange light has been reported to stimulate germination of *Physalis franchetti* (Baar, 1912). It was more effective than blue-violet light. Today all these effects can probably be related to phytochrome action which will be discussed later.

Whether or not blue light does in fact inhibit germination or stimulate it was disputed for many years, because it was contended that the light used by Flint and McAllister contained far-red, or infra-red, light as well as blue light. However, later both Wareing and Black (1957, 1958) and Evenari *et al.* (1957) showed that blue light can in fact inhibit germination. The latter showed that under certain conditions blue light may also stimulate germination. Whether germination was stimulated or inhibited depended entirely on the exact period of illumination as related to the beginning of imbibition.

The sensitivity of seeds to light increases with time of imbibition. Maximum sensitivity is reached as imbibition proceeds and does not exactly coincide with the completion of imbibition (cf. Table 3.4 and Fig. 3.5). Even storage of seeds at high relative humidities is sometimes sufficient to make them light-sensitive. If seeds are given a light stimulus while imbibed and then dried, the stimulatory effect is retained (Gassner, 1915; Kincaid, 1935). The precise time at which the seeds reach maximum sensitivity for any given species has been a matter of dispute. There have also been differences of opinion as to whether the sensitivity reaches a maximal value and then drops again or whether it remains at a high level. It appears that these differences can be related to the light intensity or amount of radiation at which sensitivity was tested on the one hand, and to the time interval which was allowed to elapse between the illumination and the determination of germination percentage on the other hand. Figure 3.5 shows some of these effects for lettuce seeds. After

10 hr or more of imbibition 30 sec of illumination were no longer enough to induce maximal germination.

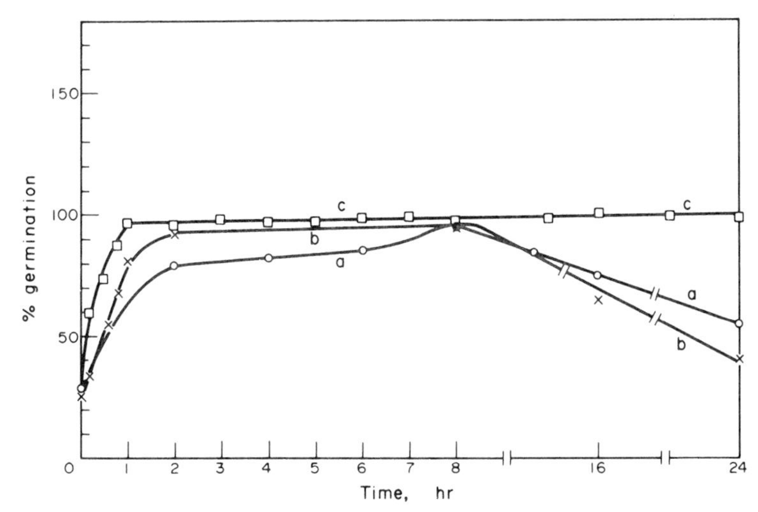

FIG. 3.5. Effect of length of imbibition period, prior to illumination, on light sensitivity of lettuce seeds, Grand Rapids, to red light at 26°C. (a) ○——○ 30 sec light, total incubation time 48 hr (reconstructed from Evenari and Neuman, 1953). (b) ×——× 30 sec light, total incubation time 72 hr. (c) □——□ 2 min light, total incubation time 72 hr. (b), (c) after Poljakoff-Mayber and Lang (unpublished).

In tobacco seeds Kincaid (1935) found that high light intensities were effective after short periods of imbibition (several hours), while low light intensities were most effective after 4 days. After 10 days of imbibition, the seeds no longer responded to illumination. In *Sinapis arvensis* sensitivity to red light had a quite sharp peak after about 4 – 8 hr imbibition at 25°C, but the peak broadened to 4 – 30 hr if the seeds were imbibed at 15°C. Clearly temperature can also play a role in the light sensitivity. In the same seeds the sensitivity to far-red light also changed during imbibition, even stimulating germination between 6 – 30 hr (Frankland, 1976). Not only is the light response changed by length of imbibition temperature and light sensitivity. It is also extremely variable between different batches of the seeds obtained from distinct parent plants. Again in the case of *Sinapis arvensis*, dark germination of between 0 – 7% in various batches resulted in germination of between 1 and 70% following the same standard red irradiation (Frankland, 1976). An unusual response to light has been found in *Bromus sterilis*, in which germination is inhibited by red light, and this inhibition is reversed by far-red (Hilton, 1984).

The effect of the mother plant has been investigated in several cases. Koller (1963) indicated that the germination of lettuce seed could be preconditioned by the light regime under which the parent plants were grown. This prompted Shropshire (1973) to study such effects more precisely. *Arabidopsis thaliana* seeds had varying dark germination depending on whether the floral primodia had been irradiated by fluorescent or incandescent light. Those illuminated with fluorescent light had about 100% dark germination when seed age was about 130 days compared to 0% after incandescent light. All seeds were viable and could be brought to germination by suitable treatment. Such effects were discovered to be common. In seeds of *Portulaca oleracea* the photoperiod to which the parent plant was exposed determined their subsequent germination in the light (Table 3.9). Light filtered through the fruit on the mother plant may affect subsequent germination of the seeds.

TABLE 3.9. *Germination of Portulaca seeds depending on treatment of parent plant (compiled from Gutterman, 1974)*

Treatment of parent plant	Germination in the light
8 hrs L, 2 hrs RL, 14 hr D	56
8 hrs L, 2 hrs FRL, 14 hr D	14
8 hrs L, 16 hrs D	30

L = white light; R = red; FR = far-red; D = dark.

The effect of light is being studied in an increasing number of plant species. Some of the earlier detailed work was concerned with lettuce (*Lactuca sativa*), *Lepidium virginicum, Nicotiana tabacum* and various *Amaranthus* species. For lettuce and *Lepidium* the precise spectral peaks for germination, stimulation and inhibition, by short illumination, have been redetermined and earlier results essentially confirmed, showing stimulation at 670 nm and inhibition at 730 nm. An important new finding was the reversibility of both germination stimulation and germination inhibition by alternating illuminations as shown by Borthwick *et al.* (1952). Germination of lettuce seeds is stimulated by red light and can be inhibited if they are subsequently illuminated with infra-red (or far-red) light. Further illumination with red light will again induce germination and always the nature of the last illumination determines the germination response. These effects are in fact very similar if not identical to those also known for flowering, etiolation and unfolding of the plumular hook of bean seedlings, as well as pigment formation in certain fruit and leaves.

All these phenomena show a very similar action spectrum. The action spectrum is that associated with the plant pigment phytochrome (P), whose existence was deduced from the biological response of the various tissues to illumination. The action spectra for the germination of lettuce and *Lepidium* seeds are shown in Fig. 3.6. Butler *et al.* (1959) were able to show that there are changes in the absorption spectrum of an intact tissue, *Zea mays* coleoptiles, as a result of illumination with red or far-red light. This is illustrated in Fig. 3.7.

As in all cases where light-effects are observed, the effect noted is dependent on both the intensity of the light, i.e. its energy, and the duration of the illumination. Provided that all the light is absorbed, then within certain ranges, the response of germination will be a function of the product of light intensity and duration of illumination, i.e. a function of the total energy of irradiation. This applies both to germination stimulation and inhibition, as well as to all the other responses to red and far-red irradiation. In order to show the reversibility of red and far-red light effects, suitable intensities and durations must be selected. The range in which these observations have been made is for total irradiations of 2×10^4 ergs/cm^2 for

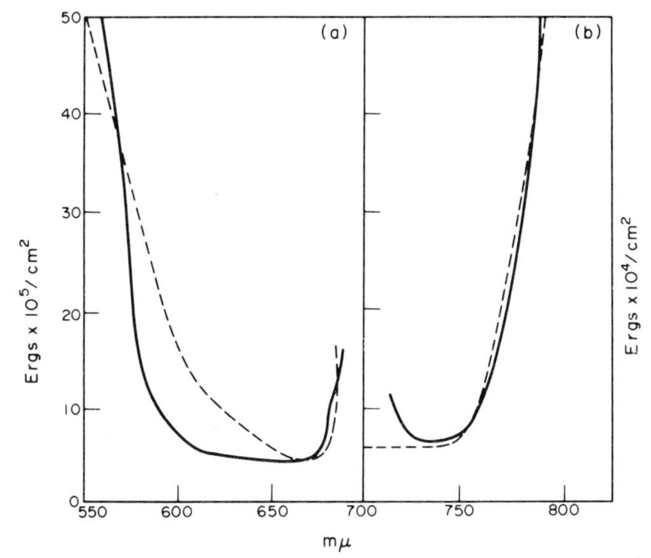

FIG. 3.6. Action spectrum for germination promotion (a) and inhibition (b) of *Lactuca sativa* and *Lepidium virginicum* seeds (after Toole *et al.*, 1956). The curve shows the amount of radiant energy at different wave lengths required to promote or inhibit germination to half its maximum value, 50%.
– – – – – Lettuce, in (a) scale × 0.04, in (b) scale × 10
——— *Lepidium.*

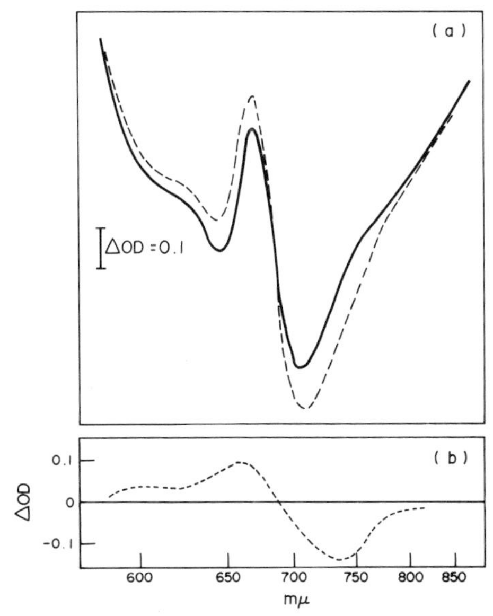

FIG. 3.7. Absorption spectrum of maize coleoptiles following red or far-red irradiation (a) as well as the difference spectrum (b) far-red irradiated spectrum minus red irradiated spectrum, (Butler *et al.*, 1959).
——— Red irradiated. – – – – – Far-red irradiated.

promotion in lettuce, 1.4×10^6 ergs/cm^2 for promotion in *Lepidium*, and 6×10^5 and 3×10^4 ergs/cm^2 for inhibition respectively in these species (see also Fig. 3.6). The period of irradiation is between a few seconds and half an hour or more.

When batches of seeds of *Sinapis arvensis* with different light requirements were germinated at different light intensities, it could be shown that the more dormant the seeds the higher the light intensity required to promote germination. Mathematical treatment by the probit transformation showed a linear relation between probit of germination and the log of the light intensity. By making various assumptions, an estimate of the photo transformation of phytochrome by the red light could be made for each seed batch. It was then found that germination percentage changed linearly with the amount of phytochrome transformed (Frankland, 1976).

In many seeds the fluence response curve of germination to light is simple. The curve is sigmoid in nature indicating that it is possible to saturate the light receptor. However, in a number of seeds such as lettuce, *Rumex* and *Arabidopsis* the fluence response curve is more complex, being biphasic. In these biphasic curves it is possible to detect one response to very low fluence (VLFR), $10^{-4} - 10^{-1}\mu$mole m^{-2} and another at higher fluence, $1 - 10^{3}\mu$mole m^{-2} (LFR) (Blaauw-Jansen and Blaauw, 1975; Kendrick and Cone, 1985). These biphasic response curves are very difficult to explain at present. It has been suggested that P_{FR} may regulate two steps in the germination process, one of which is preparatory, making ready the seed for the P_{FR} induced reaction. This preparatory process can also occur in the dark and is the very low fluence response (Blaauw-Jansen and Sewmar, 1986).

The fluence – response curves of phytochrome – mediated germination of *Kalanchoe* seeds can be modified by GA_3. GA_3 was able to induce a VLF response which indicated a 20,000 fold increase in sensitivity to P_{FR} (Rethy *et al.*, 1987). These authors indicate that this response is difficult to reconcile with the model of Blaauw-Jansen and Sewmar (1986). A second model to explain the biphasic fluence response curve has been proposed by VanderWoude (1985). According to this model phytochrome can exist as one of three dimers $P_R - P_{FR}$,$P_R - P_R$, and $P_{FR} - P_{FR}$; the dimer $P_R - P_{FR}$ leads to the VLFR, while the dimer $P_{FR} - P_{FR}$ leads to the LFR. This suggestion requires experimental study, but there is as yet no direct evidence for it.

The presence of phytochrome in seeds is now adequately documented and the properties of phytochrome, at the molecular level, are now well known (Pratt, 1982; Quail *et al.*, 1983). It is recognized that the two forms of phytochrome P_R and P_{FR} differ in their molecular configuration and it is generally assumed, although not proved unequivocally, that phytochrome at least in its P_{FR} form is associated with cellular membranes (Song, 1983; Kendrick, 1983). The transition of the phytochrome from one form to another has also been proved. An example for this is shown in Fig. 3.8. From this it can be seen that the $\delta(\delta$O.D.) at 670 nm of the intact seeds increases with length of imbibition, and a smaller but similar effect is observed at 730 nm. The detailed interpretation of the effect of light on germination is immensely complicated. In order to explain the effect of light of a given intensity and duration we must take into account the absorption spectrum of phytochrome (Fig. 3.9), which shows clearly the overlap in absorption between the R and FR forms of phytochrome as well as its absorbance in the blue. In addition we must consider the transformation which phytochrome can undergo in seeds (Fig. 3.10). These transformations are in essence the following:

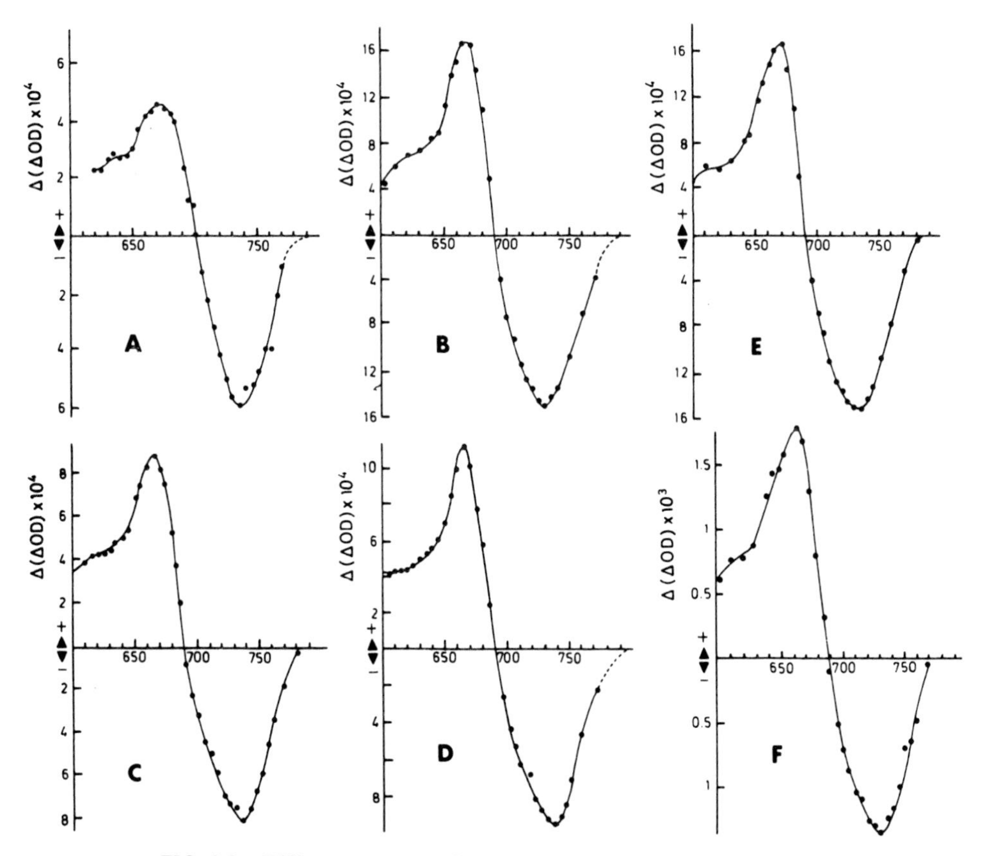

FIG. 3.8. Difference spectrum for phototransformation of phytochrome in gherkin seeds. (A) dry seeds. (B) after 24 hr imbibition. (C) after 3 hr imbibition. (D) after 8 hr imbibition. (E) after 16 hr imbibition. Sample thickness: 4 mm. Temperature of measurement: 0°C. Temperature of imbibition: 22°C. (F) Difference spectrum for phototransformation of phytochrome on a single pumpkin embryo after 24 hr imbibition at 22°C. Temperature of measurement 22°C (after Malcoste *et al.*, 1970).

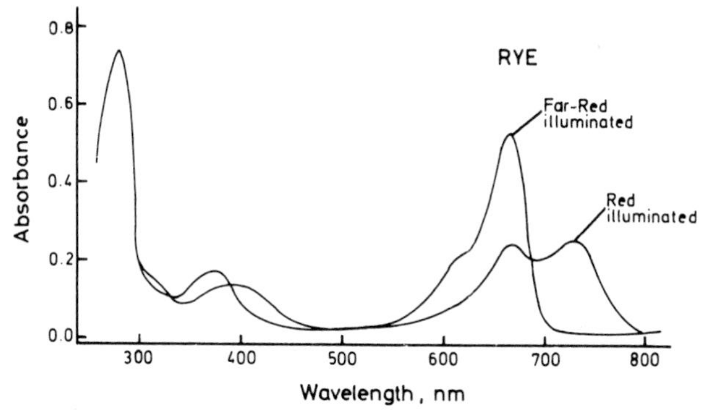

FIG. 3.9. Absorption spectrum of purified rye phytochrome, at 4°C, after 4 min R and FR irradiation. Specific activity was 0.720 and the A_{280}/A_{665} ratio was 1.27. The phytochrome as dissolved in sodium phosphate buffer, pH 7.8, containing 5% (v/v) glycerol (after Rice *et al.*, 1973).

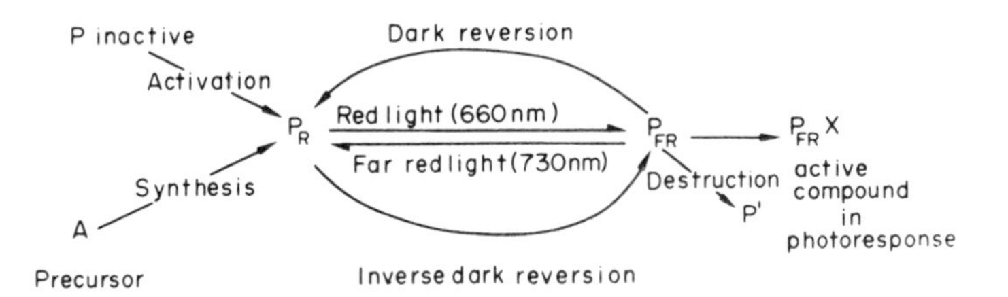

FIG. 3.10. The transformations undergone by phytochrome.

(1) activation of phytochrome from an inactive pre-existing form;
(2) synthesis of phytochrome from its precursors;
(3) transformation of P_R to P_{FR} by red light, or less effectively blue light;
(4) transformation of P_{FR} to P_R by far-red light, which requires higher energies than the reverse conversion;
(5) the reversion in the dark of P_{FR} to P_R;
(6) the inverse dark reversion, detected in dry seeds, causing a partial conversion of P_R to P_{FR}, even in the dark;
(7) the destruction, enzymatic or otherwise, of P_{FR};
(8) the reaction of P_{FR} with some unknown substance in the seed such as a chemical compound or a membrane which results in the actual effect on germination.

Transformations 5 and 6 take place via a number of intermediates, some of which are quite stable (Kendrick and Spruit, 1974). These intermediaries of phytochrome transformation are involved in the photo-inhibition of germination when dry seeds are illuminated by red light, e.g. *Sinapsis* or *Plantago* (Bartley and Frankland, 1984).

From this it is clear that phytochrome is normally in a state of equilibrium between its two forms P_R and P_{FR} and that many factors will determine the actual ratio of the two forms at a given time. Furthermore there is at least some evidence to indicate that there is more than one pool of phytochrome in plant tissues. The relative sizes of such pools and how and to what extent they are involved in the light responses in germination are not as yet known. In general it can be stated that germination is determined by the amount of P_{FR} as a percentage of the total phytochrome in seed. However, the percentage required in order to induce germination seems to be quite variable depending on the seed involved. In lettuce the amount required for germination may be as high as 30 – 40%, while in *Eragrostis* the amount required for germination seems to be much smaller. In *Nigella* the P_{FR} level required is as low as 1 – 3% of the total and in tomato seeds 22%. The ratio of $P_{FR}/(P_R + P_{FR}) = \Phi$, established when seeds are irradiated by red or far-red light, appears to be particularly important in determining germination behaviour. Yet the (Φ) required by any given species to cause germination is not fixed. Thus different populations of *Poa trivialis* from different habitats required quite different P_{FR}/P_{total} ratios, in order that germination could occur (Hilton *et al.*, 1984).

While the situation is reasonably well understood for those seeds whose germination is induced by red light and in which the red light effect is reversed by far-red light, other cases of light effects are less well understood. As already mentioned promotion of germination by blue light is one of them.

Stimulation by blue light of lettuce seeds made dormant either by exposure to high temperatures or prolonged irradiation with far-red light has recently been reinvestigated (Small *et al.*, 1979a, b). Both kinds of dormant seeds responded to blue light in a way similar to their response to red light. The blue light stimulation was reversed by far-red light. However, it was found that blue light-induced germination had an action spectrum with two peaks, at 422 and 446 nm. It is difficult to explain this action spectrum unless it is assumed that there is cooperation of blue absorbing pigments which may transfer energy to phytochrome. The studies of Small *et al.* (1979a) also bring out another important feature of the phytochrome-mediated germination. The energy of light required to induce germination depends on the way in which dormancy was induced. Seeds in which dormancy was induced by high temperatures required less light, by a factor of 4 orders of magnitude, than those made dormant by prolonged far-red light. Furthermore the amount of light energy required to induce germination by blue light was about 80 times as big as for red light.

There is at present no valid reason to doubt that blue light stimulation is mediated by phytochrome, which as already mentioned absorbs light in the blue part of the spectrum.

A further phenomenon to be considered is photo-inhibition of germination. Many seeds which germinate in the dark are inhibited by irradiation with white light, e.g. *Amaranthus caudatus, Nemophila insignis* and *Phacelia tanacetifolila* (Rollin and Maignan, 1967; Rollin, 1968). This inhibition by white light is usually due to the FR irradiation and is dependent on the fluence rate. Many species have now been shown to be inhibited by FR and this inhibition may be supposed to operate through phytochrome as is the R promotion of germination. It seems to be obtained by maintaining the P_{FR}/P total ratio at a rather low value, 0.1 or below. This is the result of the fact that when phytochrome is irradiated with FR it is converted back to P_R, but some of this is reconverted to P_{FR} due to the absorption by P_R of some far-red light (Fig. 3.9). Since the actual ratio will not be greatly affected by the fluence rate, but the degree of inhibition of germination is dependent on this rate, it cannot simply be the actual ratio which causes inhibition. It is therefore not simply the amount of P_{FR} which determines inhibition. There is some evidence supporting this view, that it is the actual rate of interconversion between the two pigment forms which is responsible for photo-inhibition (Bartley and Frankland, 1984).

It seems possible that in the inhibitory process some intermediary between P_R and P_{FR}, which is very reactive, might be the cause of inhibition. Evidence on all these points is still not entirely clear. Photo-inhibition is often referred to as being due to a high energy or high irradiation reaction.

The way in which the active form of phytochrome brings about its effect on germination is still quite unclear. The identity of the hypothetical compound *X* (Fig. 3.10), with which the active form combines, is also unknown. It must be remembered that there is as yet no biochemical reaction which can be directly linked to promotion or inhibition of germination by light. Although the evidence for phytochrome being on or near a membrane is good and although molecular models have been constructed to explain how it could change ionic channels in a membrane, the central questions remain unanswered: how does P_{FR} (or intermediaries in its formation) act, where are they located at the organ at the cellular and subcellular level and what reactions are carried out at this site (Fig. 3.11)?

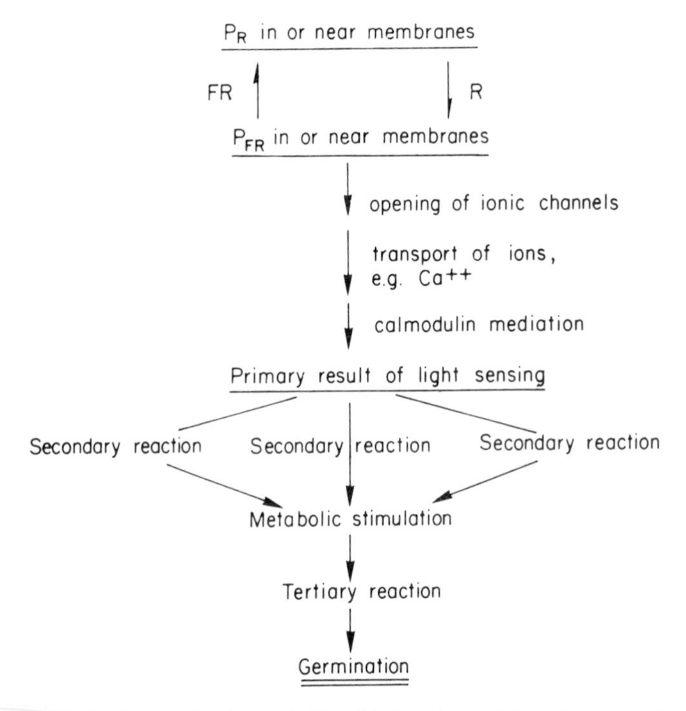

FIG. 3.11. Light sensing by seeds. Possible locations of the primary reaction and subsequent secondary reactions (Mayer, 1986).

As long as these questions remain open, most of the information now available cannot be properly interpreted or incorporated into our views on the effect of light. Furthermore, all theories, such as those locating phytochrome at a membrane, require: (1) demonstration of a specific interaction between phytochrome and its site of action, (2) binding affinity with a high affinity constant, and (3) demonstration that binding between phytochrome and its site of action is directly related to its morphogenetic expression, in our case germination. Unfortunately none of these criteria have yet been met (Pratt, 1978).

The effect of blue light in inhibiting germination is even more complex. Generally it is still supposed that here too phytochrome is involved. However, there have been suggestions that an additional pigment might be involved. This will have to be established much more firmly before it can be accepted.

It must be mentioned that the response of seeds to light is modified by various internal and external factors. Osmotic stress, the presence of growth promoters or growth inhibitors, the oxygen tension and other factors, all can change the duration and intensities of light required to evoke a certain response. Even the spectrum of radiation to which the parent plant has been exposed during seed formation, as demonstrated, for example, for seeds of *Arabidopsis thaliana*, affects the response of the seeds to light. In general it may be said that all internal and external factors affecting germination interact in some way.

Another type of radiation which affects living organisms is shortwave irradiation such as γ-rays and X-rays. This type of radiation can also affect germination and development of seedlings. Such radiation affects not only germination but also chromosomal structure and integrity and therefore also affects seedling development and even subsequent seed formation. However, a consideration of the effects of this type of radiation as well as that of α- and β-irradiation and the irradiation with other particles is outside the scope of this book.

Strictly speaking, the effects of light discussed here are effects on the breaking of dormancy or its induction. Other light effects, perhaps more directly related to the phenomenon of dormancy, are the photoperiodic response of the germination of certain seeds as well as the complex interactions between light and other factors affecting germination. These will be dealt with in the following chapter.

Bibliography

Abdul-Baki A. A. and Anderson J.D. (1972) *Seed Biology*, vol. 11, p. 283 (Ed. Kozlowski T.T.), Academic Press, New York.

Abeles F. B. and Lonski J. (1969) *Plant Physiol.* **44**, 277.

Al-Ani A., Bruzau F., Raymond P., Saint-Ges V., Leblanc J. M. and Pradet A. (1985) *Plant Physiol.* **79**, 885.

Attenberg A. (1928) *Landw. Versuchsstation* **67**, 129.

Baar H. (1912) *S. Acad. Wiss. Math.-Naturn. K. L.* **121**, 667.

Ballard L.A.T. (1967) in *Physiologie Okologie und Biochemie der Keimung*, p. 209 (Ed. H. Boriss), Ernst-Moritz-Arndt-Universitat, Greifswald.

Ballard L. A. T. and Grant-Lipp A. E. (1969) *Aust. J. Biol. Sci.* **22**, 279.

Bartley M. R. and Frankland B. (1984) *Plant Physiol.* **74**, 601.

Barton L. V. (1961) *Seed Preservation and Longevity*. Leonard Hill, London.

Beadle N. C.E. (1952) *Ecology* **33**, 49.

Beal W. J. (1914) *The Plant World* **17**, 329.

Becquerel P. (1932) *C. R. Acad. Sci. Paris* **194**, 2158.

Becquerel P. (1934) *C. R. Acad. Sci. Paris* **199**, 1662.

Bewley J. D. (1986) in *Physiology of Seed Deterioration* (Eds M. B. McDonald Jr and C. J. Nelson), Crop Science Society of America Special Publication No. 11, p. 27, Madison, Wisconsin.

Blaauw-Jansen G. and Blaauw O. H. (1975) *Acta Bot. Neerl.* **24**, 199.

Blaauw-Jansen G. and Sewmar R. (1986) *Acta Bot. Neerl.* **35**, 101.

Blacklow W. M. (1972) *Crop Sci.* **12**, 643.

Bonner F. T. (1968) *Bot. Gas.* **129**, 83.

Borthwick H. A., Hendricks S. B., Parker M. W., Toole E. H. and Toole V. K. (1952) *Proc. Nat. Acad. Sci. Wash.* **38**, 662.

Brown R. (1940) *Ann. Bot. N.S.* **4**, 379.

Butler W. L., Norris K. H., Siegelman H. W. and Hendricks S. B. (1959) *Proc. Nat. Acad. Sci. U.S.A.* **45**, 1703.

Chanoeuf-Hannel R. and Taylorson R. B. (1985) *Plant Physiol.* **78**, 228.

Ching T. M., Parker M. C. and Hill D. D. (1959) *Agron. J.* **51**, 680.

Chu G. and Tang P. S. (1959) *Acta Bot. Sinica* **8**, 212.

Cieslar A. (1883) *Forsch. Gebiete Agrikultur Physik* **6**, 270.

Connelly A., Lohman J. A. B., Loughman B. C., Quiquampoix H. and Ratcliffe R. G. (1987) *J. Exp. Bot.* **38**, 1713.

Corbineau F. and Come D. (1980/81) *Is. J. Bot.* **29**, 157.

Corbineau F. and Come D. (1988a) *Malaysian Forester*. (In press).

Corbineau F. and Come D. (1988b) *Life Sciences Advances*, series B,4 vol. 6. (In press).

Corbineau F. and Come D. (1988c) *Seed Sci. Technol.*

Crocker W. (1938) *Bot. Rev.* **4**, 235.

David R. L. (1936) *Influence des Temperatures Elevées sur la Vitalité des Graines Oleagineuses*, Imprimerie Universitaire, Aix en Provence.
Denny F. E. (1917a) *Bot. Gaz.* **63**, 373.
Denny F. E. (1917b) *Bot. Gaz.* **63**, 468.
Discussions of the Faraday Society No. 27 (1959) *Energy Transfer with Special Reference to Biological Systems*, Aberdeen University Press, Scotland.
Edwards T. I. (1932) *Q. Rev. Biol.* **7**, 428.
Egley G. H., Paul R. N. Jr and Lax A. R. (1986) *Plant Physiol.* **67**, 320.
Egley G. H., Paul R. N. Jr, Vaughn K. C. and Duke S. O. (1983) *Planta* **157**, 224.
Ellis R. H. and Roberts E. H. (1981) *Seed Sci. Technol.* **9**, 373.
Esashi Y. and Leopold A. C. (1968) *Plant Physiol.* **43**, 871.
Evenari M. and Neuman G. (1953) *Bull. Res. Council, Israel* **3**, 136.
Evenari M., Neuman G. and Stein G. (1957) *Nature Lond.* **100**, 609.
Flint H. L. and McAllister E. D. (1935) *Smithsonian Miscellaneous Collection* **94**, No. 5.
Flint H. L. and McAllister E. D. (1937) *Smithsonian Miscellaneous Collection* **96**, No. 2.
Frankland B. (1976) in *Light and Plant Development*, p. 477 (Ed. H. Smith), Butterworths, London.
Fujii T. (1963) *Plant Cell Physiol.* **4**, 357.
Gardner W. A. (1921) *Bot. Gaz.* **71**, 249.
Gassner G. (1915) *Z. Bot.* **7**, 609.
Givelberg A., Horowitz M. and Poljakoff-Mayber A. (1984) *J. Exp. Bot.* **35**, 588.
Glasstone S. (1946) *Textbook of Physical Chemistry*. Macmillan.
Godwin H. (1968) *Nature* **220**, 708.
Griffiths A. E. (1942) *Cornell Univ. Agr. Exp. Station memoir*, **245.**
Gutterman Y. (1974) *Oecologia* **17**, 27.
Hadas A. (1977a) *J. Agronomy* **69**, 582.
Hadas A. (1977b) *J. Exp. Bot.* **28**, 977.
Hadas A. (1977c) *Seed Sci. Technol.* **5**, 519.
Hadas A. (1982) in *The Physiology and Biochemistry of Seed Development Dormancy and Germination*, p. 507 (Ed. A. A. Khan), Elsevier Biomedical.
Harel E. and Mayer A. M. (1963) *Physiol. Plant.* **16**, 804.
Harper J. L. and Benton R. A. (1966) *J. Ecol.* **54**, 151.
Harrison B. J. and McLeish J. (1954) *Nature Lond.* **173**, 593.
Hasson E. and Poljakoff-Mayber A. (1980/81) *Is. J. Bot.* **29**, 98.
Haurowitz F. (1950) *Chemistry and Biology of Proteins*, Academic Press, New York.
Hegarthy T. W. and Ross H. A. (1980/81) *Is. J. Bot.* **29**, 83.
Hendricks S. B. and Taylorson R. B. (1979) *Proc. Nat. Acad. Sci. USA* **72**, 778.
Hilton J. R. (1984) *New Phytol.* **97**, 345.
Hilton J. R., Froud-Williams R. J. and Dixon J. (1984) *New Phytol.* **97**, 375.
Hodson M. J., Di Nola L. and Mayer A. M. (1987) *J. Exp. Bot.* **38**, 524.
Holden M. (1959) *J. Sci. Food Agr.* **12**, 691.
Hourmant A. and Pradet A. (1981) *Plant Physiol.* **68**, 631.
Janerette C. A. (1979) *Seed Sci. Technol.* **7**, 347.
Jirgensons B. (1958) *Organic Colloids, Elsevier,* Amsterdam.
Juntilla O. and Stushnoff C. (1977) *Nature* **269**, 325.
Kamra S. K. (1964) *Proc. Intern. Seed Test Assoc.* **29**, 519.
Kamra S. K. (1976) *Studia Forestalia Suecica*, No. 131, p. 34.
Katchalski A. (1954) *Progr. Biophysics* **14**, 1.
Kendrick R. E. (1983) *Symp. Soc. Exp. Biol.* **79**, 299.
Kendrick R. E. and Spruit C. J. B. (1974) *Planta* **120**, 265.
Kendrick R. E. and Cone J. W. (1985) *Plant Physiol.* **79**, 299.
Kennedy R. A., Barrett S. C. H., Van der Zee D. and Rumpho M. E. (1980). *Plant Cell Environ.* **3**, 243.
Kepenczynski J. and Karssen C. M. (1985) *Physiol. Plant.* **63**, 49.
Kidd F. (1914) *Proc. R. Soc. B.* **87**, 410.
Kincaid R. R. (1935) *Tech. Bull.* **277**, Univ. Florida, Agr. Exp. St.
Kinzel W. (1926) *Frost und Licht, Neue Tabellen,* Eugen Ulmer, Stuttgart.
Kivilaan A. and Bandurski R. S. (1973) *Am. J. Bot.* **60**, 140.
Kivilaan A. and Bandurski R. S. (1981) *Am. J. Bot.* **68**, 1290.

Knapp R. (1967) *Flora* **157b**, 3.
Koller D. (1962) *Am. J. Bot.* **49**, 841.
Koller D. and Hadas A. (1982) in *Encyclopedia of Plant Physiology*, New Series 12 B, p. 401, Springer-Verlag.
Lehmann E. (1913) *Biochem. Z.* **50**, 388.
Lehmann E. and Aichele F. (1931) *Keimungsphysiologie der Graser*, Verlag Ferdinand Enke, Stuttgart.
Leopold A. C. and Vertucci C. W. (1984) *Plant Physiol.* **75**, 114.
Leopold A. C. and Vertucci C. W. (1986) in *Membranes, Metabolism and Dry Organisms*, p. 22 (Ed. A. C. Leopold), Comstoc Publ. Associates, Ithaca, New York.
Lerman J. C. and Cigliano E. M. (1971) *Nature* **232**, 568.
Levari R. (1960) Ph.D. Thesis, Jerusalem.
Levitt J. (1956) *The Hardiness of Plants*, Academic Press, New York.
Maier W. (1933) *Jahrb. Wiss. Bot.* **78**, 1.
Malcoste R., Boisard J., Spruit C. J. P. and Rollin P. (1970) *Ned. Landbouwhogeschool, Wageningen* **70-16**, 1.
Marbach I. (1976) Ph.D. Thesis, Hebrew University, Jerusalem.
Marbach I. and Mayer A. M. (1974) *Plant Physiol.* **54**, 817.
Marbach I. and Mayer A. M. (1975) *Plant Physiol.* **56**, 93.
Mayer A. M. (1986) *Israel J. Bot.* **35**, 3.
Mayer A. M. and Marbach I. (1981) *Progress in Phytochemistry* **7**, 95.
McDonald Jr M. B. and Nelson C. J. (Eds) (1986) *Physiology of Seed Deterioration*, Crop Science Society of America Inc. Special Publ. 11 Madison, Wisconsin.
Morinaga T. (1926a) *Am. J. Bot.* **13**, 126.
Morinaga T. (1926b) *Am. J. Bot.* **13**, 159.
Okagami N. and Kawai M. (1982) *Bot. Mag. Tokyo* **95**, 155.
Osborne D. J., Sharon R. and Ben-Ishai R. (1980/81) *Is. J. Bot.* **29**, 259.
Osborne D. J., Della 'Aquita A. and Elder R. H. (1984) *Folia Biologica* (Praha) Special Publication, p. 155.
Owen E. Biasutti (1956) *The Storage of Seeds for Maintenance of Viability, CAB Bulletin* **43**, Commonwealth Bureau of Pastures and Field Crops.
Pappenheimer J. R. (1953) *Physiol. Rev.* **33**, 387.
Powell A. A. and Matthews S. (1978) *J. Exp. Bot.* **29**, 1215.
Pratt L. H. (1982) *Ann. Rev. Plant Physiol.* **33**, 557.
Priestley D. A. (1986) *Seed Ageing*, Cornell Univ. Press, Ithaca, New York.
Priestley D. A. and Posthumus M. (1982) *Nature* **299**, 148.
Priestley D. A., Cullinant V. I. and Wolfe J. (1985) *Plant Cell Environ.* **8**, 557.
Probert, R. J., Smith R. D. and Birch P. (1986) *New Phytol.* **102**, 133.
Quail P. H., Colbert J. T., Hershey H. P. and Viestra R. D. (1983) *Phil. Trans. Roy. Soc. London*, series B, **303**, 387.
Raymond P., Al-Ani A. and Pradet A. (1983) *Physiol. Veg.* **21**, 677.
Resuehr B. (1939) *Planta* **30**, 471.
Rethy R., Dedonder A., De Petter E., Van Wiemeersch L., Fredericq H., De Greef J., Steyaert H. and Stevens H. (1987) *Plant Physiol.* **83**, 126.
Reynolds T. and Thompson P. A. (1971) *Plant Physiol.* **24**, 544.
Rice H. V., Brigg W. R. and Jackson-White C. J. (1973) *Plant Physiol.* **51**, 917.
Roberts E. H. (Ed.) (1972) *Viability of Seeds*, Chapman & Hall, London.
Roberts E. H. (1986) in *Physiology and Seed Deterioration*, p. 101 (Eds M. B. McDonald and C. J. Nelson), Crop Science Society of America, Special Publ. No. 11, Madison, Wisconsin.
Rollin P. (1968) *Bull. Soc. Franc. Physio. Veget.* **14**, 47.
Rollin P. (1972) in *Phytochrome*, p. 230 (Ed. W. Shropshire Jr and K. Mitrakos), Acad. Press, New York.
Rollin P. and Maignan G. (1967) *Nature* **214**, 741.
Roos E. E. (1986) in *Physiology of Seed Deterioriation*, p. 1 (Eds M. B. McDonald Jr and C. J. Nelson), Crop Science Society of America Special Publ. Madison, Wisconsin.
Rupley J. A. and Siemankowski L. (1986) in *Membranes, Metabolism and Dry Organisms*, p. 259 (Ed. A. C. Leopold), Comstoc Publishing Associates, Ithaca, New York.
Shropshire W. (1973) *Sol. Energy* **15**, 99.
Shull G. H. (1914) *The Plant World* **17**, 329.
Shull C. A. (1920) *Bot. Gaz.* **69**, 361.
Siegel S. M. and Rosen L. A. (1962) *Physiol. Plant.* **15**, 437.
Small J. G. C., Spruit C. J. P., Blaauw-Jansen G. and Blaauw O. H. (1979a) *Planta* **144**, 125.

Small J. G. C., Spruit C. J. P., Blaauw-Jansen G. and Blaauw O. H. (1979b) *Planta* **144**, 133.
Song P. S. (1983) *Symp. Soc. Exp. Biol.* **36**, 181.
Spurny M. (1973) in *Seed Ecology*, p. 367 (Ed. W. Heydecker), Butterworths, London.
Tang P. S., Wang F. C. and Chih F. C. (1959) *Acta Bot. Sinica* **8**, 199.
Taylorson R. B. and Hendricks S. B. (1972) *Plant Physiol.* **50**, 645.
Thompson P. A. (1969) *J. Exp. Bot.* **20**, 1.
Thornton N. C. (1943-45) *Contr. Boyce Thompson Inst.* **13**, 355.
Toole E. H., and Brown E. (1946) *J. Agr. Res.* **72**, 201.
Toole E. H., Hendricks S. B., Borthwick H. A. and Toole V. K. (1956) *Ann. Rev. Plant Physiol.* **7**, 299.
Turner J. H. (1933) *Bull. Misc. Information* **6**, p. 257, Royal Botanic Gardens, Kew.
VanderWoude W. J. (1985) *Photochem. Photobiology* **42**, 655.
Vertucci C. W. and Leopold A. C. (1986) in *Membranes, Metabolism and Dry Organisms*, p. 35 (Ed. A. C. Leopold), Comstoc Publ. Ass. Ithaca, New York.
Vertucci C. W., Vertucci F. A. and Leopold A. C. (1987) *Plant. Physiol.* **84**, 887.
Vidaver W. E. (1972) *Soc. Exp. Biol. Symp.* **26**, 159.
Villiers T. A. (1974) *Plant Physiol.* **53**, 875.
Wareing P. F. and Black M. (1957) *Nature, London* **180**, 385.
Wareing P. F. and Black M. (1958) *Nature, London* **181**, 1420.
Went F. W. and Munz P. A. (1949) *El. Aliso* **2**, 63.

4

Dormancy, Germination Inhibition and Stimulation

Many seeds do not germinate when placed in conditions which are normally regarded as favourable to germination, namely an adequate water supply, a suitable temperature and an atmosphere of normal composition. Nevertheless seeds can be shown to be viable, as they can be induced to germinate by various special artificial treatments, or under specific external conditions. Such seeds are said to be dormant, or to be in a state of dormancy. The fact that under natural conditions the germination of the seeds is delayed, until suitable conditions for establishment and survival prevail, may be advantageous for the survival of the species. Usually not the entire crop of seeds germinates. Only a certain fraction of the population emerges from dormancy and germinates every year. For example, Moore and Moore (cited by Tran and Cavanagh, 1984), have shown that in a stock of *Albizzia julibrissin* seeds soaked for 10 yr 39% germinated in the first year, an additional 23% in the second, 11% in the third and 23.5% in the fourth. After 5 yr all the seeds in the population had emerged from dormancy. On the other hand, *Kostaletzkya virginica* seeds hardly germinated in the first year (2 – 5%), but showed 20% germination in the second, an additional 30% germinated in the third, and dormancy was lost completely in the fourth year of storage (unpublished data). Dormancy can be due to various causes. It may be due to the immaturity of the embryo, impermeability of the seed coat to water or to gases, prevention of embryo development due to mechanical causes, special requirements for temperature or light, or the presence of substances inhibiting germination. In all these cases, the phenomenon is referred to as *primary dormancy*.

Attempts have been made to classify the states of dormancy (Nikolaeva, 1969). In these attempts at classification exogenous and endogenous causes of dormancy are distinguished. These are further divided according to the supposed cause of dormancy, e.g. physical, chemical, etc. However, it is fairly obvious today that more than one factor may be responsible for the dormancy of a given seed.

It must also be remembered that dormancy is genetically controlled (Tuan and Bonner, 1964), as shown very clearly for seeds of *Avena fatua* (Naylor, 1983). We will therefore not follow a systematic classification of dormancy but try to understand some of the underlying mechanisms. Ways of breaking dormancy are often important, particularly with regard to the technology for seed banks. A handbook on ways to alleviate various forms of dormancy in many species has been published recently (Ellis *et al.*, 1985).

In many species of plants the seeds, when shed from the parent plant, will not germinate. Such seeds will germinate under natural conditions, if they are kept for a certain period of time. These seeds are said to require a period of *after-ripening*. After-ripening may be defined as any changes which occur in seeds during storage as a result of which germination becomes possible or is improved. This is the most generally used definition of this term. An alternative definition is also possible. This would define after-ripening as those processes which must occur in the embryo and which cannot be caused by any known means other than suitable storage conditions of the seeds. After-ripening is also genetically controlled and appears to be controlled by a number of genes (Naylor, 1983).

After-ripening often occurs during dry storage. In other cases seeds must be stored in the imbibed state, usually at low temperatures. This is termed *stratification* and will be dealt with later.

There is much diversity in the conditions under which after-ripening in "dry storage" occurs. The length of the storage period required is also variable. In cereals germination is often low at harvest and increases during storage. Some barley varieties after-ripen within a fortnight. Seeds of *Cyperus* after-ripen over a period of 7 yr. In lettuce, *Amaranthus* and *Rumex* the fresh seeds can germinate but the requirements for their germination are very specific. These special requirements tend to disappear during storage. Lettuce seeds 16 days after anthesis are not light sensitive and show 100% germination at 20°C. Twenty days after anthesis they become light sensitive (Globerson, personal communication) and germinate in the dark, only below 20°C. After prolonged storage they germinate even at 30°C. During prolonged dry storage their light requirement also disappears. The reverse is observed in *Amaranthus retroflexus*. Fresh seeds germinate only above 30°C, while stored seeds germinate over a wider range, down to 20°C. This type of response to storage should not, strictly speaking, be termed "after-ripening", although this term is often applied.

Seeds of *Pisum elatius*, four weeks after harvest, showed 96% germination at 10°C and 72% at 20°C but failed to germinate at 30 or 35°C. After 10 weeks they yielded 71% germination at 30°C. After one year they yielded 14% germination even at 35°C.

The necessity for a period of after-ripening may be due to a number of factors. Various kinds of change may consequently occur during this process. In the case of an immature embryo anatomical and morphological changes may occur. In other seeds chemical changes must occur in the seed before it can germinate. Frequently germination of such seeds can be forced by suitable treatment although the resulting seedlings may be abnormal.

Immature embryos are known among many families of plants, although they are most commonly associated with plants which are saprophytic, parasitic or symbiotic. Such embryos may attain full maturity either during the actual process of germination or they may mature as a preliminary to germination. In either case the changes only occur if the seeds are kept under conditions favourable to germination. Differentiation between the two types is extremely difficult. Plants in which immature embryos occur include the Orchidaceae and Orobanchaceae, as well as some *Ranuculus* species. The period required for such embryos to reach maturity varies from a few days to several months. In all these seeds, processes of differentiation at the anatomical and morphological level occur during the period of after-ripening. Apple embryos undergo clear anatomical changes during after-ripening; these include differentiation of vascular tissue in the cotyledons and starch accumulation in the zone between the radicle and hypocotyl along the embryonic axis (Dawidowicz-Grzegorzewska and Lewak, 1978).

In other seeds, such as those of *Fraxinus*, the embryo, although morphologically complete, must still increase its size before germination can take place. In seeds in which the after-ripening is due to chemical or physical changes, the composition of the storage materials present in the seeds may alter, the permeability of the seed coat may change, substances promoting germination may appear or an inhibitory one may disappear. In no case has it been possible to ascribe after-ripening to any one definite event. Perhaps an important factor may be the inability of the dormant embryo to replicate its DNA, but till now such an inability has been described only for rye (Osborne, 1983).

I. Secondary Dormancy

The seeds described above which fail to germinate when shed show primary dormancy. Other seeds will germinate readily immediately after they are shed if conditions are favourable. However, these seeds may lose their readiness to germinate. This phenomenon is called *secondary dormancy*. Secondary dormancy may develop spontaneously in seeds due to changes occurring in them, as in some species of *Taxus* and *Fraxinus*. These changes may be the reverse of those occurring during after-ripening. Sometimes secondary dormancy is induced if the seeds are given all the conditions required for germination except one. For instance, the failure to give light to light-requiring seeds, or illuminating light-inhibited seeds, such as *Nigella* and *Phacelia*, induces dormancy. *Phacelia* seeds will not germinate if kept continuously in the light. When they are returned to, and kept in, the dark for long periods they will eventually germinate provided the period in the light has not been too prolonged. If the seeds are kept for a very long time in the light they are then unable to germinate in the dark even under conditions which were previously favourable (Mato, 1924). Seeds of *Verbascum* fail to respond to light if kept in the dark at 27°C for about three weeks. If exposed to the cold (5°C) for two days, they again responded to light (Kivilaan, 1975).

Extreme high or low temperatures may also induce secondary dormancy, for example in *Ambrosia trifida* and *Lactuca sativa*. Lettuce seeds if kept imbibed in the dark at high temperatures will not germinate even if returned to lower temperatures, and no longer respond to light. Only drastic treatment such as a chilling or chemical treatment with gibberellin will induce germination.

In *Eragrostis* dormancy is induced by cycling the seeds between a low and a high temperature during imbibition, the low temperature being the determining factor. A light requirement was noted after the induction of secondary dormancy (Isikawa *et al.*, 1961). In the field secondary dormancy may be spontaneously induced and relieved in a cyclical manner. This has been observed in *Chenopodium* and *Sisymbrium* (Karssen, 1980/81a,b). Environmentally imposed secondary dormancy may be common in nature and may be considered as a response to a change in environmental conditions.

Low oxygen tension will cause dormancy in *Xanthium* (Davis, 1930). High carbon dioxide tensions may cause secondary dormancy, for example in *Brassica alba* (Kidd, 1914). Dormancy may also be induced by chemical treatment of the seeds, for example the induction of light sensitivity in light-indifferent varieties of lettuce seeds by treatment with coumarin. This latter phenomenon is closely related to the problem of germination inhibitors and will be dealt with later.

It may be assumed that the mechanisms underlying secondary dormancy are the same as those of dormancy in general. The fact that new requirements for germination arise indicates that metabolic changes occur while the seed is exposed to conditions which impose secondary dormancy. These may be changes in permeability, shifts in the balance of the forms of phytochrome, changes in inhibitor and stimulator relationships or other events.

Secondary dormancy is apparently induced in the embryo. It can be induced after primary dormancy has been broken, as is seen in apple embryos (Bulard, 1986).

II. Possible Causes of Dormancy

The possible causes of dormancy have already been listed. They will now be considered in greater detail. The means by which seed dormancy may be broken will also be discussed. Dormancy breaking is frequently of economic importance. The mechanism of artificial dormancy breaking and the natural process leading to the same effect are frequently similar.

1. Permeability of Seed Coats

A widespread cause of seed dormancy is the presence of a hard seed coat. Hard seed coats occur in many plant families and usually cause dormancy in three main ways. A hard seed coat may be impermeable to water, impermeable to gases or it may mechanically constrain the embryo.

A general discussion of coat-imposed dormancy, including the mechanical constraints caused by the seed coat, can be found in a review by Tran and Cavanagh (1984).

The impermeability of seed coats to water is most widespread in the Leguminosae but occurs also in many other families (Rolston, 1978). The seed coats of many members of the Leguminosae family are very hard, resistant to abrasion and covered with a wax-like layer. Such seed coats appear to be entirely impermeable to water. In some cases the entry of water into the seeds is controlled by a small opening in the seed coat, the strophiolar cleft. This cleft is lined with suberized cells which prevent the entry of water (Fig. 4.1A). If the cells are disturbed or the suberized layered is fractured or removed water will enter the seed. This mechanism was first described by Hamly (1932), who also investigated various artificial ways of dormancy-breaking in such seeds. He was able to show that vigorous shaking of the seeds rendered the seeds permeable to water. This treatment is frequently called impaction and has been applied to seeds of *Melilotus alba, Trigonella arabica* and *Crotalaria aegyptiaca*. Among other treatments capable of causing the cleft to become water permeable is exposure to microwave energy (2450MHz) (Tran, 1979). This appears to open up the cleft by causing a rupture of the seed coat. Such ruptures apparently also result from impaction treatment, and may be similar to changes occurring in seeds in nature in the soil. Movement of seeds from one place to another, by wind or various disturbances, may also cause this kind of rupture of the seed coat. A

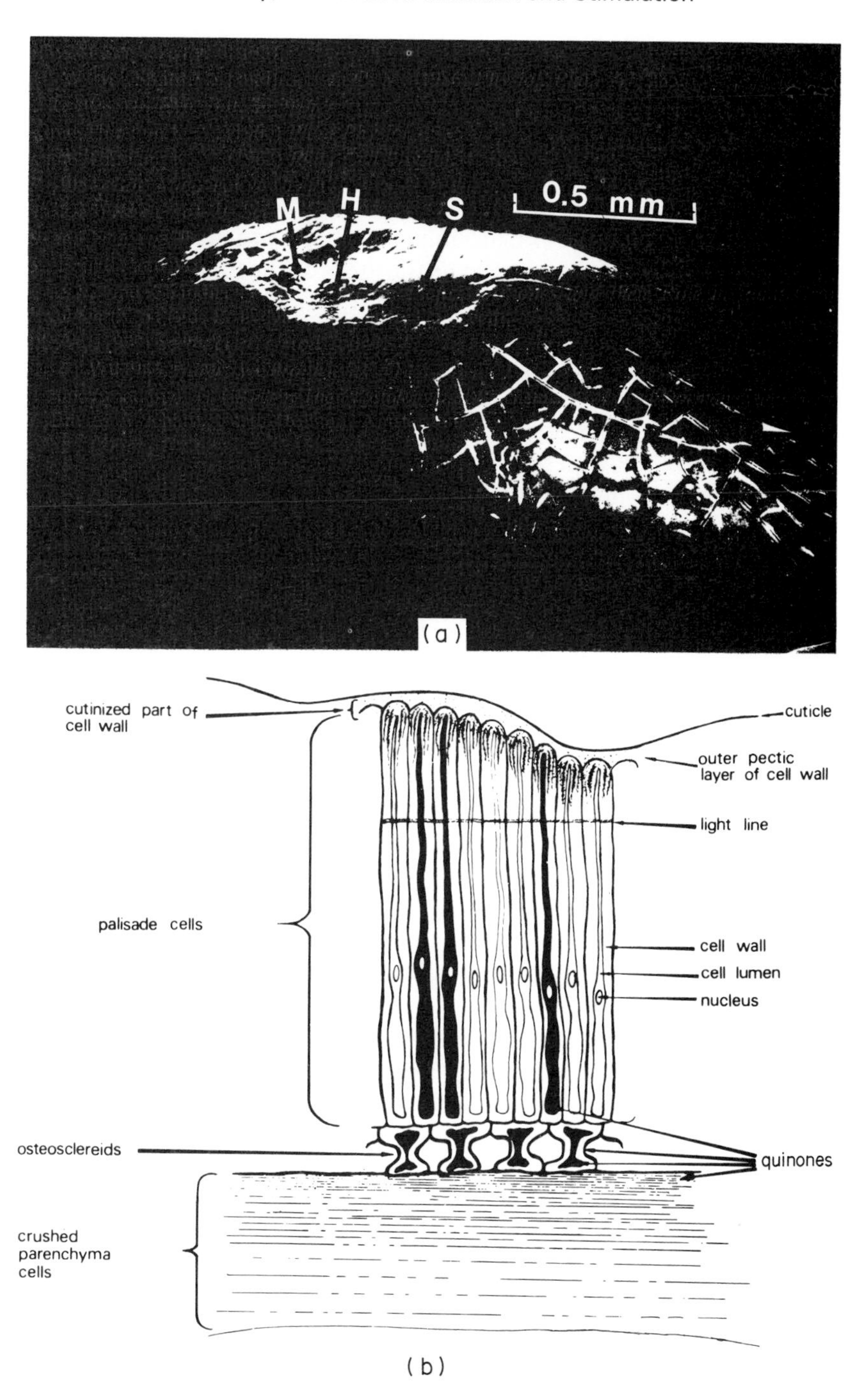

FIG. 4.1. (a) Scanning electron micrograph of the hilum region of *Acacia suavolens*. M — micropyle, H — hilum, S — strophiole (reproduced from Cavanagh, 1980 from Proceedings of the Royal Society of Victoria; by permission). (b) Section through the testa of *Pisum elatius* showing the various elements contributing to the impermeability to water (courtesy Dr E. Werker).

clearly defined strophiolar plug has been described in seeds of *Albizzia lopantha*. Heating of the seeds caused the loss of this plug by eruption and allowed the entry of water (Dell, 1980). The opening of the strophiolar gap may be reversible. The majority of seeds showing a strophiolar cleft belong to the Papilionaceae. Obviously it is not only the area of the strophiole which controls water entry. The surrounding tissues and their impregnation are also of great importance (Werker, 1980) (Fig. 4.1b). Among the substances contributing to impermeability are pectins, suberins, cutins and mucilages. In impermeable seed coats the areas of the hilum and the chalaza are usually blocked by special cells.

Other hard-coated seeds do not possess a strophiolar cleft. In the sugar maple (*Acer saccharinum*) water enters the embryo by two routes, through the chalazal area and through the micropylar region. Most of the water enters through the chalazal region as the micropylar region is apparently impregnated with more materials which have a low permeability to water (Janerette, 1979).

As mentioned above, in nature most seeds become permeable to water when the seed coat is broken down or punctured by mechanical abrasion, microbial attack, passage through the digestive tract of animals or exposure to alternating high and low temperatures which, by expanding and contracting the seed coat, cause it to crack. Under laboratory conditions and in agriculture various means of rendering the seed coat permeable have been adopted. These include either shaking with some abrasive, to cause mechanical breakage, or chemical treatment. The chemical treatment is chiefly of two kinds: removal of the waxy layer of the seed coat by some suitable solvent such as alcohol, or treatment with acids (Table 4.1).

TABLE 4.1. *Effect of acid treatment on germination of various seeds (from data of Wycherley, 1960; Khudairi, 1956)*

	% Germination	
	Untreated	Treated with sulphuric acid
Calopogonium muconoides	30	65
Centrosema pubescens	30	70
Flemingia congesta	45	60
Pueraria phaseoloides	20	70
Purshia tridentata	1	84
Cercocarpus montanus	1	52
Atriplex canescens	11	16

The mechanism by which the latter act is far from clear. Possibly chemical decomposition of seed coat components is involved, which may be analogous to the breakdown processes occurring during microbial attack or passage through the digestive tract. The treatment with alcohol is especially effective for members of the family Caesalpiniaceae. In all the cases studied, the various treatments could be shown to increase directly the water uptake of the seeds and the number of seeds which swelled (Table 4.2). The dormancy-breaking action of abrasion, alcohol or

TABLE 4.2. *Changes in permeability of seeds to water caused by various treatments. The change in permeability is expressed as the percentage of seeds which swell (from data of Barton and Crocker, 1948, and Koller, 1954)*

Species	No treatment	Alcohol treatment	Impaction	Mechanical abrasion
Trigonella arabica	5	4	100	96
Crotalaria aegyptiaca	0	20	80	95
Melilotus alba	1	0	86	100
Cassia artemisioides	2	57	3	100
Parkinsonia aculeata	2	100	8	100

sulphuric acid could therefore be directly related to an increase in the permeability of the seeds to water. Nevertheless such vigorous treatment of seeds is likely to induce other changes also, such as permeability to gases, changes in sensitivity to light or temperature and possibly even destruction or removal of inhibitory substances. Dewaxing of seeds of *Podocarpus* failed to induce germination, but removal of the epidermis did. Probably under natural conditions peristomatal fissures in the cuticle are induced by seed shrinkage and these permit water uptake (Noel and van Staden, 1976). Thus, although there is no doubt that increase in water permeability is an essential part of the dormancy-breaking action, it is by no means certain that it is the only result of the treatments. Effects on membranes may be involved and will be discussed later.

The germination of seeds is determined not only by their ability to imbibe water, but also by the conditions during imbibition. Initially seeds are leaky and when placed in water lose many substances including electrolytes, low molecular weight compounds and even proteins. This leakage gradually ceases, apparently due to repair of cellular membranes (Murphy and Noland, 1982; Marbach and Mayer, 1985).

Water inflow into the seed during imbibition may cause damage to the seed tissues. Thus, in barley, water uptake at low temperatures from the liquid phase leads to subsequent damage to the seedling, however this is not the case if the seeds are exposed to water vapour. Excess water often leads to dormancy or poor germination. Integrity of the epidermal layer of the seed coat, at least in soybean seeds, plays an important role in protecting the seeds from imbibition injury (Duke *et al.*, 1986), but protection against seed injury need not necessarily be due to seed coat effects on water permeability. In the case of *Blepharis*, mucilages in the seed coat play a vital role. In the presence of excess water the mucilage becomes a diffusion barrier to oxygen (Gutterman *et al.*, 1967). Excess water may also encourage the development of large mixed populations of micro-organisms in and around the seed envelope, which compete with the embryo for available oxygen.

Frequently seed coats are impermeable to gases despite the fact that the seeds are permeable to water. This impermeability may be either towards carbon dioxide or oxygen, or both. That such differential permeability exists seems fairly well established, despite the apparently small differences in molecular diameter of the substances involved.

The most frequently cited case of impermeability to oxygen is provided by

Xanthium. Crocker (1906) showed that the upper and lower seed in the fruit differ in their ability to germinate. Later both Shull (1911 and 1914) and Thornton (1935) showed that they differ in their requirement for external oxygen pressure to give 100% germination. The upper seed gives 100% germination at 21°C only in pure oxygen, while the lower one requires only 6% oxygen for full germination. The oxygen requirement of both seeds is reduced by higher temperatures (Thornton, 1935). Intact seeds needed higher oxygen concentrations for germination than excised embryos from both upper and lower seeds. The excised embryos gave 100% germination at 1.5% for upper and 0.6% oxygen for the lower seed. This indicates that the seed envelope is impermeable to oxygen. It is not clear whether the impermeability of the seed envelope restricts the internal oxygen supply to such an extent that oxidative respiratory mechanisms are reduced or whether some other mechanism is in operation. The latter has been suggested by Wareing and Foda (1957) who indicated that the high oxygen requirement of the upper seeds is due to the presence of an inhibitor which has to be destroyed by oxidation before germination can ccur. Porter and Wareing (1974) showed that there is no difference in the oxygen permeability of the seed coat of imbibed upper and lower seeds of *Xanthium* although there is a three-fold difference in the permeability of the dry seed. Even under these conditions diffusion of oxygen is adequate for respiration. They suggest that the high oxygen requirement is caused by the oxidation of an inhibitor.

Since in some instances anaerobic conditions could induce germination, it would appear that the oxidation of an inhibitor may not be the mechanism by which germination is regulated in such seeds (Esashi *et al.*, 1976; Katoh and Esashi, 1975a, b). However, it has been shown that the upper seed of *Xanthium* can also be made to germinate by very small amounts of ethylene. It appears that the action of ethylene and that of oxygen are independent of each other. More than one mechanism of producing ethylene appears to exist. Thus the regulatory mechanism of dormancy in a species investigated for over 70 yr is still essentially unsolved.

Another seed showing physiological and morphological dimorphism is *Bidens pilosa*. While the elongated central achenes germinated readily, the outer shorter achenes were sensitive both to red light and to higher oxygen tensions, both of which increased their germination (Forsyth and Brown, 1982). However, even the long seeds responded to increased oxygen by more rapid germination. Whether a simple restriction of oxygen supply to the embryo is the cause of this effect is still uncertain.

Improved germination by increased oxygen tension is also shown by wild oats, indicating restricted permeability to oxygen. Spaeth (1932) reported that the nucellar membrane restricted the oxygen supply of seeds of *Tilia americana* and also provided evidence that the testa is impermeable to moisture. In neither case is the mechanism clear. A striking instance of the complexity of the function of a permeability barrier to oxygen is that of *Sinapis arvense* (Edwards, 1969). Seed coats of *S. arvense* are permeable to water but less so to oxygen. Removal of the seed coat allows an adequate supply of oxygen to reach the embryo. However, it was found that the requirement for oxygen by the embryo did not exceed the supply. In the presence of low oxygen tensions there was production of a germination inhibitor which tended to accumulate within the seed and which might prevent germination. Raised oxygen tensions slowed down inhibitor formation and permitted normal germination (Table 4.3).

TABLE 4.3. *Effect of O_2 concentration on O_2 uptake and growth inhibitor production by Sinapis embryos (after Edwards, 1969)*

O_2 concn (atmosp.)	O_2 uptake (ml/g/4 hr)	Inhibitor content (units)
0	0	11·8
0·05	1·0	10·5
0·1	2·0	9·2
0·2	2·5	8·8
1·0	2·5	8·5

Coat permeability to oxygen has been directly measured in apple seeds. Imbibition of the seeds at 20°C resulted in a rapid decrease in permeability to oxygen, while at 4°C oxygen permeated the seed coat quite readily. It seems possible that the seed integuments actually used up oxygen and thereby limited its supply to the embryo (Come, 1968). The dormancy of these apple seeds can be broken by placing them at 30–35°C in a moist atmosphere. Thevenot and Come (1978) suggest that this creates conditions of anaerobiosis of the embryo. In this case then a decreased availability of oxygen for the embryo is necessary to break dormancy. One must be careful to distinguish between conditions required for breaking dormancy and those needed for the subsequent germination of the seed.

The permeability of seed coats can be changed by the presence of external factors. For example saponins, which arise in the soil from decomposition of plant material, decrease the permeability of the seed coats of lucerne to oxygen (Marchaim *et al.*, 1972). It seems quite likely that this is not an isolated instance.

In barley dormancy is caused chiefly by the glumellae. The glumellae of barley take up oxygen, probably through the action of polyphenol oxidase. A distinct difference exists in the oxygen uptake of glumellae from dormant and non-dormant seeds. In the dormant seeds oxygen uptake starts at once, with imbibition, while in the non-dormant ones it only begins after a delay of about 12 hr (Lenoir *et al.*, 1986). This delay could be related to emergence from dormancy, but no causal relation between dormancy and oxygen uptake was detected. A similar case is that of *Avena sativa*, whose glumules absorb very considerable amounts of oxygen ($60\mu l/100mg.hr^{-1}$). Here too, glumellae from freshly harvested seeds begin to take up oxygen immediately, reaching maximal absorption after 5 hr. As the seeds emerge from dormancy the oxygen uptake by the glumellae starts later and its peak occurs later. However, in this case it is clear that this oxygen uptake is not the cause of dormancy, since peeling the seeds does not induce germination (Corbineaux *et al.*, 1986). Possibly the oxidative processes in the glumellae have a secondary effect, weakening of the resistance to the penetration of the embryonic root.

Changes in water permeability of seed coats were induced in *Pisum elatius* by changing the conditions of drying the seeds after harvest. In the absence of oxygen while drying, fully permeable seeds were obtained. In the presence of oxygen the seeds were impermeable to water (Marbach and Mayer, 1974). These results indicate that the permeability of seed coats can and does change in response to environmental factors. The possible mechanisms by which such impermeability is brought about include deposition of lignin or callose, or for example in *Sida*, cellulose and hemicellulose (Egley *et al.*, 1986; see also Chapter 1). Depending on the species, different cell layers may become impermeable.

Instances are known where seed coats are differentially permeable to oxygen and to carbon dioxide. Brown (1940) demonstrated that the nucellar membrane of *Cucurbita pepo* shows such differential permeability. The isolated inner membrane is more permeable to carbon dioxide (15.5 ml/cm^2hr) than to oxygen (4.3 ml/cm^2hr). The inner membrane is more permeable to gases than the outer one and it is the inner membrane which in fact controls permeability. The permeability of the membrane to carbon dioxide is increased by treatment with chloroform or heat. This treatment is supposed to kill the living cells and thus to change the structure and permeability of the inner membrane. It did not affect the permeability to oxygen, indicating that the two gases followed different paths of diffusion through the membrane. The significance of this for the germination of the Cucurbitaceae is not clear.

In *Xanthium*, carbon dioxide at high concentrations cannot induce dormancy of the intact seeds in the presence of oxygen. As little as 1% oxygen was sufficient to reduce the effectiveness of carbon dioxide in inducing dormancy. The induction process was temperature-dependent (Thornton, 1935).

Carbon dioxide does not invariably induce dormancy. In Chapter 3, a number of instances of improved germination have already been mentioned which are due to the presence of low concentrations of carbon dioxide. A more extreme case is that of the dormancy-breaking action of carbon dioxide on *Trifolium subterraneum* (Ballard, 1958). Carbon dioxide at concentrations between 0.3 and 4.5% was effective. Carbon dioxide above 5% was found to have an inhibitory effect. Treatment of the seed with activated carbon also stimulated germination and the results were consistent with the supposition that this treatment resulted in a raising of the carbon dioxide concentration. These findings have been extended to other species of *Trifolium*, as well as to *Medicago* and *Trigonella* species. Carbon dioxide treatment caused breaking of dormancy in those cases where cold treatment was effective and even in some cases where cold treatment was ineffective. Neither carbon dioxide nor cold broke the dormancy of the freshly-harvested seeds of *Trifolium* (Grant-Lipp and Ballard, 1959). The germination of *Phleum pratense* is stimulated by raised concentrations of carbon dioxide, both in the light and in the dark, even after removal of the seed coat as an obstacle (Maier, 1933).

Axentjev (1930) studied the effect of seed coats on the germination of seeds of many species. In several seeds dormancy was partly caused by seed coat effects, probably due to the impermeability of the seed coat, and partly by a light requirement, and these factors are to some extent additive. Thus, *Cucumis melo* is light-inhibited, but its germination in the dark is further improved by removal of the seed coat.

Seed coats may also exert a physical restraint on the developing embryo. If the thrust developed during imbibition and growth is inadequate, the seed will not rupture the seed coat and will fail to germinate. Thus certain kinds of dormancy are caused by the inability of the embryo to develop the necessary thrust. This appears to be the case for the dormant seeds of *Xanthium* (Esashi and Leopold, 1968). In such cases the mere mechanical weakening of the seed coat will relieve dormancy. In other cases it is possible that there is a requirement for chemical dissolution of the seed coat by enzymes produced by the radicle (Ikuma and Thimann, 1963). The endosperm of lettuce is said to be a mechanical barrier to germination and its dissolution has been invoked as a dormancy-breaking mechanism. However, this has not been adequately proven (Bewley and Halmer, 1980; Mayer, 1977). In seeds of *Sorghum halepensis* the glumes adhere firmly to the seed after seed shed. These glumes mechanically prevent protrusion of the radicle unless they are fractured or abraded.

The seed coat in *Acer* and its function in causing dormancy has been extensively investigated (Pinfield and Dungey, 1985). The seed coat seems to have multiple functions independent of the actual dormancy of the embryo itself. It exerts a mechanical restraint on the embryo, it can to some extent restrict water uptake and it may prevent the loss of an endogenous inhibitor. In addition, the seed coat can restrict oxygen supply to the embryo. In this one species, the seed coat imposes dormancy by the combination of a number of mechanisms, and in addition there is embryo dormancy.

The ability of the embryo to take up water may be of importance in dormancy of some seeds. Water uptake may be hormonally regulated and there is at least some indication that abscisic acid (ABA) may lower the ability of the embryo to take up water under osmotic stress (Schopfer and Plachy, 1984; McIntyre and Hsiao, 1985). Water is probably a general environmental signal, particularly as there is probably differential uptake in different parts of a seed (Mayer, 1986).

2. Temperature Requirements

The effect of temperature on germination as such has already been discussed. An additional effect of temperature must be considered. In order to germinate many seeds require an exposure to some definite temperature prior to being placed at the temperature which is favourable to germination. This treatment involves exposure of fully imbibed seeds to either high or low temperatures, at which germination does not occur. Seeds requiring such temperature treatment must be considered as being dormant and events occurring in them are akin to after-ripening in wet storage. The most commonly known and used procedure is exposure of seeds to low temperature under moist conditions and is termed *stratification*, i.e. layering the seeds in moist soil at low temperatures. Evidently, during stratification changes take place within the seeds. Embryo growth has been described in cherry seeds by Pollock and Olney (1959), in which, during moist storage at 5°C, the embryonic axis increased in cell number, dry weight and total length. An increased oxygen uptake was also noted in the embryonic axis and leaf primordium, but not in the cotyledons. On a cellular basis, an increase of oxygen uptake was especially marked in the embryonic axis. It appears that during after-ripening the energy supply to the embryo was increased. Investigations of the changes in nitrogen and phosphate content of cherry seeds during stratification were also carried out. Nitrogen content rose in the embryonic axis and in the leaf primordium but the nitrogen content per cell remained constant. Phosphorus content, however, increased both on the basis of the total per organ and per cell (Olney and Pollock, 1960). The phosphorus and nitrogen which appear in the embryo presumably originate in the storage organs. Some of these results are shown in Table 4.4. All the changes were either absent or much less marked if the seeds were held at 25°C instead of at 5°C. Some of the changes observed in the phosphate content of the embryo are presumably related to its greater rate of respiration.

TABLE 4.4. *Changes occurring in cherry seeds during stratification (compiled from graphs of Pollock and Olney, 1959; Olney and Pollock, 1960)*

	Axis length (mm)		No. of cells/axis		Dry wt/axis (μg)		O_2 uptake of axis as % of non-after-ripened		Total N/axis (μg)		Total P/axis (μg)	
Temp (°C)	5°	25°	5°	25°	5°	25°	5°	25°	5°	25°	5°	25°
Initial	1·6	1·6	180	180	400	400	100	100	23	23	2·0	2·0
After 8 weeks	1·75	1·6	220	180	400	320	300	100	21	19	2·9	2·0
After 16 weeks	2·3	1·6	265	210	600	320	550	100	29	20	3·6	1·5

The metabolism of peach seeds during stratification has also been investigated (Flemion and de Silva, 1960). Marked changes in amino acid, organic acid, and phosphate composition were noted. However, it was impossible to establish a definite relationship between these changes and the actual ending of dormancy.

A number of enzymes have also been shown to change during stratification. Catalase and peroxidase were shown to increase in *Sorbus aucuparia, Rhodotypos kerrioides* and *Crategus* (Flemion, 1933; Eckerson, 1913). These changes in the enzymes may be the direct cause of emergence from dormancy but it seems much more likely that they are the secondary result of other changes in the seeds. Barton (1934) was able to show a complete absence of a correlation between increase in catalase activity during after-ripening and the completion of the after-ripening of *Tilia* seeds.

In apple embryos protein bodies in the embryonic axis were broken down during stratification at 4°C but not at 25°C. This was accompanied by increases in the nucleolar volume and with RNA formation in cells of the hypocotyl of the intact embryos (Dawidowicz-Grzegorzewska and Zarska-Maciejewska, 1979). Even if such changes are closely related to dormancy breaking, this still fails to suggest a mechanism.

Many Rosaceous seeds contain cyanogenic glycosides such as amygdalin. These are often broken down during stratification, liberating hydrocyanic acid. There is some coincidence in time between the cessation of the liberation of hydrogen cyanide and the completion of after-ripening during stratification, but any causal relationship between the two processes seems unlikely.

In many cases no changes have as yet been observed during chilling treatment, but the evidence that before treatment the seeds failed to germinate and after treatment they did, indicates that internal changes nevertheless took place. In general it appears that most phases of metabolism can be affected during stratification. From the numerous investigations which have been carried out it is not possible to point to any one metabolic event which is directly responsible for dormancy breaking by stratification.

It is sometimes possible to force seeds requiring stratification to germinate by other means, for example by the complete removal of the seed coats or by removal of the cotyledons (Flemion and Prober, 1960). Thus chilling is not absolutely essential for seedling establishment in such embryos, particularly in Rosaceous seed. Seeds forced to germinate in this way do not form normal seedlings, the seedlings are either dwarfed or otherwise deformed (Barton and Crocker, 1948). However, cold treatment of the seedlings will again cause normal growth. Cold treatment apparently increases the growth capacity of the embryo and seedling.

During stratification the growth substance balance in seeds is changed considerably. The presence of growth inhibitors, such as abscisic acid (ABA), drops, for example in the seed of *Fraxinus americana, Juglans regia* and *Corylus avellana*. In *Acer plantanoides* acid inhibitors decreased, and gibberellic acid increased, well before germination began (Pinfield and Davies, 1978). In this case there is a good correlation in time between the events occurring during stratification and subsequent germination. Such correlations have not always been observed. In hazel nuts, for instance, the increase in gibberellic acid (GA) during stratification was negligible, the main increase occurring on transfer to the germination temperature. In many species gibberellins increased during stratification but decreased again to the initial level towards the end of stratification. It seems that in these cases gibberellic acid is not a limiting factor in breaking dormancy. It is more likely that the sensitivity of the embryo to the hormone changes (Ross, 1984). If this is the case, then quantitative measurements of hormone levels during stratification, or after-ripening, cannot contribute much to the understanding of the problem. Hopefully the attempts to localize the hormones in the embryo (using immunological methods) may supply more relevant information. In some cases, as in *Betula pubescens* seeds, stratification can be replaced by treatment of the seeds with gibberellic acid. It appears therefore that stratification does remove some block to gibberellin synthesis and its amount in the seeds does increase. It is evident therefore that during stratification the hormonal balance is changed. However, we do not yet understand how such changes bring about the germination and growth that result. Moreover, great care must be taken to distinguish between cause and effect, i.e. whether the changed pattern is the result of stratification, or whether it is due to other processes which occur during stratification, but which are not directly involved in dormancy breaking.

The conditions required for effective stratification are often similar to the natural conditions to which seeds are exposed. Many seeds are shed in autumn and then subjected to low temperatures during the winter months while they are exposed to moist conditions under leaf-litter or in the soil. Although this would lead one to expect that stratification requirements are similar in many plants, they in fact differ greatly. This difference exists both with regard to the length of the cold treatment and also with regard to the actual temperature during the treatment. Even in one family, e.g. the Rosaceae, and within one genus, marked differences exist. For example, Barton and Crocker (1948) cite experiments to show that while *Rosa multiflora* has a requirement of 2 months at 5 – 10°C, *Rosa rubiginosa* requires 6 months stratification at 5°C, and will not respond to treatment at either 10°C or 0.5°C. However, such a precise temperature requirement is rare and usually stratification can be carried out within a fairly wide temperature range. The best known cases of seeds requiring low temperature treatment for germination are found among the Rosaceae and among various conifers. However, in other families as well, such requirements are known; for example in *Juglans nigra* (Juglandaceae), *Aster* (Compositae), *Adonis* and *Anemone* (Ranunculaceae), *Iris* (Iridaceae) as well as many others including aquatic plants. In the latter the temperature treatment must be given while the seeds are actually immersed in water. Fuller lists of plants requiring stratification, giving details of time and temperatures used, are given by Crocker and Barton (1953) and Nikolaeva (1969). Crocker and Barton (1953) state that in the seeds of *Paeonia suffruticosa* germinated at normal temperatures, root protrusion occurs but the shoot fails to develop and no further development occurs. However, if the germinated seeds are placed at a low temperature (1 – 10°C) for 2 to 3 months and kept moist, the bud of the epicotyl shoot develops normally when the seedlings are again transferred to normal temperatures. This phenomenon is called epicotyl dormancy by these authors. Similar phenomena occur in some species of *Lilium*.

There is very little evidence in the literature to indicate that high temperatures in themselves break dormancy. Cases are known, as already mentioned, where high temperatures cause a change in the structure of the seed coats, thus causing a change in permeability. In contrast, many instances are known where there is interaction between the effects of low temperatures and somewhat raised ones. This probably involves a differential temperature requirement for growth of different parts of the embryo.

Sometimes brief exposure to slightly elevated temperatures promotes subsequent germination at lower temperatures after exposure to light, for example in *Poa pratensis* and *Lepidium virginicum*. The mechanism of the effect of the raised temperatures is in no way clear. In many desert seeds, storage at 50°C promotes germination, while after storage at low temperatures germination and seedling growth is poor. The same seeds if stored at 20°C required prolonged storage and maturation, of up to 5 months, for normal germination. In this case high temperatures may be regarded as breaking dormancy. However, if the temperatures were raised to 75°C the seeds failed to germinate (Capon and Van Asdall, 1967).

The importance of the seed coat in thermoinhibition is apparent from the effect of hypochlorite, which by the action of chlorine, seems to weaken the cell walls in the endosperm and seed coat of lettuce, so that germination can occur (Drew and Brocklehurst, 1984).

In many seeds germination is promoted by alternating temperature changes, which may be either diurnal or seasonal. Diurnal temperature changes have already been discussed in Chapter 3 (see also Table 4.6). A clear-cut effect is observable for *Nicotiana* when alternated between 20 and 30°C diurnally in the dark. Smaller effects are observable for some of the other seeds when the germination at 20° or 30°C is compared with that when they are alternated between 20° and 30°C. Seasonal changes have been reported for seeds of *Aphanes arvensis* (Roberts and Neilson, 1982). The temperature range which permits germination is extended to higher values during late spring and summer and narrows in autumn and winter. Dormancy was greatest in April (Fig. 4.2). Germination responded both to constant and alternating temperatures. The mechanism which is responsible for this kind of cyclic change in germination response is still unknown.

The temperature requirements for proper development of seedlings of *Convallaria majalis* are extremely complicated. Root protrusion occurs at 25°C, but is greatly

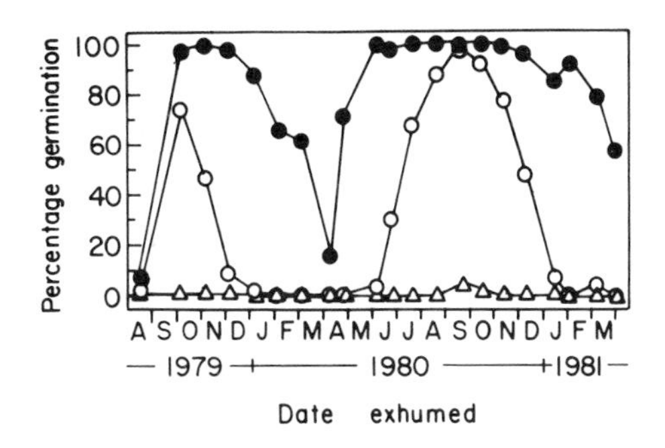

FIG. 4.2. Seasonal changes in the germinability of seeds of *Aphanes arvensis* (from data of Roberts and Neilson, 1982). Freshly collected seeds were buried outdoors in the soil (1977) and samples periodically removed and tested for germination at different temperatures. ●——● germination at 4°C, ○——○ germination at 15°C, △——△ germination at 25°C.

promoted if the seeds are first exposed to a low temperature, followed by a raised temperature that permits root growth. The first leaf enters dormancy after it has broken through the cotyledonous sheath and at this stage requires low temperature treatment at 5°C for its further development. Treatment at an earlier stage is ineffective. Further growth of the first leaf proceeds at a moderate temperature. The second leaf apparently also requires low temperature treatment for its development. Under normal conditions, in the natural habitat, such a cycle is not attained in less than 9 months and may take up to 14 months (Barton and Schroeder, 1942).

The literature on germination is full of examples where the temperature requirement is altered by exposure of the seeds to high, low or alternating temperatures. It seems clear that the changes occurring are extremely complex and affect primarily the rate of germination at the various temperatures. We have previously defined the optimal temperature of germination as involving both a rate factor and the actual germination percentage attained. Treatments with high, low or alternating temperatures can apparently widen the temperature range in which germination occurs; whether the optimal temperature is changed is not clear. This is illustrated by the behaviour of a number of seeds. *Betula lenta* seeds normally germinate only around 30°C. If the seeds are exposed to low temperature during stratification the temperature at which germination occurs is reduced to 0°C. *Festuca* seeds will not germinate at 30°C when freshly harvested, but storage at 20 – 30°C for 1 year will cause them to germinate at 30°C. Storage at low temperature does not have this effect; the seeds still fail to germinate at 30°C. A similar behaviour was observed in *Avena sativa* (Corbineaux *et al.*, 1986). In *Avena*, the changes occuring during storage at high temperature do not involve the embryo as the isolated embryos of dormant seeds germinate readily at 30°C.

Another instance is provided by annual *Delphinium* seeds, which will germinate at temperatures up to 30°C, following stratification at 5 – 10°C or even 15° C for 2 months, but not without this treatment.

In the preceding discussion it has been shown that temperature has a profound effect on dormant seeds yet there seems to be very little in common in the very varied responses.

As already discussed in Chapter 3, the effects of temperature may be sensed by the cellular membranes. Hendricks and Taylorson (1979) brought direct evidence for changes occurring in the properties of those membranes in seeds of *Barbarea verna* and lettuce at well-defined, critical temperatures. Membrane fractions were prepared, labelled with a temperature sensitive probe and the effect of temperature on the membranes was then determined. The critical temperatures at which the fluidity of the membrane occurred coincided with the temperatures at which physiological response of the seeds also occurred. For further discussion of temperature effects see Mayer and Marbach (1981) and Mayer (1977, 1986).

Other indications of the significance of membranes in dormancy breaking, come from the effect of alcohols, ether and chloroform on dormant seeds. In many cases exposure of seeds to alcohols or other solvents either in solution or to the vapour of the solvent relieves dormancy. The amounts required are frequently large, e.g. 100 – 500 mM in various species of grasses. Solvents are known to interact with lipid components of membranes. One way of interpreting the effect of solvent-dormancy breaking is to relate this to their anaesthetic effect (Taylorson and Hendricks, 1980/81). Other interpretations are also possible but there is little doubt that such effects do point to the membrane as one of the sites which controls or modifies seed dormancy.

3. *Light Requirements and their Interaction with Temperature*

The fact that light can act as a dormancy-breaking agent has already been mentioned. The question of the mechanism of light stimulation of germination has also been discussed in Chapter 3. In addition to those seeds which require light for their germination, there are many species whose germination is inhibited by light (see Table 4.5 and Table 3.8).

TABLE 4.5. *Effect of light and temperature on the germination of various seeds (compiled from data of Toole et al., 1955, and Koller, 1954)*

		% Germination at temperature stated					
Seed	Light condition	15°	20°	25°	30°	35°	20 – 30°C
Brassica juncea	Red light	90	48	18	2	—	80
	Dark	53	20	6	0	—	34
Lepidium virginicum	Red light	21	32	33	0	0	29
	Dark	0	0	0	0	0	1
Nicotiana tabacum	Red light	94	96	94	84	8	97
	Dark	2	5	6	0	0	97
Zygophyllum dumosum	White light	77	84	72	12	—	—
	Dark	80	82	82	16	—	—
Calligonum comosum	White light	0	4	8	0	—	—
	Dark	0	64	80	12	—	—
Juncus maritimus	White light	62	74	—	70	—	—
	Dark	0	0	—	0	—	—

Light does not only affect the absolute germination percentage but also the rate of germination. Thus in *Agrostis* seeds the final germination percentage is directly related to the light intensity. However, at high light intensities it takes much longer to reach maximal germination than at low ones (Leggatt, 1946).

In all these cases, no matter whether light inhibits or promotes, there is a complex interaction between light and other external conditions as well as with the age of the seeds. For example, seeds of *Salvia pratensis, S. sylvestris, Epilobium angustifolium, Echium vulgare* and others lose their light sensitivity very shortly after harvest. In *Salvia verticillata, Epilobium parviflorum* and *Rumex acetosella* sensitivity is retained at least for a year. In other seeds sensitivity is retained for much longer periods (Niethammer, 1927).

An example of the complexity of the situation is provided by lettuce seeds of the variety Grand Rapids, which are sensitive to light. These seeds, immediately after harvest, hardly germinate at all in the dark at 26°C. After a certain period of storage they germinate in the dark at 18°C, but require a light stimulus for germination at 30°C. When returned to 26°C they only germinate at about 1 – 4%. Irradiation of such seeds with red light, following the treatment at 30°C, raises the germination to 12 – 17%, if the seeds are again returned to 26°C. Controls not given treatment at 30°C will germinate 95% after this irradiation.

Induction of a light requirement was demonstrated by Toole (1959) for lettuce seeds of the variety Great Lakes. These seeds germinated 95% at 20°C. If kept for four days at 35°C they will only germinate 11% at 20°C in the dark. However, irradiation with red light before returning them to 20°C restored germination to 95%. The light requirement slowly decreases as the length of the storage period increases, and eventually, after several years of storage, germination in the dark at 26°C reaches 60 – 80%. It is possible to induce a light requirement in lettuce seeds which do not normally show it, by treating the seeds with coumarin or with high temperatures or with solutions of high osmotic pressure. Peeling the seeds, including removal of the endosperm, or merely puncturing the endosperm, which encloses the embryo, is sufficient to abolish their light requirement at 26°C.

Positional effects of seeds may determine the interaction of light and other dormancy-breaking factors such as potassium nitrate (KNO_3). In *Avena fatua* the proximal seeds in the spikelet responded differently, than the adjacent ones, to potassium nitrate in the presence of far-red radiation, designed to give a defined P_{FR}/P_{total} ratio (Hilton, 1985).

Interaction of the light response of seeds with temperature was clearly shown by Toole (1959) for seeds of *Lepidium virginicum*. Such seeds will not germinate in the dark if kept at 15°C or 25°C, or if transferred from 15 to 25°C or vice versa. If illuminated with red light, two days after sowing they will germinate to about 30 – 40%, if kept throughout at either 15 or 25°C or on transfer from 25 to 15°C. However, the reverse transfer, from 15 to 25°C, when the light is given after the first two days, causes full germination. From this work and also from the examples quoted in Chapter 3, it is quite clear that the photochemical reaction in germination, involving phytochrome, is very closely linked to other chemical reactions which will determine whether and to what extent there will be a response to illumination.

The interaction of light and alternating temperatures in dormancy breaking of *Dactylis glomerata* illustrates how complicated the systems are. It was shown that during storage a quantitative increase in the response to one factor (light or alternating temperature) was met by an identical increase in the response to the other. In this species there are subtle differences in dormancy between seeds of different populations of plants. Within each population there is a normal distribution in the length of the period of storage required to break dormancy. In addition, during storage, while primary dormancy is lost, secondary dormancy may be induced (Probert *et al.*, 1985).

Complex interaction between light and temperatures has been observed in many seed species. *Tagetes* requires light for germination, but light cannot overcome the inhibitory effect of elevated temperatures. However, light becomes effective again once the seeds are shifted to a lower temperature at which they still do not germinate in the dark (Forsyth and Van Staden, 1983).

In many seeds the light effect is related to the presence of the seed coat. The inhibition caused by light to germination of *Cucumis melo* is reduced by removal of the seed coat. In *Agropyron smithii* removal of the seed coat does not abolish the inhibitory effect of light on germination. An unusual example is provided by *Oenothera biennis* where the intact seeds of the plant do not germinate in light or dark. However, if the seed coat is pierced, the seeds will germinate in light. Germination in the dark is only slightly promoted.

An entirely different light response occurs in seeds whose germination is affected, not by short illumination, but by alternations between dark and light, i.e. by photoperiod. Photoperiodic responses of seeds were first shown by Isikawa (1954, 1955). He suggests that both light-stimulated and light-inhibited seeds may be considered to show short day behaviour. Thus in the light-requiring seeds of *Patrinia* and *Epilobium*, which germinate under illumination of between 2 and 21 hr, germination is promoted by interruption of the light period by darkness. Light-inhibited seeds such as *Nigella damascena* and *Silena armeria* germinate if illuminated for 1 min or 3 hr daily respectively with low light intensities. The only instances quoted by Isikawa indicating a long day requirement are seeds of *Leptandra* and *Spiraea*. A true long day requirement for germination appears to exist in *Begonia* seeds. The seeds germinate if given at least three cycles of long days, the critical day length being 8 hr. A break in the dark period, if illumination is less than 8 hr, is also effective in causing germination. Furthermore, the response to light is increased in the presence of gibberellin, the critical day length being greatly shortened (Nagao *et al.*, 1959).

Other more complex examples are provided by the work on *Betula* (Black and Wareing, 1955), on *Tsuga canadensis* (Stearns and Olson, 1958) and on *Escholtzia argyi* (Isikawa and Ishikawa, 1960). *Betula* seeds will germinate at 15°C only under long day conditions, eight cycles being required to induce germination. However, at 20 – 25°C germination will occur following a single exposure to light for 8 – 12 hr. *Escholtzia* seeds do not germinate either in the light or the dark at constant temperatures. The seeds will, however, germinate if the light treatment is combined with a thermoperiodic treatment. This is illustrated by Table 4.6. From this it appears that highest germination is obtained if a 6 hr photoperiod at 25°C is followed by an 18 hr thermoperiod at 5°C. The results shown in Table 4.6 are difficult to analyse as both photo- and thermoperiod were varied simultaneously. It is never clear to which change increased or decreased germination must be ascribed. *Citrullus colocynthis* seeds provide another example of photoperiodic response in germination.

TABLE 4.6. *Percentage of germination of seeds of Escholtzia with daily photoperiodic and thermoperiodic treatments (after Isikawa and Ishikawa, 1960). Each temperature and light treatment was repeated eight times; the initial treatment was done in light*

Temperature and period of light exposure	Temperature and period of darkness	Germination (%)
15°, 6 hr	5°, 18hr	60
15°, 18 hr	5°, 6 hr	24
25°, 6 hr	5°, 18 hr	80
25°, 18 hr	5°, 6 hr	25
25°, 6 hr	15°, 18 hr	40
25°, 18 hr	15°, 6 hr	10
35°, 6 hr	5°, 18 hr	63
35°, 18 hr	5°, 6 hr	28
35°, 6 hr	15°, 18 hr	66
35°, 18 hr	15°, 6 hr	0

Seeds of this plant germinate in the dark at 20°C. However, exposure to long days, i.e. 12 hr of light per day, inhibits their germination. Short day treatment, eight hours of illumination, does not inhibit.

These phenomena show a marked similarity to photoperiodism in flowering as in many cases induction phenomenon can be identified. Seeds of *Begonia* germinate when they have been given six cycles of a nine hour (or more) photoperiod, followed by seven days of darkness (Nagao *et al.*, 1959). *Eragrostis* seeds require a dark period following imbibition otherwise the light treatment is not fully effective. Interruption of the dark period reduces its promotive effect. In *Hypercicum*, short light periods separated by dark periods have been shown to be cumulative in their effect. In *Atriplex dimorphostegia* short periods of illumination promote, while long periods inhibit germination. Germination of *Citrullus lanatus* is inhibited by irradiation with far-red light, and this inhibition may be reversed by irradiation with red light at a suitable time. If the time interval between red and far-red irradiations is increased beyond a certain length the photoreversibility is lost (Botha *et al.*, 1982). Botha *et al.* (1982) also showed that induced secondary dormancy is partially released by some hormones. Ethylene was particularly effective and its effect of releasing dormancy was concentration dependent. In all the above cases the photoperiodic effects of light appear to be mediated by the phytochrome system. It should be recalled that phytochrome interconversion can occur even in dry seeds. Therefore full imbibition is not necessarily required for perception of the light effect. Furthermore, repeated cycling between P_R and P_{FR} can eventually lead to the formation of inactive phytochrome. Some of these effects were discussed by Koller (1972). It seems that the general principles relating to photoperiod responses also apply to germination.

In recent years there has been a fair amount of evidence indicating that phytochrome is located on or near membranes in the cell. As a result it is of course tempting to relate the dormancy-breaking action of light to changes in properties of such membranes. However, one must be careful in this interpretation tempting as it is until rigorous proof becomes available (see also Chapter 3).

4. Germination Inhibitors

A large number of substances can inhibit germination. All those compounds which are toxic generally to living organisms will also, at toxic concentrations, prevent germination, simply by killing the seed. Far more interesting is the action of those substances which prevent germination without affecting the seeds irreversibly. The simplest case, and one perhaps most frequently met with in nature, is that of osmotic inhibition. It is possible to prevent the germination of seeds simply by placing them in a solution of low osmotic potential. When the seeds are removed from such an environment and placed in water, they germinate. Such a situation appears to exist in fruits and in the case of seeds of plants from saline habitats. The substances responsible for the low osmotic potential may be sugars, inorganic salts such as sodium chloride, or other substances. The threshold of osmotic potential at which germination is prevented differs widely for different seeds. A few examples of those differences are shown in Fig. 4.3.

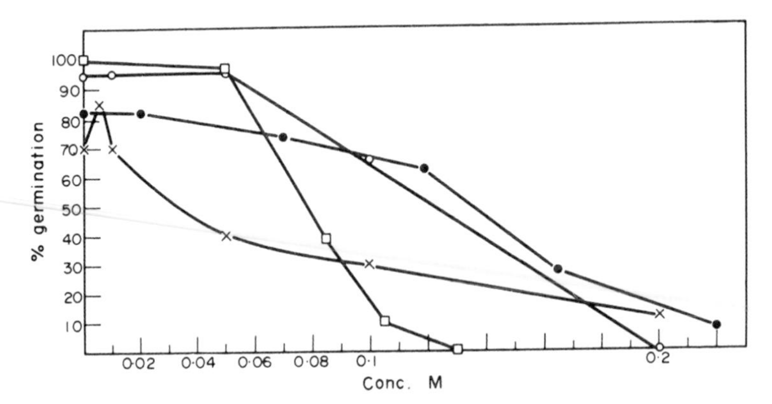

FIG. 4.3. The effect of sodium chloride concentration on the germination of various seeds (compiled from data of Uhvits, 1946; Lerner *et al.*, 1959; and Poljakoff-Mayber, unpublished). ×——× *Atrilex halimus*, ●——● Alfalfa, ○——○ Tomato, □——□ Lettuce. Germination percentages determined after 8 days for tomatoes, 4 days for *Atriplex* and 2 days for lettuce and alfalfa.

In the laboratory, convenient compounds for obtaining osmotic inhibition are PEG (polyethylene glycol) or mannitol. The use of these osmotica frequently gives results somewhat different from those obtained with sodium chloride. Mannitol and other non-ionic osmotica usually allow a higher percentage germination than sodium chloride (Fig. 4.4). This difference is ascribed to the ion toxicity of the salts. It appears therefore that the germination percentage obtained is not strictly a function of the availability of water to the seed. The inhibitory effect of salt does not persist after the salt is removed and washed off the seeds (Kurth *et al.*, 1986). The tolerance of seeds to salt, during germination, shows great variability. It is not clear whether the toxic effect of salts affects the initial stages of germination, during imbibition,

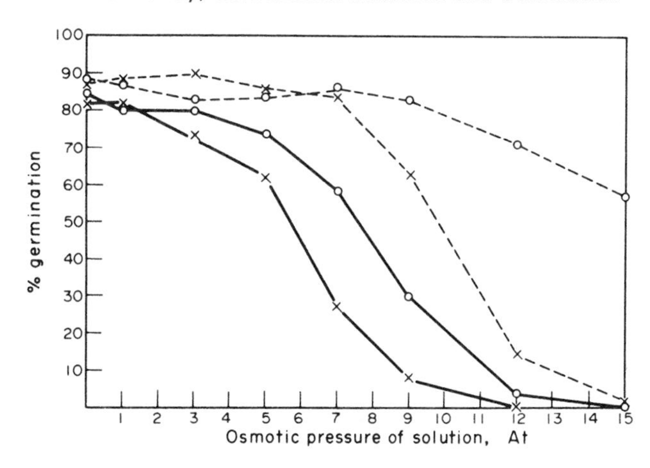

FIG. 4.4. The effect of osmotic potential and ionic toxicity on the germination of alfalfa seeds (from data of Uhvits, 1946). o——o Germinated in mannitol for 2 days, o-----o germinated in mannitol for 10 days, ×-----× germinated in sodium chloride for 2 days, ×——× germinated in sodium chloride for 10 days.

or whether it affects the seed only after it is already fully hydrated as was shown for barley (Bliss *et al.*, 1986).

Transient exposure to low osmotic potential is not always toxic, sometimes it can promote the subsequent germination, after the osmoticum is removed. This treatment is now used in agriculture and referred to as priming or osmoconditioning.

Osmoconditioning shortens the spread of germination without affecting the actual percentage germination, e.g. in carrots, celery and onions (Brocklehurst and Dearman, 1983). In other cases germination increased. Osmoconditioning also improves vigour in seed lots with low vigour, and enables formation of uniform stands in the field. As it is well known that during storage various deteriorative processes occur in seeds, the controlled hydration maintained during osmoconditioning may permit repair processes to take place (Burgass and Powell, 1984). The repair processes may be of membranes or even of DNA or RNA. This treatment is now used in order to speed up germination of seeds.

Another type of inhibition is that caused by substances which are known to interfere with certain metabolic pathways. As germination is closely associated with active metabolism, all compounds which interfere with normal metabolism are likely to inhibit germination. An example of this type of interference is provided by various respiratory inhibitors. Compounds such as cyanide, dinitrophenol, azide, fluoride, hydroxylamine and others, all inhibit germination at concentrations similar to, but not always identical with, those inhibiting the metabolic processes. However, in some cases dinitrophenol and cyanide have a dormancy-breaking effect. Cyanide is actively metabolized by some seeds if applied in low concentrations (Taylorson and Hendricks, 1973). Mancinelli (1958) claims that ethionine and hydroxylamine appear to increase the light requirement of lettuce seeds, or at any rate that light is able partially to reverse the inhibition caused by these substances. While the light reversal occurred both at 22° and 26°C for ethionine, in the case of hydroxylamine the effect seemed more marked at 26°C.

Herbicides of various kinds inhibit germination to a greater or lesser extent. Many of the commonly used substances, such as 2,4-D, affect germination of wheat at comparatively low concentrations, of the order of 10^{-5} M, to cause 50% inhibition of germination. Wheat is among the least sensitive plants. To get the same extent of germination inhibition by coumarin, concentrations of the order of 10^{-3} M have to be used. Dicotyledonous plants such as cress, carrot, beetroot and others are more sensitive to coumarin than wheat and concentrations of the order of 10^{-4} M will cause 50% inhibition of germination.

The more effective of the herbicides have been used in order to prevent the germination of weed seeds in agricultural crops. A very frequent use of herbicides is as pre-emergence weed killers. The herbicide is applied in order to kill the seedling immediately after germination and before the main crop germinated. In these cases the herbicides are not in fact acting as germination inhibitors. Most growth retardants, such as Cycocel, Phosphon D and Amo 1618, if applied in sufficiently high concentrations also inhibit germination, apparently due to their being inhibitors of the gibberellic acid synthesis (see also section IV of this chapter).

Phenolic compounds of various kinds inhibit germination. Some of the selective herbicides, such as the substituted phenols and cresols, inhibit germination due to their general phytotoxicity. In addition to these compounds, many other phenolic substances also inhibit germination. A list of a few of these and the concentration at which they inhibit the germination of lettuce seeds is given in Table 4.7.

TABLE 4.7. *The inhibition of germination of lettuce seeds, Progress variety, by a number of phenolic compounds (Mayer and Evenari, 1952 and 1953)*

Compounds	Concentration causing 50% inhibition
Catechol	10^{-2} M
Resorcinol	5×10^{-3} M
Salicylic Acid	$1{\cdot}5 \times 10^{-3}$ M
Gallic Acid	5×10^{-3} M
Ferulic Acid	5×10^{-3} M
Caffeic Aicd	$> 10^{-2}$ M
Coumaric Acid	5×10^{-3} M
Pyrogallol	10^{-2} M

Many other phenolic substances at similar concentration also affect the germination of a variety of seeds. Because of the widespread occurrence and distribution of phenolic compounds in plants and fruits, it has been suggested that these substances might act as natural germination inhibitors (van Sumere, 1960). However, usually no quantitative relationship between dormancy and endogenous content of phenolic compounds has been found. Thus the polymorphic seeds of *Atriplex triangularis* differ between them in the degree of dormancy. The small seeds are the most dormant, but contain the smallest amount of phenolics on a dry weight basis. Therefore if phenols are important in the induction of dormancy, in this case it must be due to the qualitative differences in phenolics composition and not due to a quantitative effect (Khan and Ungar, 1986). Such qualitative differences were indeed described.

All of the compounds mentioned so far are inhibitory but cannot be regarded as dormancy inducing. The term dormancy-induction is used to imply that the seed can again be made to germinate by one or other of the treatments already mentioned, which break natural dormancy. As already mentioned, coumarin can induce light sensitivity in varieties of lettuce seeds not requiring light for germination (Nutile, 1945).

Coumarin and its derivatives are of fairly widespread occurrence in nature. Coumarin itself (Fig. 4.5) is characterized by an aromatic ring and an unsaturated lactone structure. The effect of changes in structure of the coumarin molecule on its germination inhibitory activity has been investigated in some detail for wheat and lettuce seeds. This work showed that there was no single essential group in the coumarin molecule which was the cause of its inhibitory action. Reduction of the unsaturated lactone ring or substitution by most substituents such as hydroxyl, methyl, nitro, chloro and others in the ring system, all reduced the inhibitory activity of coumarin. A certain variation in the response of the two species tested was observed, however it did not materially alter the general picture (Mayer and Evenari, 1952). Opening of the lactone rings also considerably reduced the inhibitory action of coumarin.

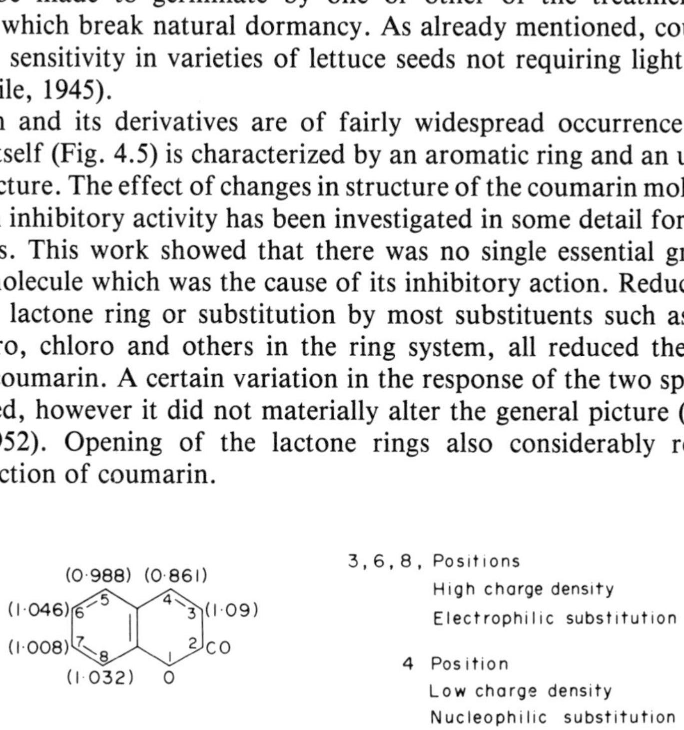

(0·988) (0·861)
(1·046) 6 5 4 3 (1·09)
(1·008) 7 8 1 2 CO
(1·032) O

3, 6, 8, Positions
High charge density
Electrophilic substitution

4 Position
Low charge density
Nucleophilic substitution

FIG. 4.5. Structure of coumarin with ring numbering and showing charge density.

Because of its widespread distribution in plants and due to its strong inhibitory action, coumarin is considered to be one of the substances which may function as a natural germination inhibitor. This has been frequently proposed, but the occurrence of coumarin itself in seeds at inhibitory concentrations has been proved only in *Trigonella arabica* (Lerner *et al.*, 1959). However, coumarin derivatives such as the glycosides of the lactone or substituted coumarins have been shown to occur in many fruits. It is not unlikely therefore that such compounds act as natural germination inhibitors. Coumarin is actively metabolized in germinating seeds.

The most important other germination inhibitor is undoubtedly abscisic acid (Fig. 4.6), which will be discussed together with other hormones active in germinating seeds (see section IV of this chapter).

The idea that fruits and seeds contain natural germination inhibitors is not new. Wiesner (1897) suggested that seeds of *Viscum* contain a germination inhibitor which prevents germination within the fruit. This idea was further developed by

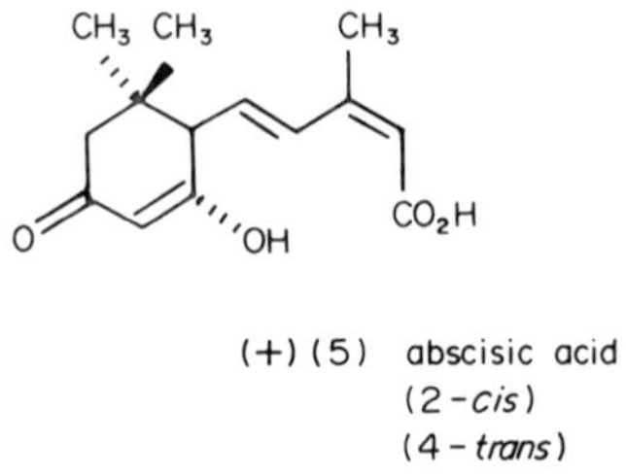

(+) (5) abscisic acid
(2-*cis*)
(4-*trans*)

FIG. 4.6. Structure of (+) (S) abscisic acid – (2 – *cis*;4 – *trans*).

other workers including Molisch. It appears that Oppenheimer (1922) was among the first to study this problem experimentally. He tested the juice of tomatoes to determine whether it contained a germination inhibitor. He concluded that it did contain an inhibitory substance. This conclusion has since been disputed. Some authors claim that the inhibitory action is solely due to osmotic effects while others claim to have isolated an inhibiting substance. Akkerman and Veldstra (1947) claim to have shown that caffeic and ferulic acids are the compounds responsible for the inhibition of germination of seeds within the tomato fruit. Despite this claim it must be pointed out that these authors failed to show that caffeic and ferulic acids occur in the tomato at concentrations sufficiently high to account for the inhibition of germination of the seeds in the mature fruit. It seems likely that the explanation of the failure of tomato seeds to germinate in the fruit can be ascribed to the combined action of osmotic inhibition together with a specific or non-specific germination inhibitor, such as ABA, whose presence in tomato juice has been demonstrated (Dorffling, 1970), or coumarin. Such interaction between the osmotic effect and the effect of an inhibitor has been shown to exist for example in the case of lettuce seeds (Lerner *et al.*, 1959). It seems very probable that this situation exists in many other cases (Evenari, 1949), for example in grapes. The same author reviewed the older literature regarding germination inhibitors. Numerous substances have been identified in seeds which were able to inhibit germination but their biological role is often in doubt. Coumarin and hydroxycinnamic acid and their derivatives, as well as vanillic acid, have been shown to occur in barley husks (van Sumere *et al.*, 1958). Both Varga (1957a,b,c,d) and Ferenczy (1957) studied the inhibiting compounds occurring in lemons, strawberries and apricots. Coumarin and derivatives of cinnamic and benzoic acids were identified as the compounds responsible for inhibitory action. The same type of compounds have also been identified in clusters of sugarbeet seeds (Roubaix and Lazar, 1957; Massart, 1957). In sugarbeet fruit *cis* – 4 – cyclohexene – 1,2 – dicarboximide was identified, which has inhibitory properties (Mitchell and Tolbert, 1968). The fruit of *Gardenia jasminoides* contains genipin (or its derivatives — see Fig. 4.7). When the seeds retain part of the flesh of the fruit they fail to germinate. However, after they pass through the alimentary tract of a bird the seeds do germinate — apparently the inhibitor is digested (Shimomura *et al.*, 1983). Since the compound constitutes about 4% of the dry weight of the fruit and is active at a concentration of 5×10^{-4} M, it might have some role in inducing dormancy in the seed. In all these cases, especially when more than one substance

KNO_3

I

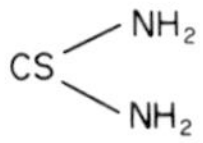

III

C_2H_4

II

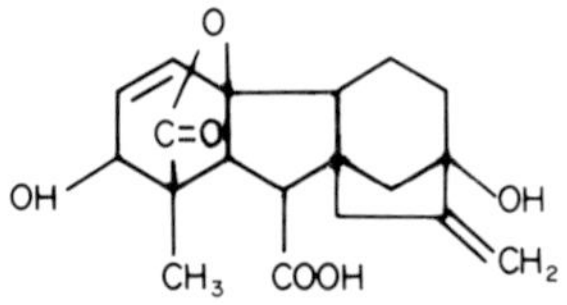

IV

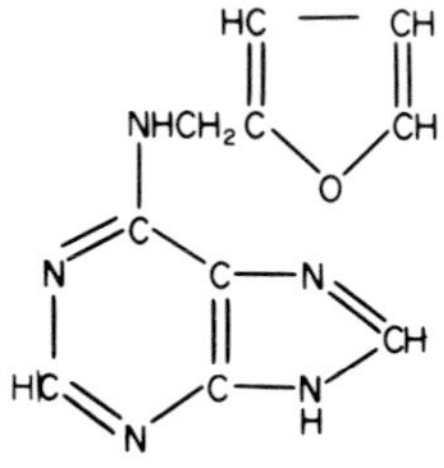

V

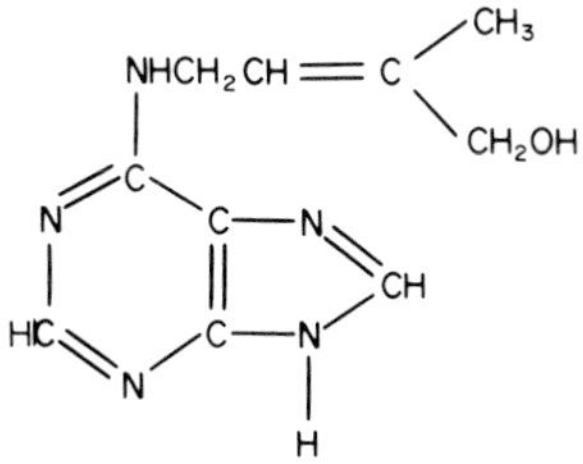

VI

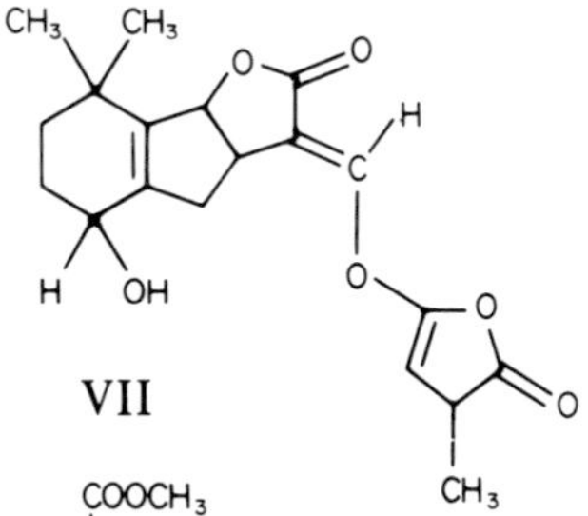

VII

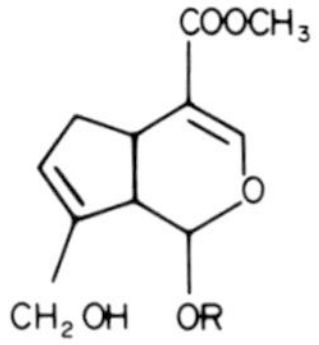

VIII

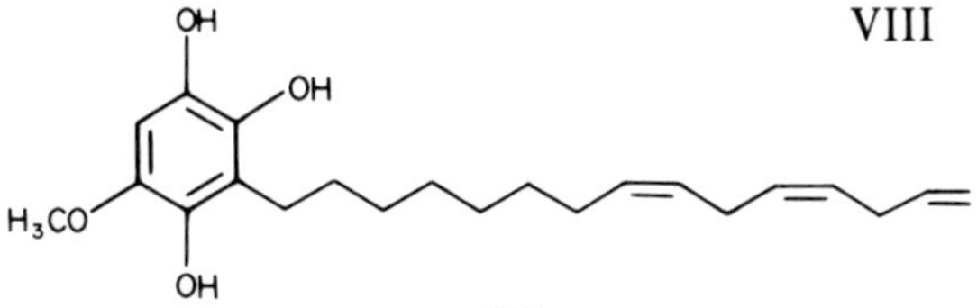

IX

FIG. 4.7. Structure of some dormancy-breaking compounds.
(I) Potassium nitrate, (II) ethylene, (III) thiourea, (IV) gibberellin A_3,
(V) kinetin, (VI) zeatin, (VII) strigol, (VIII) genipin, (IX) hydroquinone derivative:
2 – hydroxy – 5 – methoxy – 3 – [(8'Z,11'Z) – 8',11',14' – pentadecatriene] – *p* – hydroquinone.

was isolated from the fruit or the seed, it seems likely that dormancy, or inhibition of germination, is due to the combined effect of several compounds.

Evenari (1949) in his review attempts to classify the naturally occurring inhibitors found in seeds and fruits. He mentions cyanide-releasing complexes especially in Rosaceous seeds, ammonia-releasing substances, mustard oils, mainly in Cruciferae, various organic acids, unsaturated lactones, especially coumarin, parasorbic acid and protoanemonin, aldehydes, essential oils, alkaloids and phenols. From a glance at this list it can be seen that these compounds are not in any way restricted in occurrence to seeds and fruits. On the contrary substances of this general type occur in leaves, roots and other parts of plants as well as in the fruits and seeds. For this reason it is easy to understand why germination inhibitory substances have been isolated from a variety of plant tissues. Unfortunately, frequently the precise chemical nature of the inhibitor has not been established and it is often not clear what, if any, the biological importance of these substances is. This aspect of the problem will be dealt with in Chapter 7.

III. Germination Stimulators

Various chemical substances can completely or partially substitute for light or temperature in breaking dormancy. These substances are very varied in chemical nature. Some of them are simple compounds, such as potassium nitrate and thiourea, while others are complicated molecules such as the hormones, gibberellins and cytokinins which will be dealt with in section IV. The structure of a few of these dormancy-breaking substances is shown in Fig. 4.7.

The effect of potassium nitrate on germination was discovered when it was noted that Knop's solution promoted the germination of some seeds. Further experiments showed that potassium nitrate was responsible for this stimulation. Potassium nitrate promotes the germination of a number of seeds, for example *Lepidium virginicum, Eragrostis curvula, Polypogon monspelliensis*, various species of *Agrostis* and *Sorghum halepense*. The effect on some other species is shown in Table 4.8. The stimulation obtained by potassium nitrate is dependent on its concentration, showing an optimum, (Fig. 4.8). As with light, potassium nitrate stimulation shows interaction with temperature. The germination of *Eragrostis curvula* is stimulated between 15° and 30°C in the dark by 0.2% potassium nitrate. At higher temperatures or alternating temperatures there was no effect. In contrast the germination of *Polypogon* is promoted by potassium nitrate only at alternating temperatures (Toole, 1938).

TABLE 4.8. *Effect of KNO_3 on germination of various seeds (after Hesse, 1924) Concentration of KNO_3 0.01M*

Seed	Temp.	Incubation time in days	% Germination	
			Water	KNO_3
Veronica longifolia	16–20°	20	3.5	45
Hypericum perforatum	17–22°	20	28.0	57
H. hirsutum	15–20°	28	20.0	43
Epilobium hirsutum	16–20°	14	23.0	40
E. montanum	16–20°	14	6.5	89
Verbascum hypsiforme	18–19°	14	9.0	71

Interactions between nitrate and other growth promoting factors exist. Seeds of *Sisymbrium officinale* undergo seasonal changes in dormancy in the presence of nitrate in the soil. Nitrate was required in combination with red light in order to permit germination, while if either of these is absent the seeds enter secondary dormancy (Hillhorst *et al.*, 1986).

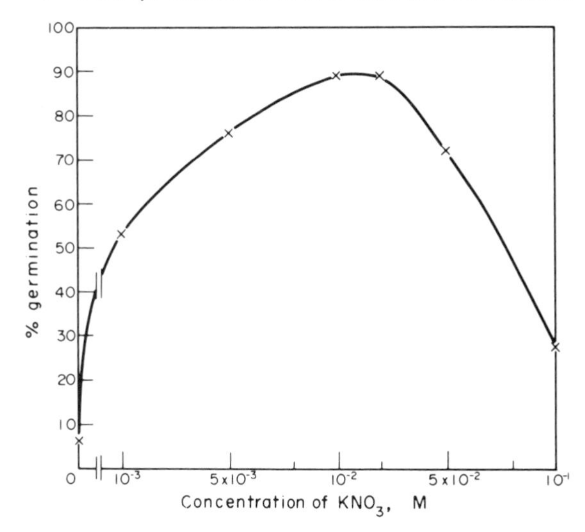

FIG. 4.8. Relation between germination percentage of seeds of *Epilobium montanum* and the potassium nitrate concentration. Germination in the dark for 14 days at 15 – 20°C (after Hesse, 1924).

Interactions between light and nitrate have often been reported for other seeds, for example, *Poa, Capsella* and *Chenopodium*. In the case of *Sisybrium* the effect of nitrate appears to be directly related to phytochrome, nitrate apparently acting as some kind of cofactor to phytochrome action. It is possible that the combined actions of the two lead to gibberellin synthesis. Indeed, in the same seed GA_{4+7} could substitute for nitrate and red light, but at the same time red light alone could reduce the level of GA required to induce germination. Various suggestions have been made with regard to the exact site at which nitrate acts, but although action at the membrane seems quite probable, no direct evidence is at present available.

The dark germination of many seeds is stimulated by thiourea. For lettuce seeds the effective concentration is of the order of 10^{-2} to 10^{-3} M. These concentrations apply if the seeds are actually germinated in the solution of thiourea. A procedure frequently adopted involves soaking the seeds in a concentrated solution of thiourea, $0.6 - 4 \times 10^{-2}$ M, for a short time and then transferring them to water. Thiourea has been shown to stimulate the germination of seeds of *Cichorium, Gladiolus* (Shieri, 1941) and *Quercus* . In oak seed, as well as in other tree species such as *Larix* and *Picea*, thiourea was substituted for the cold temperature treatment (Deubner, 1932; Johnson, 1946).

The effect of thiourea on the germination of lettuce was first described by Thompson and Kosar (1939) and its interaction with light was studied by Evenari *et al.* (1954). Both in lettuce and endive, thiourea abolished the inhibitory effect of high temperature; in lettuce thiourea also abolished the light requirements of the seeds. Thiourea, in addition to stimulating germination, inhibits growth. Prolonged treatment of seeds with high concentrations of thiourea is therefore liable to cause an apparent inhibition of germination, because the emergence of the root is prevented. It was found that germination of lettuce above 90% was obtained if the seeds were kept for 24 hr in 5×10^{-2} M thiourea and then transferred to water.

Water controls germinated at only 40%. Keeping the seeds longer in thiourea reduced the germination below 90%. As in many other effects on germination, here also there exist interactions between the effect of thiourea and that of temperature. Moreover, the effects observed are a function of the thiourea concentrations (Table 4.9). It is clear that at different temperatures the concentration of thiourea effective in causing germination is different. The effects of thiourea on the metabolism of seeds, which have been studied in some detail, will be discussed in Chapter 6. Thiourea substitutes for the natural germination stimulator which develops in *Fraxinus* seeds during chilling (Villiers and Wareing, 1960). In some varieties of peach seeds, thiourea could substitute for after-ripening in promoting germination but subsequent seedling growth was abnormal (Tukey and Carlson, 1945).

TABLE 4.9. *The effect of thiourea on the germination of lettuce, variety Grand Rapids, at different concentrations and temperatures (Poljakoff-Mayber et al., 1958)*

Thiourea	% Germination	
	20°C	26°C
0	26.8	5.8
$1{\cdot}0 \times 10^{-2}$ M	67·4	21·4
$2{\cdot}5 \times 10^{-2}$ M	76·5	16·1
$5{\cdot}0 \times 10^{-2}$ M	56·3	12·4

The action of derivatives of thiourea on the germination of lettuce seeds has also been studied. All modifications of the structure of thiourea, e.g. ethyl, methyl and phenyl substitution converted it from a germination stimulator to a substance inhibiting germination (Mayer, 1956). Thiourea has been shown to be present in the seeds of at least one plant species, *Laburnum anagyroides* (Klein and Farkass, 1930). The effect on germination of a large number of other thiol compounds was studied by Reynolds (1974), who found that most of them had some stimulatory effect on germination.

It seems likely that other compounds of stimulating activity are of widespread occurrence in nature. Libbert and Lubke (1957) showed that the widely distributed scopoletin can promote germination in old seeds of *Sinapis alba* at very low concentrations. Ruge (1939) indicated that in *Helianthus annuus* seeds, there is a balance between a germination-stimulating and an inhibiting factor, however, neither was identified. The best established cases of natural stimulators are those of the factors responsible for stimulating the germination of the seeds of the parasitic plants *Striga* spp. and *Orobanche* spp. Brown *et al.* (1952) showed that seeds of both these plants germinate in the vicinity of the host only after the latter secretes a germination-stimulating substance. One of the stimulating substances appears to be xylulose. A more active compound is strigol (Fig. 4.7, VII), which causes stimulation of the germination of *Striga* at extremely low concentrations (Cook *et al.*, 1972). Strigol was isolated from a non-host plant of *Striga*, cotton. Recently a germination stimulant for *Striga* has been isolated from the exudate of a host, *Sorghum*. This compound has been characterized as a hydroquinone derivative, which is active at 10^{-7} M. The structure of this hydroquinone is shown in Fig. 4.7, IX (Chang *et al.*, 1986). The germination of *Orobanche* is stimulated by root exudates and this stimulation is enhanced by gibberellins (Garas *et al.*, 1974). A number of additional substances have been partially isolated from exudates of host plants of *Orobanche* and *Striga* (Edwards, 1972).

Volatile compounds are often produced by plants, often following injury, and after infection by fungi. It is not surprising that some of these compounds are able to stimulate germination. Among the more active compounds found were nonanenitrile, 2 – nonanone and octylthiocyanate (French, 1985). The latter compound stimulated germination of *Rumex* at a concentration of 1 ppm by volume and the other two at 5 – 10 ppm/vol. These are low concentrations and might, under very specific conditions, arise in the soil or near plants and affect germination.

Lastly the stimulating action of the group of compounds, such as ethanol, ether and chloroform, generally regarded as anaesthetics, must be mentioned. These compounds act if applied to the seed during imbibition and then can be removed and yet stimulate germination (Taylorson, 1982). Many species are stimulated to some extent and it has been suggested, but not proven, that they stimulate germination and break dormancy by acting at cellular membranes.

IV. Hormones in Germination Stimulation and Inhibition

Plant hormones are known to affect germination. The term hormone is here used in the widest sense of the word, including various growth substances and natural compounds having a regulatory function in the plant.

Lona (1956) and Kahn *et al.* (1956) were among the first to show that gibberellic acid stimulates the germination of *Lactuca scariola* and *Lactuca sativa* in the dark. In these cases gibberellic acid substituted for light in promoting germination. Other cases of substitution of gibberellic acid for red light are known in the case of seeds of *Arabidopsis, Kalanchoe* and *Salsola volkensii*. However, in a number of seeds whose germination is either promoted or inhibited by light, e.g. *Juncus maritimus, Oryzopsis miliacea* and others, treatment with gibberellic acid was not effective in promoting germination (Leizorowitz, 1959). A number of species of plants whose germination is not affected by light have been shown to be promoted by gibberellic acid, as shown in Table 4.10. The sensitivity of some of the seeds depended on the time of harvest. Although the effect of gibberellic acid seems to be similar to that of light (Table 4.11) gibberellic acid is far more effective than red light in reversing the high temperature inhibition of germination in lettuce. Thus 100 ppm gibberellic acid stimulated the germination of lettuce seeds, at 30°C, from 2% in water to 33%. In contrast red light had no effect. When both gibberellic acid and light were given a germination percentage of 50 was obtained. Gibberellic acid can also reverse the inhibition of germination caused by high osmotic pressure. Khan (1960a,b) showed that lettuce seeds germinating to 82% in the dark, in water, gave only 22% germination in 0.15 M mannitol. However, addition of 35 ppm gibberellic acid to the mannitol resulted in germination of 61%, showing a reversal of osmotic inhibition.

TABLE 4.10. *Effect of gibberellic acid on the germination of various seeds (compiled from data of Kallio and Piiroinen, 1959; Corns, 1960)*

		% Germination	
Plant	Treatment with GA	Treated	Water control
Avena fatua	500 ppm	57	26
Sinapis arvensis	500 ppm	89	9
Thlapsi arvense	500 ppm	90 – 100	0
Gentiana nivali	1000 ppm (24 hr)	90	0
Bartschia alpina	1000 ppm (24 hr)	90	5
Draba hirta	1000 ppm (24 hr)	95	0

TABLE 4.11. *Effect of light and gibberellic acid on the germination of lettuce at 26°C (Evenari et al., 1958)*

	% Germination	
Treatment	In water	In gibberellic acid $2{\cdot}9 \times 10^{-5}$ M
Dark	12·0	39·0
Red	44·0	66·5
Far-red	5·0	25·0
Red and far-red	11·0	35·0

Similar reversals of osmotic inhibition were obtained using red light. More detailed investigation along the lines mentioned, studying the interaction of gibberellic acid with far-red light and with high temperatures, have resulted in the general conclusion that gibberellic acid and red light act only partially in the same way and that their mode of action is not identical (Evenari *et al.*, 1958; Ikuma and Thimann, 1960). That gibberellic acid might be of importance in determining the germination of seeds in nature is indicated by the isolation from *Phaseolus*, lettuce and many other seeds of gibberellin-like compounds (Phinney and West, 1960).

For a long time the accepted view of the mode of action of the gibberellins in the germination of cereals was that these compounds are transported from the scutellum or embryonic axis to the aleurone where they induce synthesis of specific enzyme proteins active during germination. The enzyme most studied was α-amylase. Lately, there are some indications that gibberellin is also synthesized in the aleurone in response to some other factor originating in the embryo (Jacobsen and Chandler, 1987).

The level of gibberellin (GA) changes during seed development and shows fluctuations, so that it is difficult to relate development directly to a given level. During dormancy breaking, rises in the level of gibberellin have frequently been reported (Frankland and Wareing, 1966). Such increases in the GA level may be due to its liberation from bound forms, its transport from one part of the seed to another or due to synthesis. There is evidence for each of these processes (Lang, 1970; Jones and Stoddart, 1977).

In considering the role of GA in dormancy and dormancy breaking one is again confronted with a major problem, namely whether the observed correlations between GA level and dormancy breaking are relations between cause and effect or whether they merely accompany the relief of dormancy. The wealth of data on this crucial question is discussed by Black (1980/81) but no clear-cut conclusion is reached. It is important to distinguish between the effect of externally applied GA, or any other germination promoter or inhibitor, and the changes in the endogenous level of the compound under investigation. The endogenous effect of externally applied compounds or their lack of effect is never convincing proof of their involvement in regulatory mechanisms.

Some answers to these questions will perhaps be provided by continued work on mutants of plants deficient in, or insensitive to, GA. One of the species studied is *Arabidopsis*, some of whose mutants have an absolute requirement for GA_{4+7} in germination (Koornneef and Van der Zeen, 1980). The seeds of such mutants were normal in many respects and had normal weights and size. The requirements of the mutants for GA for germination were lower than those for growth, and GA biosynthesis appeared to be affected. Another mutant studied in greater detail did not germinate in the dark even in the presence of GA_{4+7}, but did so in the light. In this case light seemed to increase the sensitivity to GA. The GA requirement in the dark was apparently so high that it could not be fulfilled by an exogenous supply of GA (Koornneef *et al.*, 1985). Changed sensitivity to GA is clearly an additional factor which must be considered when studying the role of GA in dormancy. The known effects of the gibberellins on metabolism during germination will be discussed in Chapter 6.

Anothei plant hormone, cytokinin, also affects germination of seeds. Miller (1958) showed that kinetin promotes the germination of seeds. Later the natural cytokinin, zeatin, was isolated and characterized (Fig. 4.7) (Letham *et al.*, 1964). This natural cytokinin acts in the same way as kinetin. Zeatin ribotide and riboside, which also are present in seeds, are less effective in stimulating germination than zeatin itself (Van Staden, 1973). The cytokinins are actively metabolized in germinating seeds. A large number of derivatives of kinetin, where the furfuryl group is replaced by other groupings, also stimulate germination, for example benzyladenine. While originally it was thought that kinetin substitutes for red light in germination, it was later shown that the seeds were sensitized by kinetin so that a smaller dose of light would induce their germination. Weiss (1960) has shown that while in water lettuce seed required 3600 f.c. sec. light to promote their germination, in the presence of kinetin 720 f.c. sec. light were sufficient to bring about the same effect. The seeds need not be kept continuously in a solution of kinetin. As little as a few hours in the solution, at a suitable period, is sufficient to cause the sensitization. That kinetin does not substitute for light is also indicated by the fact that light-inhibited seeds such as *Amaranthus* are not affected by kinetin at all.

It seems likely that the requirement of light for the full expression of the stimulatory effect of kinetin depends on the phytochrome reaction (Reynolds and Thompson, 1973).

In addition to their interaction with light, cytokinins also interact with other exogenously applied compounds, such as ABA and GA, and also with temperature. The involvement of cytokinins in dormancy breaking has been investigated repeatedly, but the results are rather confusing. During stratification the cytokinin level rises only after the ABA level has already declined. This suggests the possibility that cytokinins are part of a much more complicated system in regulating dormancy if they have a role at all. There is not enough evidence about the quantitative changes of cytokinins during emergence from dormancy and about the possible role of such changes (Thomas, 1977; Ross, 1984).

Further evidence for the complexity of hormonal effects may be brought from the interrelation between osmotic stress and hormones. Exposure of seeds to low water potentials using osmotic treatment with, for example, polyethylene glycol, prevents their germination (Khan *et al.*, 1980/81). When, however, the seeds are treated with one hormone or with combinations of several hormones during the exposure to osmotic stress their dormancy, e.g. their light requirement on subsequent transfer to normal germination conditions, is broken. Application of hormones during osmotic conditioning can also prevent the imposition of secondary dormancy. Osmoconditioning allows a controlled hydration and solute uptake by the seeds which allows various processes in the seed to continue, while radicle elongation is prevented due to insufficient yield pressure in the root cells (Carpita and Nabors, 1981; Bradford, 1986). If during osmoconditioning the seeds are also exposed to the effect of hormones their response seems to be the result of an interaction between the two treatments (Fig. 4.9).

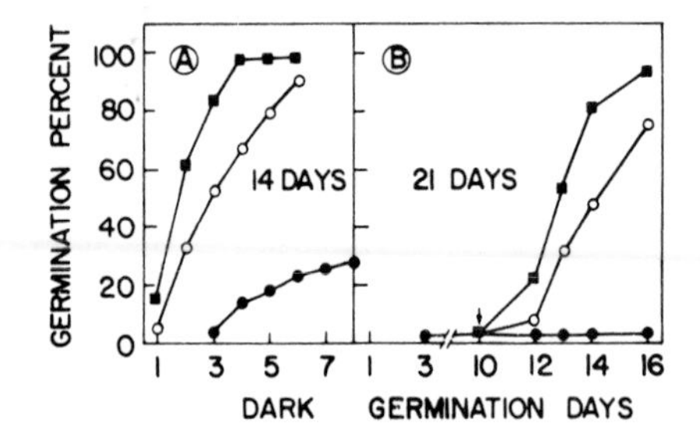

FIG. 4.9. Interaction of osmotic treatment with growth regulators. Seeds of *Chenopodium* were treated (A) for 14 days with polyethylene glycol (PEG) ●——● ; or PEG + 4 mM GA_{4+7} ○——○ ; or PEG + 40μM kinetin +3.5 mM ethephon + 1mM GA_{4+7} ■——■ and dark germination was then determined or (B) The seeds were treated with PEG for 21 days, soaked at alternating temperatures, 22°/12°C for 10 days, then transferred to growth regulators as above for 10 days and then germination determined. Germination in both treatments was at alternating temperatures of 22°/12°C and treatment at 15°C (after Khan and Karssen, 1980).

The effect of hormones of the indolylacetic acid (IAA) type on germination has long been in dispute. Numerous workers have investigated the effect of IAA and similar substances on the germination of a variety of seeds, and have obtained conflicting results, stimulation or inhibition being obtained, depending on the concentration of IAA and the type of seed used. However, the most general effect is an absence of response of the seeds to physiological concentrations of IAA. Soeding and Wagner (1955) attempted to relate divergent observations to the state of dormancy of the seeds. However, they failed experimentally to prove such a relation for seeds of *Poa*. A relation between depth of dormancy and response to IAA was, however, established for lettuce seeds. Even in this case the observed absolute effects were usually very small. In seeds with a dark germination of the order of 4 – 10%, IAA at a concentration of 10^{-7} M raised the dark germination to 20 – 30%, while on seeds germinating in the dark at 60 – 80% IAA had no effect. In other experiments the response of the seeds to IAA was found to be temperature-dependent. At 20°C, 10^{-7} M IAA raised the germination of lettuce seeds from 27 to 47%, while at 26°C the germination percentage was 8, both in water and in IAA (Poljakoff-Mayber, 1958).

The effect of IAA on wheat has been reinvestigated by Radley (1979). Externally applied IAA had no effect on isolated embryos or on grains whose pericarp had been removed. However, when the auxin level in the intact grains was followed it appeared that, in the developing grain, capability to germinate is found about a week after the free auxin content of the seeds begins to fall. This might indicate an inhibitory action of free auxin. The level of IAA in seeds of *Phaseolus vulgaris, Zea mays* and *Pinus silvestris* rose quite sharply during imbibition, especially during the early hours (Tillberg, 1977). However, the level of IAA, in contrast to reports for other hormones, showed wide divergence. Thus after 24 hr it was 27 ng/g in *Phaseolus*, 182 in *Pinus* and 4505 in *Zea*. Such wide differences would seem to suggest that one may be dealing with the liberation of IAA from a bound form, perhaps in storage organs, and that such rises are the result of germination. In this respect it must be recalled that there is now good evidence that IAA occurs as esters at least in seeds of *Zea mays* and *Oryza sativa* (Epstein *et al.*, 1980; Hall and Bandurski, 1979; Nowacki and Bandurski, 1980). The amounts of the IAA esters are appreciable, e.g. 2700 ng/g in rice, part of which is the inositol derivative of IAA. It was shown that IAA inositol is the form in which IAA is transported from the endosperm of *Zea* to the shoot and that IAA and IAA inositol are freely inter-convertible. These data indicate the need for a re-examination of previous results on the function of IAA in germination. It is not unlikely that if IAA inositol, and not its free form, reaches its site of action then the application of free IAA to seeds may lead to erroneous conclusions. Equally the fact that the bulk of the IAA is esterified makes data on free IAA levels dubious. Of course the experiments quoted apply only to two species of cereal seeds. This does not mean that IAA is metabolized in this fashion in other seeds.

As mentioned earlier ethylene can induce germination in seeds of *Xanthium*. This stimulatory effect of ethylene is more general (Ketring, 1977) in seeds showing some degree of dormancy. Ethylene is effective when applied exogenously at quite low concentrations of a few microlitres per litre. Ethylene can be applied to seeds as 2 – chloroethane phosphoric acid (Ethrel or Ethepon) which liberates ethylene in the presence of water above pH 4.0. Ethylene chlorhydrin can in some cases also stimulate germination, in a manner similar to thiourea, but it is less effective.

There is no doubt today that seeds are capable of producing ethylene and the internal concentration has been measured in seeds of *Arachis* and found to be of the order of 0.01 – 0.05 μl/l. The product of ethylene by seeds rises during imbibition. As is the case with other plant hormones, the question of interaction again arises. IAA is known to enhance ethylene production (Lieberman, 1979), yet auxin itself does not normally stimulate germination. Ethylene can also affect the level of other hormones and its production can be affected by them. There exists a very intricate series of interactions between ethylene, hormones and environmental factors. Seeds of *Amaranthus retroflexus* respond to ethylene, and under conditions of water stress their sensitivity to ethylene actually increases, recalling osmoconditioning. Temperature increases resulted in an increased response to ethylene (Schonbeck and Egley, 1980). Ethylene at a concentration of 0.1μl/l can induce the germination of *Striga* very readily (Egley and Dale, 1970). In lettuce, ethylene at high concentrations, 1 – 100 μl/l, overcame thermo-inhibition between 25 and 30°C. The seeds actually produced ethylene both at 25°C and at 30°C and inhibition was not due to blockage of ethylene production. The presumed target site for ethylene was probably the hypocotyl (Abeles, 1986).

Generally it has been observed that ethylene enhances the effect of light in stimulating germination (Speer *et al.*, 1974), but as little as 0.2 μl/l of ethylene inhibited the P_{FR} stimulation of germination of *Potentilla norvegica* (Suzuki and Taylorson, 1980). As far as ethylene is concerned we are unable to pinpoint its primary action. We do not yet know whether its increased production precedes, or results from, dormancy breaking and we are left with a state of confusion similar to that previously noted for the other plant hormones.

The presence of abscisic acid has been reported in very many seeds and fruits (Addicott and Lyon, 1969; Milborrow, 1974). To what extent abscisic acid is responsible for dormancy of seeds is still debated (Walton, 1980/81). There is a lot of evidence that the ABA level in seeds drops during stratification. However, it does not always do so. Even if the ABA level does drop this does not necessarily mean that the decrease is the cause of the alleviation of dormancy. It may equally be its outcome.

A close correlation between ABA concentration in the embryo and ability to germinate was found in soybeans. The seeds will not germinate until 21 days after anthesis, and at this time the ABA level dropped very considerably. Also when ABA is removed, by washing, germination occurs (Quatrano, 1987). ABA induces the synthesis of storage protein, while it inhibits the synthesis of carboxypeptidase C which is considered as a "specific enzyme" for germination (Dure, 1985). However, it is not clear how these events, at the molecular level, can lead to induction and maintenance of dormancy or its termination. The effect of ABA might be that of a dormancy-inducing compound, or it may be acting by preventing embryo growth, for example in *Chenopodium album* (Karssen, 1968).

A role has been suggested for ABA in preventing the premature germination of embryos during seed formation (Dure, 1975), and at least for cotton seeds the evidence is quite convincing. Nevertheless, this is probably not the main or only mechanism by which premature germination is prevented.

Exogenously applied ABA prevents the germination of seeds of many species. The inhibition gradually disappears on removal of the seeds from the ABA solution and by washing them. Exposure of intact seeds to osmotic or ionic stress during germination induces an increase in the level of free and bound ABA (Hasson and Poljakoff-Mayber, 1980/81). However, this increase is followed by a decrease. In tomato juice, which has been shown to prevent germination of the seeds inside the fruit, the level of ABA present cannot account for the inhibitory activity. ABA did, however, increase osmotic inhibition (Dorffling, 1970).

Perhaps the most important evidence for the role of ABA, in inducing primary dormancy, comes from the studies using single gene mutants of *Arabidopsis thaliana* (Koornneef *et al.*, 1984; Karssen, 1982; Karssen *et al.*, 1983). The onset of dormancy of seeds in *Arabidopsis* was found to be correlated with the ABA present in the embryo and not with ABA derived from the mother plant. Mutants with reduced embryonic ABA had a reduced level of dormancy (Karssen, 1982; Karssen *et al.*, 1983), $< 10 ng.g^{-1}$ compared with $200 - 500$ $ng.g^{-1}$ fresh weight in the wild type. In another mutant, the endogenous ABA was equal to or higher than in the wild type. This mutant had a reduced sensitivity to ABA again resulting in seeds less dormant than those of the wild type (Koornneef *et al.*, 1984). The concentration of ABA required to inhibit germination varies for different seeds from micromolar in some to millimolar in others. The inhibitory effect of ABA on germination, like many other effects of ABA, can be reversed by GA and sometimes also by cytokinins.

The information brought above illustrates the difficulties encountered in interpreting the function of ABA in inducing dormancy. ABA appears to be an important natural factor in inducing primary dormancy in some seeds and in inhibiting germination, but it is not the only factor. Induction of dormancy, and its termination, is apparently a result of interaction between ABA and GA and possibly also with other plant hormones or other factors. When ABA was first identified, it was thought that it could account for all the dormancy phenomena, but it is by now clear that this view is no longer tenable. Clearly many problems remain in interpreting results of studies on the function of ABA in dormancy.

ABA is readily metabolized in seed tissue (Walton, 1980/81) and some of the products of its metabolism are biologically active. This aspect of the problem has been inadequately studied and is particularly important for interpretation of the effect of ABA applied exogenously.

A more precise localization of ABA and other plant hormones in the cells and the tissues might be of great importance for interpretation of their effect and the interaction between them. Such localization can be achieved nowadays with the use of the immuno-chemical and radio-immuno assays developed recently.

From the above discussion of inhibitors and stimulators it may be concluded that the germination of seeds is controlled by a variety of external and internal factors. These factors include, in addition to simple environmental conditions, the presence of external and internal germination-inhibiting and stimulating substances. One of the controlling factors is the balance between stimulatory and inhibitory concentrations of the compounds at their site of action.

A clarification of the role of hormones and inhibitors in germination requires a great deal of further research. It is necessary to know how the level of all the compounds change during dormancy breaking and germination. Much more information is needed about the transition between bound and free forms of the hormones and whether the bound forms are active. Furthermore, knowledge about the sites in the seeds where the hormones are present is needed and when and where they are formed and how they are translocated. Additional questions relate to the interaction of these compounds with environmental factors such as the ambient atmosphere, light, temperature and water potential. Only when much more information becomes available will it be possible to interpret the vast amount of data on plant hormones and their action in germinating seeds.

The entire topic of primary, or secondary, dormancy and of germination inhibition and stimulation, is extremely complex. At this stage it is impossible to assign dormancy to any single cause. Indeed dormancy can be the result of several interacting factors operating via the seed coats or other enveloping structures on the one hand and via the embryo itself on the other hand. Dormancy may be due to physical constraints of some kind, permeability blocks or due to metabolic blocks. The latter appear to be possible at any one of a number of pathways and may involve a variety of mechanisms. Changes in levels of inhibitors or stimulators may be metabolically regulated or they may be the result of physical events due to leakage, or combinations of the two. At present it is not possible to pinpoint a single seed coat effect as being responsible for seed dormancy and it is also quite clear that dormancy and its control differs in different species. It may even show subtle differences in the different cultivars or populations of the same species. While differences between species are due to different combinations of control mechanisms it is probable that differences within a given species are quantitative ones and are genetically controlled. We do not know enough about the genetic control of dormancy and of germination to be able to reach any more definite conclusions and many of the answers may come from further genetic studies.

Bibliography

Abeles F. B. (1986) *Plant Physiol.* **81**, 780.
Addicott F. and Lyon J. L. (1969) *Ann. Rev. Plant Physiol.* **20**, 139.
Akkerman A. M. and Veldstra H. (1947) *Rec. Trav. Chim.* **66**, 411.
Axentjev B. N. (1930) *B.B.C.* **46**, 119.
Ballard L. A. T. (1958) *Austr. J. Biol. Sci.* **11**, 246.
Barton L. V. (1934) *Contr. Boyce Thompson Inst.* **6**, 69.
Barton L. V. and Crocker W. (1948) *Twenty Years of Seed Research*, Faber and Faber, London.
Barton L. V. and Schroeder E. M. (1942) *Contr. Boyce Thompson Inst.* **12**, 277.
Bewley J. D. and Halmer P. (1980/81) *Is. J. Bot.* **29**, 118.
Black M. (1980/81) *Is. J. Bot.* **29**, 161.
Black M. and Wareing P. F. (1955) *Physiol. Plant.* **8**, 200.
Bliss R. D., Platt-Aloia K. A. and Thomson W. W. (1986) *Plant Cell Environment* **9**, 727.
Botha F. C., Small J. G. C. and Grobbelaar N. (1982) *South African J. Bot.* **1**, 131.
Bradford, K. J. (1986) *Hort. Sci.* **21**, 1105.
Brocklehurst P. A. and Dearman J. (1983) *Ann. Appl. Biol.* **102**, 577.
Brown R. (1940) *Ann. Bot. N.S.* **4**, 379.
Brown R., Johnson A. W., Robinson E. and Ryler G. (1952) *Biochem. J.* **50**, 596.
Brown R., Greenwood A. D., Johnson A. W., Landsdown A. R., Long A. G. and Sunderland N. (1952) *Biochemistry* **52**, 571.
Bulard C. (1986) *Physiol. Plant.* **67**, 279.
Burgass R. W. and Powell A. A. (1984) *Ann. Bot.* **53**, 753.
Capon D. and Van Asdall W. (1967) *Ecology* **48**, 305.
Carpita N. C. and Nabors M. W. (1981) *Planta* **152**, 131.
Cavanagh A. K. (1980) *Proc. R. Soc. Victoria* **91**, 161.
Chang M., Netzly D. H., Butler L. G. and Lynn G. L. (1986) *J. Am. Chem. Soc.* **108**, 7858.
Come D. (1968) *Bull. Soc. Franc. Physiol. Veg.* **14**, 31.
Come D. (1980/81) *Is. J. Bot.* **29**, 145.
Cook C. E., Whichard L. P., Wall M. E., Egley G. H., Coggan P., Luhan P. A. and McPhail A. P. (1972) *J. Am. Chem. Soc.* **94**, 6198.
Corbineaux F., Lecat S. and Come D. (1986) *Seed Sci. Technol.* **14**, 725.
Corns W. G. (1960) *Can. J. Plant Sci.* **40**, 47.
Crocker W. (1906) *Bot. Gaz.* **42**, 265.

Crocker W. and Barton L. V. (1953) *Physiology of Seeds, Chronica Botanica.*
Davis W. E. (1930) *Am. J. Bot.* **17**, 58.
Dawidowicz-Grzegorzewska A. and Lewak St. (1978) *New Phyt.* **81**, 99.
Dawidowicz-Grzegorzewska A. and Zarska-Maciejewska B. (1979) *New Phyt.* **83**, 385.
Dell B. (1980) *Am. J. Bot.* **67**, 556.
Deubner C. G. (1932) *J. Forestry* **30**, 672.
Dorffling K. (1970) *Planta* **93**, 243.
Drew R. L. K. and Brocklehurst P. A. (1984) *J. Exp. Bot.* **35**, 986.
Duke S. H., Kafekuda G., Henson C. A., Loeffler N. L. and van Hulle N. M. (1986) *Physiol. Plant.* **68**, 525.
Dure L. S. (1975) *Ann. Rev. Plant Physiol.* **26**, 259.
Dure L. S. (1985) in *Oxford Survey of Plant Molecular Biology*, Vol. 2, p. 179 (Ed. B. J. Miflin), Oxford University Press, Oxford.
Eckerson S. (1913) *Bot. Gaz.* **55**, 286.
Edwards M. M. (1969) *J. Exp. Bot.* **20**, 876.
Edwards W. G. H. (1972) in *Phytochemical Ecology*, p. 235 (Ed. J. B. Harborne), Academic Press, London.
Egley G. H. and Dale J. E. (1970) *Weed Science* **18**, 586.
Egley G. H., Paul R. N. and Lax A. R. (1986) *Physiol. Plant.* **67**, 320.
Ellis R. H., Hong T. D. and Roberts E. H. (1985) *Handbook of Seed Technology for Genebanks*, vol. 2, International Board for Plant Genetic Resources, Rome.
Epstein E., Cohen J. D. and Bandurski R. S. (1980) *Plant Physiol.* **65**, 415.
Esashi Y. and Leopold A. C. (1968) *Plant Physiol.* **43**, 871.
Esashi Y., Kotaki K. and Ohhara Y. (1976) *Planta* **129**, 109.
Evenari M. (1949) *Bot. Rev.* **15**, 143.
Evenari M., Stein G. and Neuman G. (1954) *Proc. Ist Int. Photobiological Congress*, p. 82. Amsterdam.
Evenari M., Neuman G., Blumenthal-Goldschmidt S., Mayer A. M. and Poljakoff-Mayber A. (1958) *Bull. Res. Council Israel* **6D**, 65.
Ferenczy I. (1957) *Acta Biol. Hung.* **8**, 31.
Flemion F. (1933) *Contr. Boyce Thompson Inst.* **5**, 143.
Flemion F. and Prober P. L. (1960) *Contr. Boyce Thompson Inst.* **20**, 409.
Flemion F. and De Silva D. A. (1960) *Contr. Boyce Thompson Inst.* **20**, 365.
Forsyth C. and Brown N. A. C. (1982) *New Phytol.* **90**, 151.
Forsyth C. and van Staden J. (1983) *Ann. Bot.* **52**, 659.
Frankland B. and Wareing P. F. (1966) *J. Exp. Bot.* **17**, 596.
French R. C. (1985) *Ann. Rev. Phytopathology* **23**, 173.
Garas N. A., Karssen C. M. and Bruinsma J. (1974) *Z. Pflanzenphysiol.* **71**, 108.
Grant Lipp A. E. and Ballard L. A. T. (1959) *Austr. J. Agr. Res.* **10**, 495.
Gutterman T., Witztum A. and Evenari M. (1967) *Is. J. Bot.* **16**, 213.
Hall P. J. and Bandurski R. S. (1979) *Plant Physiol.* **63**, Suppl. p. 50, Abstr. 278.
Hamly D. H. (1932) *Bot. Gaz.* **93**, 345.
Hasson E. and Poljakoff-Mayber A. (1980/81) *Is. J. Bot.* **29**, 98.
Hendricks S. B. and Taylorson R. B. (1979) *Proc. Nat. Acad. Sci. USA* **76**, 778.
Hesse O. (1924) *Bot. Archiv.* **5**, 133.
Hilhorst H. W. M., Smitt A. I. and Karssen C. M. (1986) *Physiol. Plant.* **67**, 285.
Hilton, J. R. (1985) *J. Exp. Bot.* **36**, 974.
Ikuma G. and Thimann K. V. (1960) *Plant Physiol.* **35**, 557.
Ikuma H. and Thimann K. V. (1963) *Plant Cell Physiol.* **4**, 169.
Isikawa S. (1954) *Bot. Mag. Tokyo* **67**, 51.
Isikawa S. (1955) *Kumamoto J. Sci.* Ser. **B2**, 97.
Isikawa S. and Ishikawa T. (1960) *Plant Cell Physiol.* **1**, 143.
Isikawa S., Fujii T. and Yokohama Y. (1961) *Bot. Mag. Tokyo* **74**, 13.
Jacobsen, J. V. and Chandler, P. M. (1987) in *Plant Hormones and their Role in Plant Growth and Development* p. 64 (Ed. P. L. Davies) Nyhoft, Dordrecht.
Janerette C. A. (1979) *Seed Sci. Technol.* **7**, 347.
Johnson L. P. V. (1946) *Forestry Chronicle* **22**, 17.
Jones R. L. and Stoddart J. L. (1977) in *The Physiology and Biochemistry of Seed Dormancy and Germination*, p. 77 (Ed. A. A. Khan), North-Holland, Amsterdam.

Kahn A. (1960a) *Plant Physiol.* **35**, 1.

Kahn A. (1960b) *Plant Physiol.* **35**, 333.

Kahn A., Goss J. A. and Smith D. E. (1956) *Plant Physiol.* **31**, Suppl. 37.

Kallio P. and Piiroinen P. (1959) *Nature, Lond.* **183**, 1930.

Karssen C. M. (1968) *Acta Bot. Neerl.* **17**, 293.

Karssen C. M. (1980/81a) *Is. J. Bot.* **29**, 45.

Karssen C. M. (1980/81b) *Is. J. Bot.* **29**, 65.

Karssen C. M. (1982) in *Plant Growth Substances*, p. 623. (Ed. P. F. Wareing), Academic Press, London.

Karssen, C. M., Brinkhorst-Vander Swan D.L.C., Breekland, A.E. and Koorneef, M. (1983) *Planta,* **157**, 158.

Katoh H. and Esashi Y. (1975a) *Plant Cell Physiol.* **16**, 687.

Katoh H. and Esashi Y. (1975b) *Plant Cell Physiol.* **16**, 697.

Ketring D. L. (1977) in *The Physiology and Biochemistry of Seed Dormancy and Germination*, p. 157 (Ed. A. A. Kahn),North-Holland, Amsterdam.

Khan A. A. and Karssen C. M. (1980) *Plant Physiol.* **66**, 175.

Khan M. A. and Ungar I. A. (1986) *Bot. Gaz.* **147**, 148.

Khan A. A., Peck N. H. and Samimy C. (1980/81) *Is. J. Bot.* **29**, 133.

Khudairi A. K. (1956) *Physiol. Plant.* **9**, 452.

Kidd F. (1914) *Proc. R. Soc.* **B87**, 609.

Kivilaan A. (1975) *Flora* **164**, 1.

Klein G. and Farkass E. (1930) *Ost. Bot. Z.* **79**, 107.

Koller D. (1954) Ph.D. Thesis Hebrew University, Jerusalem (in Hebrew).

Koller D. (1972) in *Seed Biology*, vol. 2, p. 2 (Ed. T. T. Kozlowski), Academic Press, New York.

Koorneef M. and van der Zeen J. H. (1980) *Ther. Appl. Biol.* **102**, 57.

Koorneef M., Engelsma A., Hanhart C. J., Van Leenen-Martinet E. P., Van Rijn L. and Zeevart J. A. P. (1985) *Physiol. Plant.* **65**, 33.

Koorneef, M., Reuling, G. and Karssen C. M. (1984) *Physiol. Plant.* **61**, 377.

Kurth E., Jensen A. and Epstein E. (1986) *Plant Cell Environ.* **9**, 667.

Lang A. (1970) *Ann. Rev. Plant Physiol.* **21**, 537.

Leggatt C. W. (1946) *Can. J. Res.* **C24**, 7.

Leizorowitz R. (1959) M. Sc. Thesis, Jerusalem (in Hebrew).

Lenoir C., Corbineau F. and Come D. (1986) *Physiol. Plant.* **68**, 301.

Lerner H. R., Mayer A. M. and Evenari M. (1959) *Physiol. Plant.* **12**, 245.

Letham D. S., Shannon J. S. and McDonald I. R. (1964) *Proc. Chem. Soc.* p. 230.

Libbert E. and Lubke H. (1957) *Flora* **145**, 256.

Libbert E. (1957) *Phyton* **9**, 81.

Lieberman M. (1979) *Ann. Rev. Plant Physiol.* **30**, 533.

Lona F. (1956) *Ateneo Parmense* **27**, 641.

Maier W. (1933) *Jahr. Wiss. Bot.* **78**, 1.

Mancinelli A. (1958) *Ann. di Botanica* **26**, 56.

Marbach I. and Mayer A. M. (1974) *Plant Physiol.* **54**, 817.

Marbach I. and Mayer A. M. (1985) *J. Exp. Bot.* **36**, 353.

Marchaim U., Birk Y., Dovrat A. and Berman T. (1972) *J. Exp. Bot.* **23**, 302.

Massart L. (1957) *Biochimia (USSR)* **22**, 117.

Mato N. (1924) *Sitz. Acad. Wiss. Matt. Natur.* K. L. **133**, 625.

Mayer A. M. (1977) in *The Physiology and Biochemistry of Seed Dormancy and Germination*, p. 357, (Ed. A. A. Khan), North-Holland, Amsterdam.

Mayer A. M. (1986) *Is. J. Bot.* **35**, 3.

Mayer A. M. and Evenari M. (1952) *J. Exp. Bot.* **3**, 246.

Mayer A. M. and Evenari M. (1953) *J. Exp. Bot.* **4**, 257.

Mayer A. M. and Marbach I. (1981) *Progress Phytochem.* **7**, 95.

McIntyre G. I. and Hsiao A. J. (1985) *Bot. Gaz.* **146**, 347.

Millborrow B. V. (1974) *Ann. Rev. Plant Physiol.* **25**, 259.

Miller C. O. (1958) *Plant Physiol.* **33**, 115.

Mitchell E. D. and Tolbert N. E. (1968) *Biochemistry* **7**, 1019.

Murphy J. B. and Noland T. L. (1982) *Plant Physiol.* **69**, 428.

Nagao M., Esashi Y., Tanaka T., Kumaigai T. and Fukumoto S. (1959) *Plant Cell Phys.* **1**, 39.

Naylor J. M. (1983) *Can. J. Bot.* **61**, 3561.
Niethammer A. (1927) *Biochem. Z.* **185**, 205.
Nikolaeva M. G. (1969) *Physiology of Deep Dormancy in Seeds,* Israel Program Sci. Trans. Jerusalem.
Noel A. R. A. and van Staden J. (1976) *Z. Pflanzensphysiol.* **65**, 422.
Nowacki J. and Bandurski R. S. (1980) *Plant Physiol.* **65**, 422.
Nutile G. E. (1945) *Plant Physiol.* **20**, 433.
Olney H. O. and Pollock B. M. (1960) *Plant Physiol.* **35**, 970.
Oppenheimer H. (1922) *Sitz. Akad. Wiss. Wien Abt.* 1 **131**, 279.
Osborne D. J. (1983) *Can. J. Bot.* **61**, 3568.
Phinney B. O. and West C. A. (1960) *Ann. Rev. Plant Physiol.* **11**, 411.
Pinfield N. J. and Davies H. V. (1978) *Z. Pflanzenphysiol.* **90**, 171.
Pinfield N. J. and Dungey N. O. (1985) *J. Plant Physiol.* **120**, 65.
Poljakoff-Mayber A. (1958) *Bull. Res. Council Israel.* **6D**, 78.
Poljakoff-Mayber A., Mayer A. M. and Zachs S. (1958) *Ann. Bot. N. S.* **22**, 175.
Pollock B. M. and Olney H. O. (1957) *Plant Physiol.* **34**, 131.
Porter N. G. and Wareing P. F. (1974) *J. Exp. Bot.* **25**, 583.
Probert R. J., Smith R. P. and Birch B. (1985) *New Phytol.* **101**, 521.
Quatrano, R. S. (1987) in *Plant Hormones and their Role in Plant Growth and Development* p. 494 (Ed. P. J. Davies), Nyhoft, Dordrecht.
Radley M. (1979) *Sprouting Symposium,* England, p. 131.
Reynolds T. (1974) *J. Exp. Bot.* **25**, 375.
Reynolds T. and Thompson P. A. (1973) *Physiol. Plant.* **28**, 516.
Roberts H. A. and Neilson J. E. (1982) *New Phytol.* **92**, 159.
Rolston M. P. (1978) *Bot. Rev.* **44**, 365.
Roubaix J. de and Lazar O. (1957) *La Sucrerie Belge* **5**, 185.
Ross J. D. (1984) in *Seed Physiology* vol. 2, p. 45 (Ed. D. R. Murray), Academic Press, New York.
Ruge U. (1939) *Z. f. Bot.* **33**, 529.
Schonbeck M. W. and Egley G. H. (1980) *Plant Physiol.* **65**, 1149.
Schopfer P. and Plachy C. (1984) *Plant Physiol.* **76**, 155.
Shieri H. B. (1941) *The Gladiolus* **16**, 100.
Shimomura A., Sashida Y., Nakata H., Yamamoto A., Kawakubo Y. and Kawasaki J. (1983) *Plant Cell Physiol.* **24**, 123.
Shull C. A. (1911) *Bot. Gaz.* **52**, 455.
Shull C. A. (1914) *Bot. Gaz.* **57**, 64.
Soeding H. and Wagner M. (1955) *Planta* **45**, 557.
Spaeth J. N. (1932) *Am. J. Bot.* **19**, 835.
Speer H. L., Hsiao A. and Vidaver W. (1974) *Plant Physiol.* **54**, 852.
Staden, van J. (1973) *Physiol. Plant.* **28**, 222.
Stearns F. and Olsen J. (1958) *Am. J. Bot.* **45**, 53.
Sumere C. F. van (1960) in *Phenolics in Plants in Health and Disease*, p. 25, Pergamon Press, Oxford.
Sumere C. F. van, Hilderson H. and Massart L. (1958) *Naturwiss.* **45**, 292.
Suzuki S. and Taylorson R. B. (1980) *Plant Physiol.* **65**, Suppl. p. 102, Abstract 59.
Taylorson R. B. (1982) in *The Physiology and Biochemistry of Seed Development, Dormancy and Germination,* p. 323 (Ed. A. A. Khan), Elsevier Biomedical Press, Amsterdam.
Taylorson R. B. and Hendricks S. B. (1973) *Plant Physiol.* **52**, 23.
Taylorson R. B. and Hendricks S. B. (1980/81) *Is. J. Bot.* **29**, 273.
Thevenot C. and Come D. (1978) *C. R. Acad. Sci. Paris* **287D**, 1127.
Thomas T. H. (1977) in *The Physiology and Biochemistry of Seed Dormancy and Germination* , p. 111 (Ed. A. A. Khan), North-Holland, Amsterdam.
Thompson R. C. and Kosar W. F. (1939) *Plant Physiol.* **14**, 561.
Thornton N. C. (1935) *Contr. Boyce Thompson Inst.* **7**, 477.
Tillberg E. (1977) *Plant Physiol.* **60**, 317.
Toole E. H. (1959) in *Photoperiodism and Related Phenomena in Plants and Animals*, AAAS Publ. No. 55 (Ed. R. B. Withrow).
Toole E. H., Toole V. K., Borthwick H. A. and Hendricks S. B. (1955) *Plant Physiol.* **30**, 473.
Toole V. K. (1938) *Proc. Ass. Off. Seed Analyst N. America*, p. 227, 30th Ann. Meeting.

Tran V. N. (1979) *Aust. J. Plant Physiol.* **6**277.
Tran V. N. and Cavanagh A. K. (1984) in *Seed Physiology* vol. 2, p. 1 (Ed. D. R. Murray), Academic Press, New York.
Tuan D. Y. H. and Bonner J. (1964) *Plant Physiol.* **39**. 768.
Tukey H. B. and Carlson R. F. (1945) *Plant Physiol.* **20**. 505.
Uhvits R. (1946) *Am. J. Bot.* **33**, 278.
Varga, M. (1957a) *Acta Biol. Hung.* **7**, 39.
Varga M. (1957b) *Acta Biol. Szeged.* **3**, 213.
Varga M. (1957c) *Acta Biol. Szeged.* **3**, 225.
Varga M. (1957d) *Acta Biol. Szeged.* **3**, 233.
Villiers T. A. and Wareing P. F. (1960) *Nature, Lond.* **185**, 112.
Walton, D. C. (1980/81) *Is. J. Bot.* **29**, 168.
Wareing P. F. and Foda H. A. (1957) *Physiol. Plant.* **10**, 266.
Weiss J. (1960) *C. R. Acad. Sci. Paris* **251**, 125.
Werker E. (1980/81) *Is. J. Bot.* **29**, 22.
Wiesner J. (1897) *Ber. Deut. Bot. Ges.* **15**, 503.
Wycherley, P. R. (1960) *J. Rubber Res. Institute Malaya,* **16**, 99.

5

Metabolism of Germinating Seeds

The dry seed is characterized by a remarkably low rate of metabolism. This is probably a direct result of the very low level of hydration of the seed, whose water content is of the order of 5 – 10%. Despite this almost complete absence of metabolism in the seed, it cannot be assumed that it lacks the potentiality for metabolism. As already shown in Chapter 1 a few enzymes are active even in the desiccated seed. Moreover when dry seeds are broken up by grinding in a suitable aqueous medium it is possible to show in the extract the presence of a considerable number of enzyme systems. Thus it can be concluded that the dry seed is a well equipped functional unit which can carry out a large number of biochemical reactions, provided that the initial hydration of the proteins and more specifically of the enzyme proteins, has taken place.

One of the most frequently used criteria for evaluating the rate of metabolism is the rate of respiration. Dry seeds show a very low rate of respiration, which begins to rise rapidly when the seeds are placed in water. This is the first easily observable metabolic change in the seed and occurs well before germination. The second change which is also very easily observed, is the breakdown of reserve materials in the seed. This reaction is also initiated by the hydration of the proteins. As long as the seeds are dry very few and small changes are observed in their chemical composition. As soon as the seed is hydrated very marked changes of composition in its various parts occur. The chemical changes which occur are complex in nature. They consist of three main types: the breakdown of certain reserve materials in the seed; the transport of the breakdown products from one part of the seed to another, especially from the endosperm to the embryo, or from the cotyledons to the growing parts; and the synthesis of new materials from the breakdown products formed. Of special importance is the initiation of protein synthesis during germination and its relation to nucleic acid metabolism.

At the ultrastructural level many changes occur during germination and such changes must in some way be related to the overall metabolism of the germinating seed. The only substances normally taken up by seeds during germination are water and oxygen. In many instances substances are lost from the seed during the initial stages of germination. These initial stages are consequently accompanied by a net loss of dry weight due to the oxidation of substances on the one hand and leakage out of the seed on the other hand. Only when the seedling has formed, i.e. when a root has emerged and takes up minerals, and the cotyledons or first leaves are exposed to light and are capable of photosynthesis, does an increase in dry weight begin.

In this chapter we will discuss the breakdown of reserve materials in seeds, respiration, protein and nucleic acid metabolism and the general metabolism of seed components. In considering the different metabolic pathways attention will also be paid to the subcellular organelles in which they occur, their development and their specific function. The complex effect of growth substances, germination inhibitors and stimulators and their role in the control of the initiation of germination will be discussed in a separate chapter, Chapter 6.

I. Changes in Storage Products During Germination

As already mentioned in Chapter 2, the chief types of seeds are those containing lipids and those containing carbohydrates as a major reserve material. A variety of other materials are also present in the dry seeds as already discussed in Chapter 2. Although some of these compounds are present only in small amounts, their turnover is nevertheless of considerable importance during germination. A number of other compounds do not seem to be metabolized to any appreciable extent.

Changes in the composition of seeds during germination have been investigated in detail in a number of species. Among them are *Vigna sesquipedalis* (Oota *et al.*, 1953), rice (Palmiano and Juliano, 1972), corn (Ingle *et al.*, 1964, 1965) and Douglas fir (Ching, 1966).

A very clear illustration of the changes in composition in different parts of a seed of *Vigna sesquipedalis* during germination is given in Fig. 5.1 (Oota *et al.*, 1953). This seed is characterized by large cotyledons and epigeal germination. The cotyledons show a loss of all compounds studied, while all the other organs show an increase in the various seed constituents. Particularly striking is the increase of materials in the rapidly growing hypocotyl, which ceases as the hypocotyl ceases to grow. The seed studied by Oota lends itself particularly to such detailed analysis as it is technically feasible to differentiate between the hypocotyl, root, cotyledons, plumule and epicotyl, even in the dry seed.

The most striking change which can be observed after the onset of germination is growth — an increase in the size and dry weight of the hypocotyl, which is accompanied by an almost similar loss of weight in the cotyledons. However, the cotyledons continue to lose dry weight for some 30 hr after the growth of the hypocotyl has slowed down (Fig. 5.1,I). The increase in weight of the other parts of the embryo, epicotyl, plumule and radicle, begins later and takes place at slower rate. Various substances such as soluble sugars (Fig. 5.1,II) insoluble polysaccharides (Fig. 5.1,III) soluble and protein nitrogen (Fig. 5.1,IV, V) as well as pentose nucleic acid phosphorus (Fig. 5.1,VI) move out of the cotyledons and are transferred to other parts of the growing embryo. The transient nature of the growth of the hypocotyl is evident specially if metabolism of soluble sugars and nucleic acid phosphorus is followed (Fig. 5.1,II and VI). In the period up to 24 – 48 hr of germination the hypocotyl is the main sink for these compounds. Apparently when the cotyledons are pushed above the ground, the plumule and radicle take over as the main sinks. The large amount of soluble sugars present in the hypocotyl is apparently not only a result of transport from the cotyledons but also the result of the breakdown of reserve carbohydrates, such as starch and oligosaccharides. Protein synthesis begins in the various embryonic organs immediately with the beginning of their growth. However, soluble nitrogen which disappears from the cotyledons appears during the first 48 hr in larger amounts in the radicle than in the hypocotyl. In the cotyledon there is apparently an initial breakdown of proteins which results in a transient

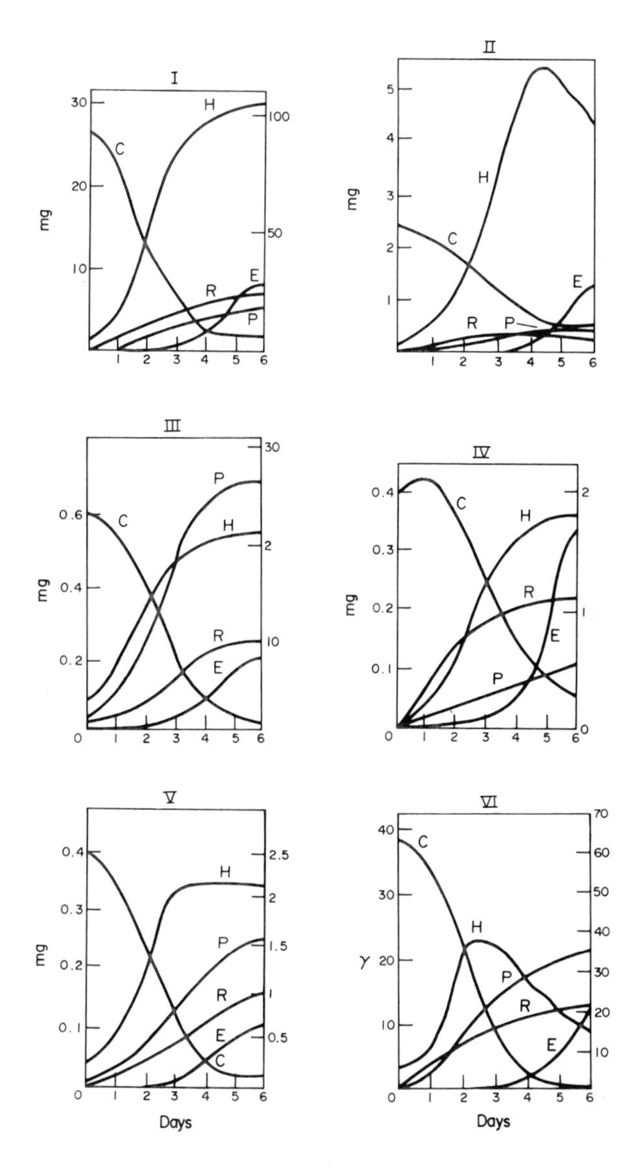

FIG. 5.1. Changes in dry weight and amount of various constituents in the different organs of *Vigna sesquipedalis* seeds during germination. Values per individual seed part (after Oota *et al.*, 1953). P — plumule, E — epicotyl, R — radicle, C — a pair of cotyledons. (I) Dry weight: P, E, H and R on the left ordinate, C on the right. (II) Soluble sugars as glucose equivalents. (III) Insoluble polysaccharides as glucose equivalents: P, E, H and R on the left ordinate, C on the right. (IV) Soluble nitrogen: P, E, R and C on the left ordinate, H on the right. (V) Protein nitrogen: P, E, H and R on the left ordinate, C on the right. (VI) Pentose nucleic acid phosphorus: P, E, H and R on the left ordinate, C on the right.

accumulation of soluble nitrogen. After 24 hr the flow of soluble nitrogen out of the cotyledons increases when the radicle and the hypocotyl act as sinks.

The data above show a marked mobility of the various components of the seed. The function of the cotyledons as storage organs, which empty out as other parts of the embryo develop, is clearly demonstrated.

In corn, dry weight and total nitrogen of the whole seedling decrease during the first 120 hr of germination (Fig. 5.2) (Ingle *et al.*, 1964).

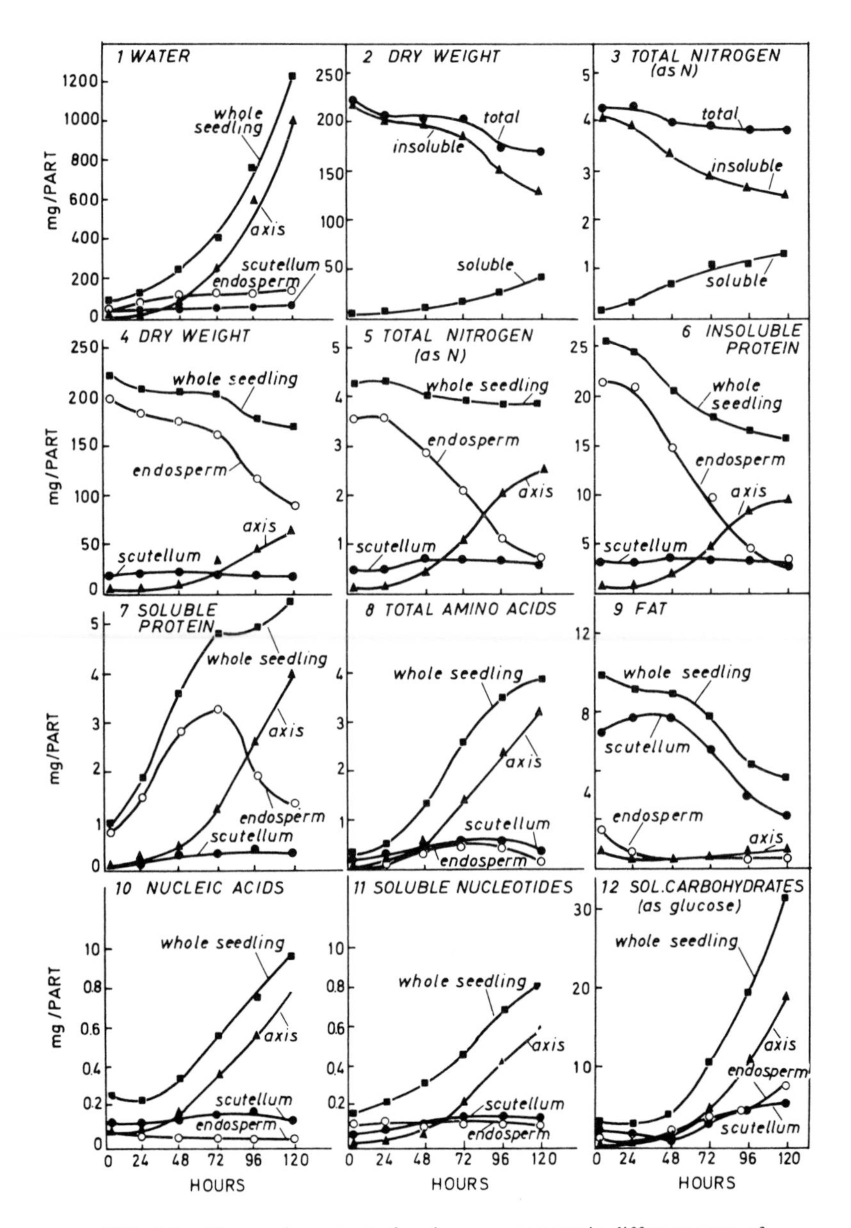

FIG. 5.2. Changes in content of various components in different parts of *Zea mays* seeds during germination (after Ingle *et al.*, 1964).

The drop is primarily in the endosperm while total nitrogen and dry weight increase in the axis. Insoluble protein shows similar changes. However, soluble protein and total amino acids rise in the whole seedling and the axis, while soluble protein in the endosperm shows a peak after 3 days. Nucleic acids and soluble nucleotides rise both in the whole seedling and the axis during the same time period, but remain at a constant level in the endosperm and scutellum. A more detailed investigation showed that the overall RNA and DNA content rises for 4 days of germination (Ingle and Hageman, 1965). This rise occurs in the axis and the scutellum, but a decrease was noted in the endosperm. Soluble nucleotides increased in all parts of the corn seedling. Generally, changes in the scutellum of corn are small (Fig. 5.2). Its various components decrease after 48 hr of germination except for the lipids. Most of the lipids of the seed are localized in the scutellum. The lipid content of the endosperm and axis is quite small and it changes very little during germination (Fig. 5.2). The overall trend is therefore similar to that in *Vigna*, movement from the storage organ, the endosperm, to the embryonic axis. Soluble carbohydrates increase in all parts of the seedling after about 48 hr of germination.

It should be remembered that the analysis of the data on the changes in content of the various substances in the seeds of *Vigna* and of corn is expressed on a "per part" basis. The different parts of the seed differ greatly in size and relative dry weight. Had the changes been expressed on a dry weight basis, they would have been quite dramatic for the small seed parts such as the radicle or epicotyl of *Vigna* and the embryonic axis of corn.

The overall changes in the composition of rice seeds during germination is illustrated in Table 5.1 (Palmiano and Juliano, 1972). Again a loss of starch is observed together with increases in free sugars, soluble amino nitrogen and soluble protein and a decrease in RNA. The loss in dry weight begins after 2 – 4 days, indicating an initial slow rate of metabolism. Other investigations indicate that the loss in starch occurred in the endosperm, but no starch accumulated in the root or shoot. Dry weight increase was more rapid in the shoot than in the root, since in rice the coleoptile grows out before the roots (Fukui and Nikuni, 1956).

TABLE 5.1. *Changes in composition of rice during germination in the dark (from data of Palmiano and Juliano, 1972)*

Period of germination days	Dry weight mg	Starch mg	Free sugars mg	Crude protein mg	Soluble amino N μg	RNA μg	Soluble protein μg
0	18·4	16·2	0·15	1·36	2·18	26	258
2	19·6	—	—	0·83	—	14	276
3	17·0	13·9	0·37	0·82	9·79	11	268
4	17·0	12·4	0·77	0·70	14·25	11	296
5	12·6	10·8	1·14	0·64	15·80	11	304

Yocum (1925) grew wheat in soil and analysed the composition of both the entire seed or seedling and, after separation into different organs was possible, also the composition of the endosperm, root and plumule. He found that starch decreased continuously in the endosperm as well as in the whole seedling. Oligosaccharides also disappeared from the endosperm, although on the first day some were formed from other substances. Lipids also decreased during the first few days, but later they accumulated again, mainly in the plumule and in the radicle. The resynthesis of lipids and the appearance of reducing and other sugars is presumably the result of transformations of the disappearing starch. Nitrogen also leaves the endosperm and appears in other parts of the seedling.

An oil-storing seed, the castor bean has been investigated by Yamada (1955). Total lipids decreased both in the endosperm and the cotyledons, as did the total amount of fatty acids as germination proceeded. In contrast carbohydrates began to appear in the endosperm as the lipids disappeared, showing accumulation up to 4 days. After this, carbohydrates disappeared again from the endosperm and appeared in the hypocotyl. This is true both for reducing and non-reducing sugars (Fig. 5.3). It appears that lipids are converted to sugar in the endosperm itself, and later the carbohydrates are transferred to the embryo.

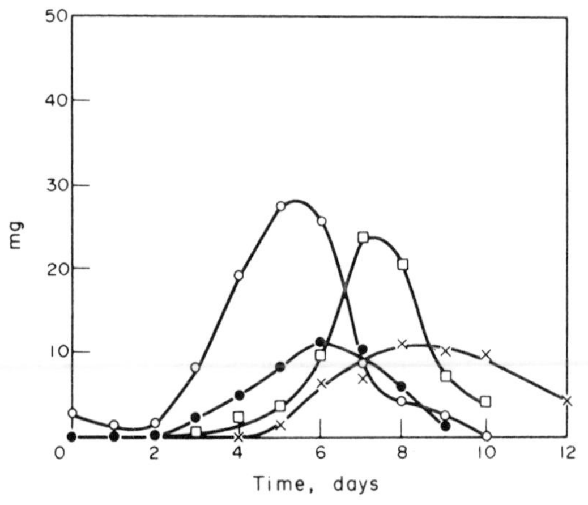

FIG. 5.3. Changes in reducing and non-reducing sugars during the germination of castor beans (after Yamada, 1955). ○——○ Non-reducing sugars in endosperm. ●——● Reducing sugars in endosperm. □——□ Non-reducing sugars in hypocotyl. ×——× Reducing sugars in hypocotyl.

A detailed study of the utilization of lipids in Douglas fir seeds during germination showed a decrease in total lipids. Glycerides were utilized while certain phospholipids increased slowly at first and then more rapidly. The relative distribution of short and long chain fatty acids changed considerably in the free fatty acid fraction (Ching, 1963, 1966). It is obvious that the various lipid fractions undergo different metabolic fates during germination. This is probably true not only for Douglas fir, but for all lipid containing seeds (Trelease and Doman, 1984).

Cell wall components can also serve as reserve materials. Galactomannans in carob seeds, mannans in various palm seeds and other polysaccharides, often referred to as hemicelluloses, may be present in various seeds. These hemicelluloses may contain in addition to galactose and mannose, arabinose, xylose, rhamnose and other sugars. These polysaccharides are usually made up of a linear β-linked backbone consisting of one or two sugars, such as mannose and galactose, and have side chains, several residues long, which often differ from those in the backbone structure. These polysaccharides usually accumulate in the thickened cell walls of the endosperm or the cotyledons. The galactomannan content of seeds of *Trigonella foenum-graecum* drops in the endosperm during germination. At the same time small amounts of sugars accumulate in the cotyledons, for example sucrose, glucose and fructose. This would indicate that the breakdown products of galactomannan, mannose and galactose are transported to the cotyledons where they are immediately metabolized further. Only small amounts of free mannose and galactose appear. If, however, endosperms are isolated then monosaccharides do accumulate as a result of hydrolysis of the galactomannan. During breakdown the galactose-mannose ratio changes. As the galactomannan is broken down the starch content of the cotyledons begins to rise sharply (Reid, 1971).

Overall increases in the RNA and DNA content have been reported during the germination of many seeds, including wheat, oats (Semenko, 1957), lettuce, rye and castor beans. Probably breakdown occurs in the storage tissue, cotyledons or endosperm and synthesis in the growing tissues, in which active cell division and elongation occur.

The changes in storage materials during germination are the result of the activity of enzymes able to break them down. The enzymes are not necessarily produced in the same cells in which the storage materials are located. Moreover signalling systems exist which regulate the production of enzymes and the interaction between different parts of the seed — embryonic axis and cotyledons, endosperm embryo and aleurone layer or cotyledon, endosperm and embryonic axis, depending on the seed, examples being pea, cereal seed and castor bean respectively. Some of these regulating mechanisms will be dealt with later on in this chapter and in Chapter 6.

The changes in storage materials are associated with quite complex changes in ultrastructure. As the storage carbohydrates are broken down the ultrastructure of the organ in which they are located changes. The gradual dissolution of starch grains is generally accompanied by vacuolization. When the carbohydrates are cell wall located, such as mannans and galactomannans, the ultrastructure changes occur in the cell wall with a gradual disintegration of wall structure. The changes are by no means uniform throughout the tissue. For example in the endosperm of lettuce most of the cell walls are degraded after radicle emergence, but cells near the micropylar end appear to be unique in structure and ultrastructure and are degraded before radicle emergence (Psaras, 1984).

A definite sequence of breakdown can be observed at the ultrastructural level, protein bodies vacuolating first and then the storage cell walls becoming electron transparent. Eventually the entire cell breaks down. The sequence of breakdown of protein bodies in endosperm shows distinct differences in different species (Pernollet, 1982) and will be discussed in section II.3.

II. Metabolism of Storage Products

From the discussion on the changes in storage products it can be concluded that some of them undergo very marked metabolic changes. Not necessarily those substances which are present in the greatest amounts are broken down first.

The metabolic changes occurring in the early stages of germination are the result of the activity of various enzymes, which are either present in the dry seed or very rapidly become active as the seed imbibes water. Generally, enzymes breaking down starch, proteins, hemicellulose, polyphosphates, lipids and other storage materials, rise in activity fairly rapidly as germination proceeds, although there is no reason to think that these changes are the direct cause of the actual germination process. Changes in the activity of various enzymes have been studied in many seeds but the same general trend is observed. In view of the importance of some of these enzymes, the reactions which are catalysed by them will be briefly discussed.

1. Carbohydrates

As already stated, the main storage carbohydrates are starch, oligo- and polysaccharides of the cell wall and soluble sugars.

Seeds usually contain both amylose and amylopectin organized in the starch grains. Starch is normally broken down by α- and β-amylases, but only some of the α-amylases are capable of attacking the native starch grains of cereals (Halmer, 1985). Dry seeds contain mainly β-amylase, as is the case for dry *Zea mays* seeds. The rise in amylase activity in the seed during germination is primarily in α-amylase which, when amylolytic activity is at its peak, accounts for 90% of total amylolytic activity of the endosperm. However, small amounts of α-amylase are present in dry wheat seeds. β-Amylase detaches units of maltose from starch. When amylopectin is attacked oligosaccharides are formed, while when amylose is attacked the molecule is broken down completely. The product maltose rarely accumulates in the seeds, it is usually broken down to glucose by the action of maltase. When both amylose and amylopectin are present, the latter is broken down preferentially. Attack of starch by α-amylase results in a mixture of sugars, maltose and glucose. The enzymes involved in breakdown and synthesis of starch in cereals have been reviewed by Marshall (1972), see also Halmer (1985). Obviously the precise course of starch breakdown in any given seed will be determined by the relative amounts of enzymes present, as well as by the ratio of the two main forms of starch.

α-Amylase originates in the scutellum and is secreted into the endosperm, while β-amylase appears to form only in the endosperm (Dure, 1960). In barley seeds it was found that the rise in α-amylase activity of the endosperm depended on the presence of the embryo. Removal of the latter resulted in a drop in amylase activity (Kirsop and Pollock, 1957).

Formation of α-amylase in the barley endosperm is controlled by gibberellic acid, which is initially secreted by the scutellum. However, the scutellar secretion is in turn controlled by the axis (Radley, 1968). Enzyme formation occurs by *de novo* synthesis in the aleurone layer (Varner *et al.*, 1965). The significance of this control will be discussed in the following chapter.

In rice seeds β-amylase is first synthesized in the scutellum and later an apparently latent form, present in the endosperm, becomes active and this latter enzyme then becomes the dominant form. The first formation of α-amylase in barley and rice is also in the scutellum and only later does enzyme formation in the aleurone layer take place (Gibbons, 1979; Okamoto and Akazawa, 1980; Akazawa and Miyata, 1982). It has been possible precisely to locate the initial formation of these enzymes by immunofluorescence and immunoprecipitation techniques. These allow the location of enzymes in the tissue by reaction with suitable antibodies prepared after isolation of the enzymes which serve as antigens. It is likely that when these techniques are applied to other enzymes, better information will become available on the precise location and sequence of formation of some of them. As germination proceeds other starch degrading enzymes appear such as debranching enzymes and phosphorylases. In some cases they are present in small amounts already in the dry seed.

The changes in the enzymes involved in starch breakdown during the germination of peas are illustrated in Fig. 5.4. Low activity of β-amylase was present in the dry seed, but no α-amylase activity could be detected. The activity of both these enzymes began to increase only 2 days after the onset of imbibition. In addition, in pea seeds, phosphorylase activity rises very rapidly in the cotyledons within 4 days of germination, so that phosphorolytic breakdown probably also occurs.

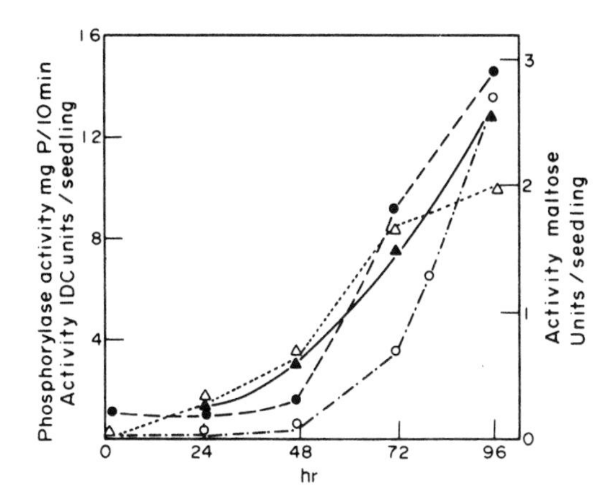

FIG. 5.4. Development of enzymes involved in starch breakdown in peas during germination (redrawn from data of Shain and Mayer, 1968 and Swain and Dekker, 1966). ● β-amylase per seedling. △ amylopectin-1, 6-glycosidase (per seedling). ○ α-amylase (per seedling). ▲ phosphorylase (per 30 cotyledons).

Breakdown of starch in peas was shown by Swain and Dekker (1966) to follow the pathway:

$$\text{starch} \xrightarrow[\alpha\text{-amylase}]{} \text{soluble oligosaccharides} \xrightarrow[\beta\text{-amylase}]{} \text{maltose} \xrightarrow[\alpha\text{-glucosidase}]{} \text{glucose}$$

At least two distinct phosphorylases, I and II were shown to be present in a number of seeds, including peas. Their level did not change at the same rate, phosphorylase I apparently increasing more rapidly. Both phosphorylases were able to catalyse glucan formation from suitable primers, as well as the phosphorolytic breakdown of starch (Matheson and Richardson, 1976). As mentioned above gibberellin is the best known growth substance controlling the initiation of synthesis of hydrolytic enzymes. It is not, however, the only substance effective. In some seeds including peas, cytokinins are clearly involved. In some cases the formation of hydrolytic enzymes appears to be a release reaction from some bound form, for example β-1,3-glucanase in barley aleurone. A starch debranching enzyme is released from a particulate fraction in peas by an activation reaction.

To what extent germination and rise in enzyme activity are causally related is at best doubtful as germination preceded the rise in enzyme activity of some of the enzymes.

The ratio between the various sugars and oligosaccharides changes as a result of the activity of the enzymes discussed above. These changes have been studied in detail in germinating barley because of their importance in the malting process. Glucose and fructose content rose very considerably up to 6 days germination at 20°C and then began to fall again. After 6 days the seedlings were 5 – 7 cm long. Other sugars also showed marked changes. Maltose rose from 1 mg per gram dry seeds to more than 55 mg after 7 days of germination. Almost as great an increase occurred in oligosaccharides containing more than three hexose residues. Sucrose showed far smaller and less regular increases, as did glucodifructose. Raffinose and maltotriose content stayed more or less steady for the first 5 days of germination and then rose steeply, increasing five-fold in the next 2 days. Apparently sucrose, raffinose, glucodifructose and fructosans are associated with respiration (McLeod *et al.*, 1953). Raffinose metabolism in barley was studied in greater detail by McLeod (1957). Raffinose was absent from the endosperm of barley, but accounted for 9% of the dry weight of the embryo. Raffinose was rapidly utilized by the embryo under normal conditions, but at the same time the sucrose content of the seedling increased. In isolated embryos raffinose metabolism was retarded. No changes in either sucrose or raffinose occurred during the first 24 hr of germination if the seeds were kept immersed in water, i.e. steeped as in the malting process. Their metabolism seems to be closely connected to aerobic processes.

The scutellum is located between the endosperm and the embryo proper and its role in the metabolism of germinating cereal seeds is important. In cereal grains such as barley, the scutellum has a role in the breakdown of starch and other components of the endosperm. The epithelial cells of the surface of the scutellum are papillar in nature and this may facilitate exchange between the embryo and the endosperm (Negbi and Sargent, 1986). These epithelial cells appear to be the source of enzymes needed for the initial breakdown of the endosperm. Later the cells in the aleurone become metabolically active, as can be seen from their ultrastructural organization, such as Golgi formation. This organization starts near the embryo. The breakdown of starch is also gradual and starts at the embryo end of the seed (Gram, 1982a,b). Ultrastructural studies therefore show a graded response in the tissue, and not a uniform activation of enzymes and disappearance of reserve materials throughout the seed. Edelman *et al.* (1959) studied the function of the scutellum in considerable detail. They showed that glucose is removed from the endosperm, converted to sucrose in the scutellum and transported to the embryo. The scutellum always has a low hexose and high sucrose content, the reverse being true for the endosperm and for the embryo. Even the isolated scutellum can readily form sucrose from hexose. Sucrose is formed by a complex mechanism. Glucose is phosphorylated in the six position in the presence of ATP. Part of the glucose-6-phosphate formed is converted to fructose-6-phosphate (F-6-P) and part to glucose-1-phosphate. The glucose-1-phosphate, in the presence of uridine triphosphate (UTP), is converted to uridine diphosphoglucose (UDPG). Sucrose is then formed by the condensation of UDPG and F-6-P. The enzymes necessary for all these reactions could also be demonstrated in the scutellum.

Martin (cited by Halmer, 1985) reported that 19 angiosperm families had non-starchy endosperms, containing mannans or galactomannans as storage materials in the cell walls, which can be mobilized, during germination. The metabolism and breakdown of these compounds has been studied in greatest detail in the Leguminosae. Thus in *Trigonella foenum-graecum* galactomannan breakdown begins 18 hr after radicle emergence. Galactomannans are broken down by the combined action of α-galactosidase, β-mannosidase, exo-β-mannanase and endo-β-mannanases. In the leguminous seeds galactomannan breakdown occurs after other existing oligosaccharides have been utilized. The change in the activity of the enzymes involved is different in various seeds. There are similarities or at least analogies between the sequence of events of galactomannan breakdown in leguminous seeds and starch breakdown in cereals (Mayer and Marbach, 1981). In *Cyamopsis* all three enzyme activities are required for galactomannan breakdown. During germination endo-β-mannanase is formed in the endosperm as is α-galactosidase. Exo-β-mannanase and α-galactosidase are present in the dry cotyledons but their level increases during germination (McCleary, 1983). Exo-β-mannanase is present in the endosperm of the dry seeds. The cotyledons were able to take up mannobiose and apparently to metabolize it. The presumed sequence in this tissue is the concerted attack of α-galactosidase and endo-β-mannanase on the galactomannan in the endosperm resulting in galactose and manno-oligosaccharide formation. The galactose is further metabolized in the cotyledon, while the oligosaccharide is first broken down by exo-β-mannanase to mannobiose and mannotriose. These are then degraded in the cotyledon to mannose by exo-β-mannase action and the mannose is metabolized.

In *Phaseolus vulgaris* seeds α-galactosidase activity is high in the embryo and low in the cotyledons of the dry seeds. During germination its activity falls in the embryonic axis and rises in the cotyledons (Lechevallier, 1960).

The cell walls of the endosperm of lettuce are heavily impregnated with galactomannans and mannans. The sequence of breakdown is similar to that in legumes and the endo-β-mannase is produced by the endosperm itself, which in lettuce has no defined aleurone layer. However, the rate of formation of the enzyme is controlled by the embryonic axis. The first product of hydrolysis, oligomannan, is further broken down by a β-mannosidase which is produced in the cell walls of the cotyledon (Bewley and Halmer, 1980). The level of this enzyme does not change with time. It acts on mannobiose and mannotriose, which are released from the endosperm and diffuse to the cotyledons, the mannosidase acting extracellularly to form mannose and galactose, which are then absorbed (Ouelette and Bewley, 1986). As in the cereal seed there exists an intricate relationship between the different parts of the seed which regulate the timing and sequence of decomposition of the storage material. It is particularly interesting that this should persist in the endosperm of lettuce which is essentially a residual tissue containing only a very small amount of the total storage products of the lettuce seed.

In the Palmae the storage material in the endosperm is mannan. In seeds such as coconut and date palm, the distal end of the single cotyledon develops into an haustorium during germination. The epithelial cells of this haustorium are rich in organelles and the endosperm cells adjacent to the haustorium are the first ones to be digested (De Mason, 1983, 1985; De Mason *et al.*, 1985). The breakdown of mannan in the endosperm appears to be regulated by the haustorium, but the hydrolytic enzymes do not originate there.

Sucrose is often present in dry seeds in small amounts or is formed as a result of raffinose breakdown. The presence of invertase has been demonstrated in a number of germinating seeds, for example barley (Prentice, 1972) and lettuce (Eldan and Mayer, 1974). It arises during germination and could at least partly account for sucrose breakdown. In addition it is probable that part of the sucrose is metabolized by glycosyl transfer reactions (Pridham *et al.*, 1969).

In *Phaseolus* seeds malonic acid is formed during germination, while it is absent or present only in trace amounts in the dry seeds. After five days of germination appreciable amounts of this acid are found in the embryonic axis (Duperon, 1960). In the same seeds other organic acids were also metabolized. Citric acid decreases during germination, while malic acid accumulates. The changes were quite marked after five days of germination. In contrast, in *Zea mays* the amounts of citric, malic and aconitic acids all increase during germination although at different rates. Other tricarboxylic acid cycle intermediates were found, in very small amounts, in a number of seeds (Duperon, 1958).

Generally speaking it can be concluded that most of the enzymes involved in the breakdown and interconversion of carbohydrates become active during germination, most by *de novo* synthesis, some by activation or release.

A special aspect of carbohydrate mobilization is the relationship between axis and cotyledons in controlling the rate of breakdown. The arguments in this respect hinge around the question of whether the axes and cotyledons simply have a sink – source relationship, or whether there exists a hormonal interaction between the two, with regulatory function. This question is still not satisfactorily resolved and some of the arguments for the different views can be found in articles by Murray (1984) and Halmer (1985).

2. Lipids

The lipids are generally present in special organelles referred to as lipid bodies or sometimes as spherosomes. The lipid bodies contain or acquire part or all of the enzymes required for the breakdown of lipids to fatty acids and glycerol. This is essentially the first step in lipid breakdown and is carried out in a stepwise manner by lipases. Normally neither of the breakdown products of hydrolysis of lipids accumulates in the seeds and are present, if at all, in small amounts. The fatty acids are processed further, chiefly by β-oxidation, while the glycerol which is formed becomes part of the general carbohydrate pool present in the seed and as such becomes available for various processes including respiration. In *Arachis* cotyledons enzyme systems have been shown which convert glycerol to glycerol phosphate which is then converted to triose phosphate. This can then be either converted to pyruvic acid or to sugars (Stumpf and Bradbeer, 1959). The bulk of the fatty acids formed following lipase action are broken down by the process of β-oxidation, resulting in the cleavage of two carbon units in the form of acetyl-CoA, which can enter the tricarboxylic acid cycle. This reaction requires both CoA and ATP. β-Oxidation was shown to occur in extracts of various seeds (Rebeiz *et al.*, 1965). The

β-oxidation activity was shown to be associated with the glyoxysomes (for discussion of glyoxysomes and their role see later). The fatty acids may also be broken down by α-oxidation. In this process the fatty acid is peroxidatively decarboxylated and carbon dioxide formed. The long chain aldehyde is oxidized to the corresponding acid by a reaction linked to NAD.

In many seeds disappearance of lipids is accompanied by the appearance of carbohydrates. This reaction apparently proceeds as follows. The fatty acids undergo β-oxidation. The acetyl CoA formed is converted to malate via the glyoxylate cycle. The malate thus formed is converted to carbohydrate by a number of reactions. All these reactions have been shown to occur in the cotyledons or endosperm of lipid-containing seeds such as soybeans, castor beans and groundnuts (*Arachis*). At the cellular level three cell compartments are involved in this process, glyoxysomes, mitochondria and the cytoplasm.

Lipoxidase is also believed to play a part in fatty acid oxidation. This enzyme, which occurs in seeds, is supposed to break the unsaturated fatty acid chain into two smaller parts by peroxidative attack at a double bond. The precise mechanism of this system is at present in doubt.

Lipids are not always converted to carbohydrates during germination. Boatman and Crombie (1958), and Crombie and Comber (1956), followed lipid breakdown in two different seeds, *Citrullus vulgaris* and the oil palm *Elaeis guineensis.* In *Citrullus* lipids are rapidly broken down both in the cotyledons and the rest of the seed and the breakdown products utilized in respiration. There does not appear to be conversion to carbohydrates. The course of fatty acid breakdown differed in the light and the dark. While in the light all fatty acids were broken down at the same rate, in the dark linolenic acid disappeared at a greater rate than in the light.

A rather different situation was found for *Elaeis* seeds. In these seeds the lipids are located in the endosperm, which is invaded by an haustorium during germination. Free fatty acids accumulate in the endosperm, but not in the haustorium. Although lipids are found in the haustorium they appear in the esterified form. It seems that free fatty acids are transferred from the endosperm to the haustorium and are immediately re-esterified there. In *Elaeis* seeds the bulk of the lipids are lost during germination, apparently as in *Citrullus* by respiration, and little or no conversion to carbohydrate occurs. All fatty acids are metabolized at about the same rate, but saturated fatty acids disappear more rapidly than unsaturated ones. Lipases are apparently absent in the haustorium and present only in the shoot of the seedling (Oo and Stumpf, 1983a,b).

The lipid bodies in which the lipids are stored are surrounded by a membrane. Since the initial breakdown of lipids is by the action of lipases, at least some lipase activity may be assumed to be associated with this membrane. However, demonstration of this association has not always been possible, especially in the lipid bodies of dry seeds (Huang and Moreau, 1978).

There are normally several lipases present in seeds, which differ in the pH optimum for their activity. Lin *et al.* (1982) showed that in homogenates of soybean cotyledons three different activities of lipase can be demonstrated — an acid lipase with a pH optimum of 4.9, a neutral lipase with an optimum between pH 6.0 and pH 7.5 and an alkaline lipase with an optimum at pH 9.0. The neutral lipase is the first to increase in activity, at the start of imbibition (Fig. 5.5), but its activity

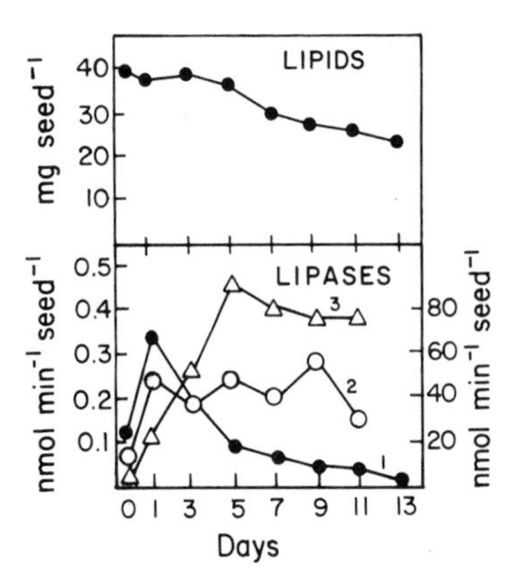

FIG. 5.5. Changes in lipid content and lipolytic activity in the cotyledons of soybean seeds and seedlings during germination. (1) Neutral lipase (right ordinate). (2) Acid lipase. (3) Alkaline lipase (left ordinates). Lipolysis was carried out using whole homogenates at pH 5.0 or 9.0 and the release of free fatty acids measured. Neutral lipase was assayed fluorimetrically at pH 6.5 using N-methyl indoxyl myristate as an artificial substrate. Total lipids were determined by drying aliquots of chloroform extracts and weighing the residue (redrawn from data of Lin *et al.*, 1982).

decreases again between the 3rd and 7th day of germination, when the lipid content begins to fall. The acid and the alkaline activities begin to increase after less than one day. The acid lipase is active during most of the period of depletion of lipids, from the 3rd day onwards, at a more or less constant level. However, the activity of both the acid and the alkaline lipases are much lower than that of the neutral lipase even when at its lowest level ($0.2-0.4$ nmol.min^{-1}.seed^{-1} as compared to $10-20$ nmol.min^{-1}.seed^{-1}). The alkaline lipase was shown to be glyoxysomal in origin. This lipase was incapable of utilizing the lipid bodies isolated from dry seeds as substrates. However, it was capable of utilizing the lipid bodies isolated after $5-7$ days of germination.

In other cases the acid lipase appeared first and seems to be associated with the spherosomes as in castor beans. An alkaline lipase appeared later, which was largely associated with the glyoxysomes (Muto and Beevers, 1974; Mayer, 1977). The alkaline lipase seems to be more specific, hydrolysing monoglycerides while the acid one also attacks di- and triglycerides.

The triglycerides are not necessarily broken down uniformly. Certain fatty acids may be released selectively by lipase attack. The lipase from *Brassica campestris* attacked triolein more slowly than either triacetin or tributyrin (Wetter, 1957). In seeds of cotton, linoleic and palmitic acid are broken down preferentially compared to stearic and oleic acid (Joshi and Doctor, 1975).

As can be seen from Fig. 5.6 lipase activity changes during germination and the course of the change differs in different parts of the seed. Figure 5.6 shows the changes in activity in the neutral lipase in the endosperm and the embryo of germinating castor beans (*Ricinus communis*) (Yamada, 1957). Lipase activity in the embryos reaches a peak after 24 hr germination, while in the endosperm it only begins to increase after about 50 hr.

In Douglas fir (Ching, 1968) lipase appears to be associated with the lipid bodies

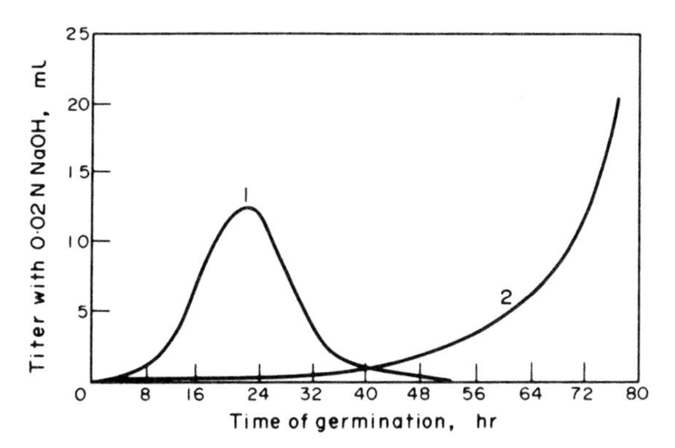

FIG. 5.6. Lipase activity in castor bean seeds during germination. (1) Neutral lipase in the embryo. (2) Neutral lipase in the endosperm (Yamada, 1957).

themselves and in jojobe an alkaline lipase was shown in the wax bodies (Huang and Moreau, 1978).

The subcellular location of lipase varies in different seeds. Not necessarily the same type of lipase, as characterized by pH optimum or specificity, will be located in the same subcellular location. In scutella 50% of the lipase having a pH optimum of 7.0 is associated with lipid bodies. In rape and mustard cotyledons it is also associated with lipid bodies (Lin and Huang, 1973; Lin *et al.*, 1983). In soybean cotyledons the alkaline lipase is apparently associated with glyoxysomes and the neutral one with the lipid bodies (Lin *et al.*, 1982). In *Brassica napus* lipase is apparently synthesized in the ribosomes and is attached to the lipid bodies by special handle-like appendices, whose formation can be demonstrated using biochemical techniques (Theimer and Rosnitschek, 1978).

There is a great amount of diversity in lipase activity from the point of view of pH optimum, substrate specificity and tissue and subcellular location. It is difficult to explain such diversity or to understand the physiological control exercised over lipase formation and lipid disappearance.

It is still unclear how the lipids stored in the lipid bodies become accessible to the action of lipase. The data brought above, showing that glyoxysomal alkaline lipase can only use lipid bodies as substrate at a certain stage of germination, points to the complexity of the system. Recently the presence of a lipase inhibitor, which is a protein, has been demonstrated in sunflower seeds (Chapman, 1987). The presence of such an inhibitor can easily interfere with assays for lipase activity. Whether it has a physiological regulatory role remains to be determined. More detailed studies on the membranes surrounding lipid bodies and their properties are clearly needed.

The fatty acids formed by hydrolysis from the glycerides are further metabolized by a complex pathway — the glyoxylate cycle — or they may be directly utilized in cell metabolism. The relation of the glyoxylate pathway to the tricarboxylic acid cycle and the conversion of lipids to carbohydrates is illustrated in Fig. 5.7. The main function of the glyoxylate cycle is to convert lipids to carbohydrates. The cycle has a number of important features. Fatty acids are oxidated by β-oxidation and acetyl CoA is formed. The acetyl CoA condenses with glyoxylate to form malate,

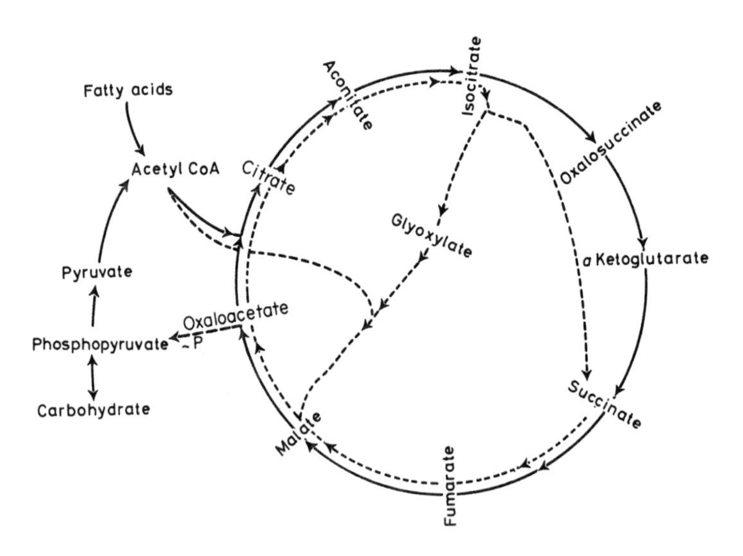

FIG. 5.7. The relationship between the tricarboxylic acid cycle (solid line) and the glyoxylate cycle (broken line) and the possible interconversion of lipid to carbohydrate.

in the presence of malate synthetase and ATP. The glyoxylate arises from isocitric acid, with the concomitant formation of succinate, by the action of isocitric lyase. The malate is converted to carbohydrates via oxaloacetate and phosphopyruvate, a process which requires ATP. Both the succinate and the malate can enter the tricarboxylic acid cycle. As a result of the glyoxylate cycle α-ketoglutaric acid in the tricarboxylic acid cycle can be completely bypassed (Marcus and Velasco, 1960). Therefore, there are two alternative points of entry of the two carbon units in the respiratory pathway.

This course of events involves glyoxysomes and mitochondria. During operation of the glyoxylate cycle NADH is formed in the glyoxysomes due to β-oxidation and condensation of acetate to succinate. The glyoxysomes are incapable of reoxidizing NADH to NAD. It has been assumed in the past that NADH leaves the glyoxysomes and is oxidized in the mitochondria. It has been shown that the oxidation of NADH is by a malate – aspartate shuttle (Mettler and Beevers, 1980). Malate is transferred from the glyoxysomes to the mitochondria where it is oxidized to oxaloacetic acid. The latter is aminated to aspartate which is returned to the glyoxysomes. In the glyoxysomes aspartate is transaminated to glutamate utilizing α-ketoglutaric acid and oxaloacetic acid is formed, which is reduced to malate using NADH. The relationship between the two cell compartments is illustrated in Fig. 5.8.

The glyoxylate cycle has been shown to operate in various seeds, for example *Arachis* and *Ricinus communis* (Kornberg and Beevers, 1957; Marcus and Velasco, 1960; Yamamoto and Beevers, 1960; Mettler and Beeevers, 1980). The activity of the enzymes involved in this cycle has been shown to increase during germination of *Arachis* seeds, as shown in Table 5.2. It will be seen that both malate synthetase and isocitrate lyase increase during germination and are completely absent in the dry seed.

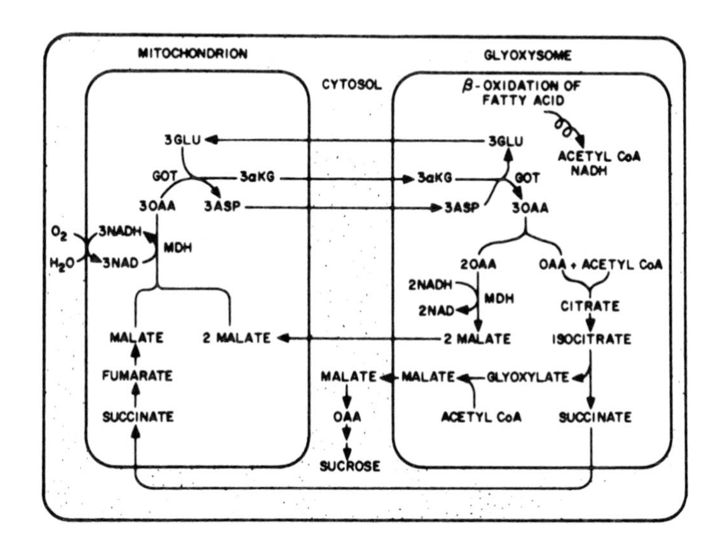

FIG. 5.8. Scheme showing interrelationship between glyoxysome and mitochondrion and the possible mechanism by which NADH produced in the glyoxysomes is oxidized by a malate aspartate shuttle (from Mettler and Beevers, 1980). (Glu = glutamate, OAA = oxaloacetic acid, α-KG = α-ketoglutarate, GOT = glutamate oxaloacetate amino transferase, MDH = malate dehydrogenase).

TABLE 5.2. *Changes in activity of the glyoxylate cycle enzymes during germination of peanuts. Enzyme activity given as μ moles glyoxylate formed (isocitritase) or removed (malate synthetase) (from Marcus and Velasco, 1960)*

Time in days	Length of radicle (mm)	Enzyme activity Isocitritase	Malate synthetase
0		—	
1	0	0	0
2	2	7·9	10·2
3	16	20·4	24·6
4	35	30·0	50·2
5	55	42·3	69·4

One of the questions relating to the operation of the glyoxylate cycle is how the level of enzymes participating in it is regulated. Using cDNA probes it was shown that the level of malate synthetase was regulated by gene transcription and not by translation as might have been thought (Smith and Leaver, 1986). Light which is known to affect the level of this enzyme exerted its effect by changing the number of gene transcripts. This kind of approach will no doubt be used more extensively in the future and help in determining levels and mechanisms underlining their control.

The glyoxysomes (Fig. 1.8D) were first described by Breidenbach and Beevers (1967). The glyoxysomes are a special class of microbodies found in cells, they have a density slightly greater than mitochondria. They can be separated from the latter by differential or density gradient centrifugation. Glyoxysomes as other microbodies have a single boundary membrane, and are characterized by the occurrence in them of catalase, uricase and glycolate oxidase. In addition to the above the glyoxysomes contain the enzymes required for the glyoxylate cycle as well as the enzymes required for β-oxidation, transamination and often also a malic dehydrogenase. Glyoxysomes are formed during germination, although some may be present already in the dry seeds or develop during stratification. Their formation is clearly linked to the onset of lipid utilization. The origin and nature of microbodies in general and glyoxysomes is reviewed by Beevers (1979) and by Huang *et al.* (1983).

It is very likely that at least the membranes of glyoxysomes are derived from the endoplasmic reticulum. Whether they arise with a full enzyme complement or whether enzymes are added post-translationally to the glyoxysomes concomitant with lipid degradation is still not resolved. However, it is clear that at least part of the enzymes are formed *de novo* and reach a peak in activity when storage lipids are degraded at maximal rate (Huang *et al.*, 1983). However, it has been demonstrated that glyoxysomal malate dehydrogenase in cotyledons of germinating pumpkin seeds is synthesized in the cytosol, and is incorporated into the organelle after processing (Yamaguchi et al., 1987).

The presence and characteristics of glyoxysomes has now been reported in many seeds (Mayer and Shain, 1979; Mayer and Marbach, 1981), although the most detailed studies are those on the glyoxysomes of the castor bean endosperm. An unusual feature of the glyoxysomes is their disappearance after the lipid reserve of the seed has been broken down. There is some evidence that glyoxysomes are converted to peroxisomes. Other explanations have also been proposed to account for the disappearance of glyoxysomes and the appearance of peroxisomes. These are: (1) destruction of glyoxysomes and the more or less simultaneous appearance by *de novo* synthesis of peroxisomes; (2) the loss of some enzymes from the glyoxysomes and the insertion into them of enzymes characteristic of the peroxisomes; and (3) the existence of a transient form of subcellular organelle showing properties between peroxisomes and glyoxysomes. Although there is some evidence for such an interconversion model, it is still not convincing and the question remains unresolved (Beevers, 1979; Huang *et al.*, 1983).

It should be remembered that other enzymes, active in some stages of germination, reach a peak in activity and then disappear again when no longer required, perhaps due to the activity of specific enzymes which break them down or inactivate them.

Membranes are composed of proteins and phospholipids. Phospholipid metabolism will be considered later in the section on phosphorus compounds.

3. Proteins

The storage proteins of most seeds are found in well-defined organelles, the protein bodies. These are small membrane bound particles between 1/2 – 10 nm in diameter which, in addition to proteins, often also contain phytin (see later). The protein bodies are apparently vesicles derived from the endosplasmic reticulum in which proteins are deposited during seed formation and maturation. During germination the storage proteins are broken down and the protein bodies empty, their membranes remaining intact. Eventually the protein bodies give rise to vacuoles (Ashton, 1976; Pernollet, 1978). The sequence of breakdown of protein bodies in some of the cereals differs from that in seeds of dicotyledonous plants (Pernollet, 1982). In maize the chief storage protein is zein but the skeleton of the protein body is made up of glutelin; in other cases, for example *Sorghum* or rice, the skeleton consists of prolamine. These proteins are located at the surface of the protein body and are the first to be degraded. Their degradation is followed by the hydrolysis of the proteins within the protein body (Harvey and Oaks, 1974; Taylor *et al.*, 1985). This mode of dissolution differs from that observed in many dicotyledonous plants. In these dissolution of the protein bodies is by autolysis from the inside, the protein body becoming vacuolated while the external membrane remains intact. Figure 1.8 (B – D) shows the gradual dissolution of the protein bodies in the embryonic radicle of a pea root together with vacuole formation (Hodson *et al.*, 1987). The protein bodies in the endosperm of some cereals apparently behave in the same way.

Breakdown of storage proteins in the cotyledons is accompanied by the appearance of new proteins in other parts of the seedling (Fig. 5.1,V). Seeds contain a variety of proteolytic enzymes some of which are present in the dry seeds while others appear during germination. The proteolytic enzymes can be divided into proteinases and peptidases depending on the size of the molecule which is attacked. Peptidases are usually divided into the endopeptidases and the exopeptidases, carboxypeptidases and aminopeptidases, depending on the site of the bond in the protein molecule which is attacked. In most cases the enzymes having proteolytic activity are soluble and are present or develop in the storage organs, i.e. in the cotyledons or the endosperm. The types of proteolytic enzymes and peptidases occurring in germinating seeds do not seem to differ fundamentally from those occurring in other plant tissues, although in a few cases enzymes with special characteristics have been reported. The development of proteolytic enzyme activity appears to be under hormonal control, the hormones originating in the axis or embryo. In barley the appearance of enzyme activity is regulated by gibberellic acid secreted by the aleurone or the embryo, while in *Cucurbita maxima* cytokinin regulates enzyme activity. In the seeds of dicotyledonous plants the hormones usually originate in the embryonic axis while the enzyme is formed in the cotyledons. In *Cucumis sativa*, the seed coat apparently also has some role in the regulatory process. Maximal breakdown of protein and accumulation of amino acids occurs in imbibing seeds when both the seed coat and the embryonic axes are present. When the seed coat is removed and only the axis is present 80% of the proteins are broken down within seven days of germination. In excised, seed coat-less cotyledons the value was only 15% and removal of the axis, with seed coat present reduced protein breakdown to only 5% (Davies and Chapman, 1979). Removal of the axis only had very little effect on trypsin and amino peptidase activities. Therefore it was suggested that the axis does not affect protein breakdown directly, but creates a sink for the breakdown products — the amino acids. The seed coat, however, does affect enzyme formation. In its presence trypsin activity develops more slowly, although after seven days of germination its level was the same as that of other enzymes, which were not affected by the seed coat. Obviously it is not simply the level of proteolytic enzyme which controls the rate and extent of breakdown of storage proteins in the cucumber seed. The interaction involved must be more complex and may well involve hormone effects as shown for other dicotyledonous seeds. Such hormonal control is not always observed and seems to be absent in pea cotyledons.

The proteolytic enzymes and peptidases of germinating seeds show great diversity with regard to their specificity for peptide linkages, their pH optima and their response to inhibitors. For example in barley eight distinct peptidases are present, as well as three different proteases. In pea seeds at least two peptidases and two proteases are present, and in lettuce four different proteolytic enzymes could be detected. Acid amino peptidases of two kinds have been reported — those sensitive to *p*-chloro mercuribenzoate indicating the presence of SH group (in barley, maize, mung beans and castor beans) and acid proteinases which are not sensitive (in *Sorghum* and in *Cannabis*). Leucine-amino-peptidases and other enzymes active towards specific synthetic substrates, as well as various carboxypeptidases have also been shown to be present in various germinating seeds. The species, the nature of the storage proteins, the part of the seed studied and of course the precise stage of germination under observation may account for the observed variability in the enzymes which have been described (Mayer and Marbach, 1981; Muntz *et al.*, 1985).

The enzymes, responsible for proteolytic breakdown, initially present in the protein bodies may be synthesized, or activated, in the protein bodies themselves or may be synthesized elsewhere in the cell and transported into the protein bodies. The protein bodies of the seeds of *Vigna radiata* (formerly referred to as *Phaseolus aureus*) contain a number of hydrolytic enzymes. However, during germination the storage protein vicilin is not broken down until an additional endopeptidase (vicilin peptidohydrolase) is synthesized *de novo* in the cytoplasm and transported to the protein bodies. Its existence can be demonstrated in the cytoplasm by immuno-fluorescence after three days of germination. On the fourth day the enzyme already appears in the protein bodies (Baumgartner *et al.*, 1978).

In *Phaseolus vulgaris* two alkaline peptidases are present in the cotyledons of the dry seed, whose activity falls during germination. However, the activity of three acid peptidases and proteinases which are also present in dry seeds rises during germination (Mikkonen, 1986). The different enzymes apparently have different functions during germination, which have not yet been fully identified. The changes in activity of the alkaline peptidases parallels the course of breakdown of proteins during germination (Fig. 5.9). The acid proteinases increase in activity much later. However, even the initial activity of the proteinases present in the dry seed may be sufficient to produce enough peptides, whose breakdown products can be transported quantitatively to the growing axis. The sharp decrease in nitrogenous reserves coincided with a sharp increase in proteinase activity, which declines again when most of the nitrogenous reserves have been depleted. It is still unclear how the polypeptides are broken down, since the measured peptidase activity in the cotyledons is low between the 10th and 20th days of germination. More data are needed to interpret fully the course of events. In the endosperm of castor bean initial hydrolysis of storage protein is due to the activity of a pre-formed protease which becomes active by hydration during germination (Gifford *et al.*, 1986).

Breakdown of storage proteins is by no means random. It involves, in its initial stages, limited proteolysis of some, but not all of these proteins. For example in

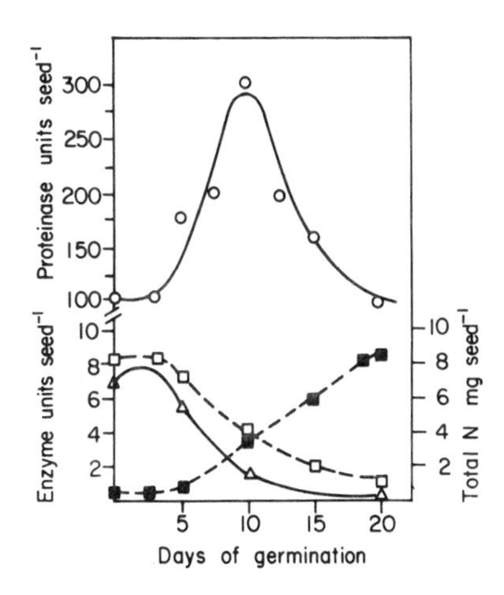

FIG. 5.9. Changes in total nitrogen and in proteinase activity of seeds of *Phaseolus vulgaris* during germination (compiled from data of Mikkonen, 1986). Total N_2 measured in both cotyledons and axis. Enzyme activities measured only in cotyledons. (1) □----□ Total nitrogen in the cotyledons. (2) ■----■ Total nitrogen in the embryonic axis. (3) △——△ Dipeptidase activity against Ala – Gly at pH 8.5. (4) ○——○ Proteinase activity at pH 3.7, using haemoglobin as substrate.

castor bean there are two storage proteins glycinin and β-conglycinin, which are broken down at quite different rates, β-conglycinin being broken down first. Moreover the acid subunits are preferentially degraded. In glycinin the acidic chains are hydrolysed before the basic ones (Wilson *et al.*, 1986).

In many seeds endogenous inhibitors of trypsin, chymotrypsin and papain are present. These inhibitors have been studied in considerable detail (Vogel *et al.*, 1968; Ryan, 1973) because of their importance when the seeds are ingested by animals. Their exact function in the germination process is still debatable. They may act by regulating proteolytic enzyme activity during germination or they may be simply a relic from the period of seed development, when they may have prevented the decomposition of newly formed storage proteins. The protease inhibitors are not equally distributed among different parts of the seed or seedling. Interaction between proteinase and proteinase inhibitor is illustrated by the course of events in germinating lettuce seeds (Fig. 5.10). As can be seen two enzymes with different pH optima increase during germination while a third enzyme disappears, due to its interaction with an endogenous trypsin inhibitor.

Although many questions are still open with regard to the detailed mechanism of protein metabolism during germination, there is some information about the fate of the breakdown products. Usually there is little change in the total nitrogen content of the seed or seedling during germination, although slight losses may occur, especially due to leaching out nitrogenous substances. Nitrogen appears to be very carefully conserved. In the place of the protein broken down free amino acids and amides appear. Table 5.3 shows some of these changes in *Phaseolus mungo* seeds.

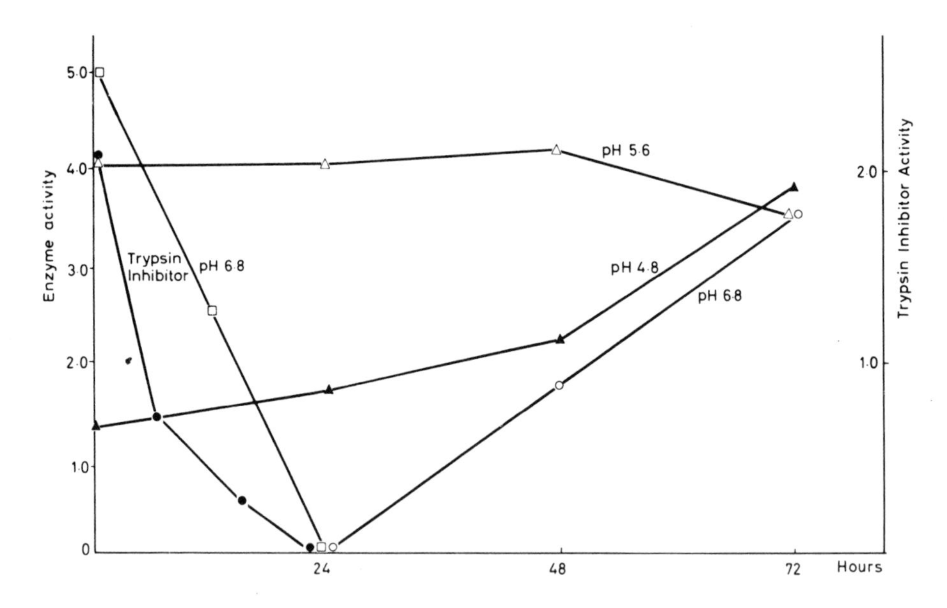

FIG. 5.10. Proteolytic activity and trypsin inhibitor of extracts of lettuce seeds as a function of pH and length of germination period. Enzyme activity as change in titer with N/50 NaOH, pectinase as substrate (left ordinate). Trypsin inhibitor, units/mg dry seeds (right ordinate) (redrawn from Shain and Mayer, 1965; 1968c).

TABLE 5.3. *Changes in protein content and other nitrogenous compounds in Phaseolus mungo seeds (from Damodaran et al., 1946). The results are given as percentage of total nitrogen*

Age of seedlings in days	1	4	7	10	13	16
Protein N (Extractable)	83·0	35·0	20·0	9·0	13·0	7·0
Ammonia N	0·3	0·9	1·2	0·8	1·1	2·5
Asparagine amide N	0·7	9·3	11·3	12·7	12·5	11·7
Glutamine amide N	0·3	1·1	1·4	2·3	2·9	2·5
Amine N	3·2	15·4	23·7	23·7	22·5	20·8
mg Total N/300 seedlings	534·5	486·2	511·2	517·2	499·6	505·5

The early work on nitrogen metabolism both by Paech (1935) and Prianishnikov (1951) and more recent work summarized by Oaks and Bidwell (1970) and Oaks (1983) showed that it is possible to induce amide formation in germinating seeds in the dark by feeding them with ammonium salts. If sugars were also fed to the seedlings this led to a marked sparing action on protein breakdown. Simultaneous feeding of both glucose and ammonia nitrogen led to greatly increased amide formation. This work has led to the view that amide formation proceeds via the following steps: during germination in the dark, proteins are broken down to amino acids, part of these amino acids are oxidatively deaminated and the carbon skeleton enters various respiratory and carbon cycles. The ammonia formed by deamination is detoxicated by the process of amide formation. The chief amides formed are glutamine and asparagine, depending on the plants (Chibnall, 1939; Lea and Joy, 1983).

Not all amino acids are deaminated. Parts of those which are not deaminated are utilized for the synthesis of proteins in the growing parts of the seedling. Soluble inorganic nitrogen already present in the seed may also be used during germination. Thus Egami *et al.* (1957) showed that the small amounts of nitrate present in *Vigna* seeds disappear as germination proceeded; they also proved the existence of a nitrate reductase system in the seedlings. Yamamoto (1955) showed, in the same seeds, that asparagine present in the cotyledons disappears and instead appears in the hypocotyl and plumule. These findings help us to understand the effect of feeding ammonia and sugars on protein metabolism. Feeding ammonia reduced protein breakdown because external nitrogen can be used to form new protein. In the case of sugars, the sparing action is chiefly due to the provision of respiratory substrates, which would otherwise be provided by the breakdown of protein and deamination of the amino acids formed.

The mechanism of glutamine formation is fairly well known. Glutamine is formed from glutamic acid and ammonia, in the presence of the enzyme glutamine synthetase and ATP, the reaction being energy-requiring. Asparagine formation is less clear (Lea and Joy, 1983). Yamamoto (1955) showed that in *Vigna* hypocotyl, asparagine is formed if the hypocotyls are fed with aspartic acid, ammonia and ATP. The enzyme responsible for the reaction has, however, never been isolated. The formation of asparagine may involve β-cyanoalanine as an intermediate.

Germinating seeds usually also contain enzymes causing hydrolysis of the amide bond, glutaminase and asparaginase, whose precise function is unknown. Asparaginase has recently been reinvestigated. The enzyme has been shown to be present in many developing seeds and previous doubts about its distribution can probably be ascribed to the failure to note the dependence of its activity on potassium ions (Sodek *et al.*, 1980). Liberation of NH_4^+ from asparagine by asparaginase can presumably lead to amino acid formation processes.

Another type of transfer enzyme concerned with amino acid metabolism is constituted by the transaminases, which transfer amino groups from amino acid to keto acid (Fowden, 1965). The presence of transaminases in a variety of seeds has been shown by Smith and Williams (1951). They found a marked increase in the activity of the transaminases transferring amino groups from alanine or aspartic acid to α-ketoglutaric acid (Table 5.4). It is probable that the same enzymes are also carrying out the reverse reactions. In most cases the activity of glutamic – aspartic transaminase increased more rapidly during germination than that of glutamic – alanine transaminase. However, in peas the reverse was the case, and in corn both enzymes seemed to increase at about the same rate during germination. These authors also showed that no fixed relation existed between increase in protein in the seeds and increase in transaminase activity, and concluded that there was no evidence to show that the two processes were directly connected.

Evidence for the presence of enzymes catalysing the reverse reaction from glutamine to pyruvic or oxaloacetic acid in wheat germination was brought by Cruickshank and Isherwood (1958).

Albaum and Cohen (1943) followed the transamination reaction glutamic – oxaloacetic acid and its reverse reaction in oat embryos. The former action took place at three times the rate of the latter. Glutamic – oxaloacetic acid transaminase activity of the embryo expressed per unit of protein increased steadily during germination.

TABLE 5.4. *Changes in glutamic – aspartic and glutamic – alanine transaminases and in protein nitrogen during germination of various plant seeds. Enzyme activity is expressed in units/embryo and protein nitrogen as mg protein/embryo. In all cases the embryo after removal of endosperm or cotyledons was used (Smith and Williams, 1951)*

Time in hours	24			48			72			96		
Seeds	Glu – Al	Glu – Asp	Protein N	Glu – Al	Glu – Asp	Protein N	Glu – Al	Glu – Asp	Protein N	Glu – Al	Glu – Asp	Protein N
Waxbean	0·03	0·10	0·11	0·11	0·16	0·16	0·15	0·28	0·26	0·17	0·48	0·32
Pea	0·03	0·06	0·09	0·17	0·13	0·12	0·28	0·18	0·15	0·37	0·23	0·16
Barley	0·08	0·09	0·03	0·14	0·26	0·04	0·24	0·56	0·08	0·42	0·88	0·11
Corn	0·09	0·10	0·08	0·17	0·17	0·11	0·20	0·28	0·16	0·44	0·53	0·32
Oats	0·05	0·14	0·02	0·10	0·18	0·02	0·17	0·44	0·04	0·20	0·67	0·07
Squash	0·06	0·04	0·08	0·07	0·06	0·08	0·08	0·13	0·08	0·13	0·43	0·09
Pumpkin	0·04	0·06	0·04	0·07	0·09	0·06	0·16	0·27	0·08	0·28	0·43	0·10

If however it was expressed per unit of dry weight it decreased, probably because of the large increase in dry weight of the embryos due to non-protein structural components. In contrast to the results of Smith and Williams (1951), Albaum and Cohen found fairly good correlation between increase in protein, transaminase activity and soluble nitrogen content especially during seedling growth. The changes are illustrated in Fig. 5.18. The recent discovery that glutamic acid arises in plants not by amination of oxoglutaric acid by ammonia but due to the action of the so-called GOGAT enzyme (glutamine – oxyglutarate aminotransferase) is no doubt of significance also in germinating seeds (Miflin and Lea, 1977). This enzyme catalyses the reaction:

$$\text{glutamine} + \alpha\text{-oxoglutarate} + \text{NAD(P)H} + \text{H}^+ \longrightarrow 2\ \text{glutamate} + \text{NAD(P)}^+$$

This enzyme seems to be the key to glutamic acid formation and all transamination reactions are consequently dependent on its functioning. Therefore glutamine itself has a key role in amino acid metabolism.

Glutamine must now be regarded as the main intermediary in metabolism. This may be related to the observation that glutamine does not accumulate in seeds germinated in the light. When seeds are germinated in the soil or in nutrient solution, in the light, protein synthesis is from external nitrogen taken up by the seedling and from the carbon skeleton formed during photosynthesis. Undoubtedly the key processes in such seedlings are the *de novo* synthesis of amino acids and proteins and not protein breakdown and amide formation. In amino acid synthesis transamination reactions probably play an important role.

Dry seeds contain very little free amino acids. The growth of the embryo in the germinating seed is dependent on a supply of amino acids for its protein synthesis. The amino acid pool increases during germination, for example in lettuce (Table 5.5). The main source for these amino acids is the storage protein, but its amino acid ratio need not necessarily be the same as that of the newly synthesized seedling protein and apparently interconversion of the amino acids occurs. The main pathways for such interconversion are the transamination and deamination reactions mentioned above. The amides and particularly glutamine have a special role in this. Virtanen *et al.* (1953) have already shown that in germinating pea seeds homoserine, which is absent from the dry seeds both in its free form and in protein, is synthesized during the first 24 hr of germination. Arginine is also synthesized *de novo* in germinating peas (Shargoal and Cossins, 1968). The synthesis and interconversion of amino acids are apparently the same as in other plant tissues and proceed via the same pathways and amino acid families (Forest and Wightman, 1971, 1972; Miflin and Lea, 1982; Lea and Joy, 1983).

In cereal seeds the scutellum plays a special role in nitrogen metabolism. It acts as an intermediary between the endosperm and the growing embryo. The scutellum is capable of hydrolysing peptides, which it takes up, by means of peptidases present in it. The peptides and amino acids are taken up by carrier-mediated, energy dependent processes. The ability of the scutellum to absorb peptides and amino acids may have a regulatory role in mobilization of endospermal protein for growth and development of the embryo (Sopanen, 1980).

Changes in the free amino acid composition in the germinating seed may be indicative of developmental changes. For example, Fine and Barton (1958) showed

TABLE 5.5. *Changes in the amino acid content of lettuce seeds during germination (Klein, 1955). The amino acid content is given as γ-amino N per gram initial dry seeds*

Amino acid	Time of germination in days 0	1	2	3
Alanine	5	30	80	220
Threonine	5	20	40	190
Leucine	20	20	60	280
Serine	30	30	60	250
γ-amino butyric acid	5	5	15	25
Lysine	15	5	20	40
Tryptophan	5	5	2	—
Glutathione	10	0	0	20
Aspartic acid	40	35	35	40
Glutamic acid	60	80	110	160
Asparagine	30	40	60	240
Glutamine	60	40	360	700

that both the ratio of amino acids and the absolute amounts change during after-ripening of tree peony seeds.

Protein synthesis and its association with nucleic acid metabolism will be discussed in sections III.1 and III.2.

4. Metabolism of Phosphorus-containing Compounds

Phosphorus appears in seeds primarily in the organic form and very little seems to be present as inorganic orthophosphate. Among the phosphorus-containing compounds are the nucleic acids, phospholipids, phosphate esters of sugars, nucleotides and phytin, which is the main form in which phosphorus is stored. Some of the phosphorus-containing compounds occurring in cotton seed and the changes in them during germination are shown in Table 5.6. Phytin is usually present in the form of the calcium and magnesium salt of inositol-hexaphosphoric acid, and sometimes it occurs as the potassium or manganese salt.

TABLE 5.6. *Changes in composition of the various phosphorus fractions in cotton seeds, c.u. Paymaster, during germination. The results are given as mg phosphorus per gram dry weight (from Ergle and Guinn, 1959)*

Time of germination in days	Dry seeds	1	2	4	6
Phytin	8·61	8·49	7·15	4·00	1·97
Inorganic	0·44	0·29	1·87	4·77	7·02
Total lipid	0·71	0·81	0·87	0·50	0·85
Ester	0·32	0·41	0·40	0·58	0·42
RNA	0·12	0·11	0·15	0·25	0·39
DNA	0·11	0·11	0·12	0·21	0·44
Protein	0·11	0·10	0·16	0·28	0·26

Phytin is present in most seeds and may constitute up to 80% of their total phosphorus content (Reddy *et al.*, 1982). The calcium and magnesium content of the phytin molecule is variable. Thus wheat phytin contains 12% calcium and 1.5% magnesium, while oat phytin contains 8.3% calcium and 15% magnesium as well as 5.7% manganese. The absolute amount of phytin is also very variable and varies not only between species but even in different varieties of the same species (Ashton and Williams, 1958). Because most of the phosphate is present in the seed in the bound form, orthophosphate may well be a limiting factor in many reactions in which phosphate participates, such as phospholipid synthesis, nucleic acid and protein synthesis and energy generating processes. The large amounts of phytin, which decrease during germination with a concomitant increase of inorganic phosphate, suggest that phytin acts as a phosphate store. The phosphate is liberated from it by enzymic hydrolysis. The enzyme involved is a phosphatase known as phytase and is probably not entirely specific for phytin and is capable of hydrolysing other phosphate ester linkages as well. Phytase hydrolyses the phosphate bonds of phytin one by one. It has been claimed that more than one phytase exists and that the different phytases attack the phosphate bond at different positions of the myoinositol ring. As a result of the attack of phytase on phytin different myoinositol phosphates can be detected in the seeds, from the penta to the monophosphates. The monophosphate is said to be hydrolysed by a separate enzyme (Tomlinson and Ballou, 1962; Loewus and Loewus, 1982). If more than one enzyme is involved in the hydrolysis of phytin, problems of the regulation of the sequential attack on the substrate must arise, which have not yet been addressed. As can be seen in Table 5.6, the amount of phytin in cotton seed drops quite quickly during germination, so that after six days most of the phytin has disappeared. Similar rapid disappearance of phytin from germinating seeds has been observed in wheat, oats, peas and lettuce, as well as other seeds. In cotton seeds, all the phytin is located in the cotyledons (Ergle and Guinn, 1959). It is also probable that in other seeds a large amount of phytin is located in the storage tissues. However, Albaum and Umbreit (1943) showed that phytin is present in the embryo of oats, disappearing rapidly during germination. Usually there is a good correlation between the rapidity of phytin breakdown and phytase activity of the seed or seedling (Fig. 5.11). In oats, phytase activity in the entire seeds seems to develop rather slowly and the total phytin content decreases to about half in the course of a week. In wheat, on the other hand, most of the phytin has disappeared after a similar period. Peers (1953) showed that in the dry wheat seed about 80% of the phytase is present in the endosperm and only 1% in the embryo. The remainder is distributed between the other parts of the seed. The enzyme prepared from wheat was characterized by stability to fairly high temperatures. The purified enzyme was less heat resistant than the crude preparation. The optimum temperature for enzyme activity was around 50°C and the pH optimum was around 5.4, which seems to be characteristic for phytase prepared from seeds. The mode of action of phytase is quite complex as described above and the various inositol phosphates are not attacked at equal rates.

Recently a previously unknown property of phytin, its action as an anti-oxidant, has been reported (Graf *et al.*, 1987). Phytin, by chelating most of the iron in the seeds, would block iron driven hydroxyl radical generation. This might result in protection of, for example, lipids in membranes. Phytin also inhibits various oxidative

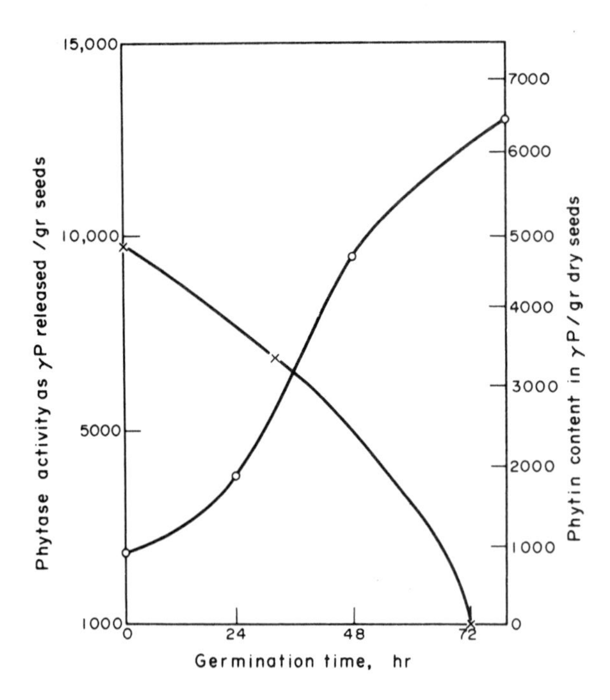

FIG. 5.11. Phytin content and phytase activity in germinating lettuce seeds. Phytin content ×——× as μg phytin phosphorus per gram dry seeds. Phytase activity ○——○ as μg phosphorus released by gram seeds (from Mayer, 1958).

enzymes e.g. polyphenol oxidase. Thus an additional previously unsuspected role might be assigned to phytin, but it must be remembered that most of the phytin is sequestered in protein bodies. It may therefore not be available *in vivo* for this supposed new function.

At one time it was thought that phytin could serve as an energy source and that transphosphorylation reactions between phytin and dinucleotides could occur. However, for thermodynamic reasons this seems very unlikely. Moreover, numerous experiments designed to show the existence of such transphosphorylation reactions were unsuccessful. Thus although the literature still contains reports on phytin as an energy source, these must be treated with great reserve (Mayer, 1973, 1977).

It is significant that the germinating seeds contain not only phytase but also the enzyme systems necessary for phytin synthesis. In fact some phytin appears to be formed during germination (Majunder *et al.*, 1972).

Phytate also acts as a source of myoinositol in the germinating seeds, as has been shown for wheat. The myoinositol liberated is channelled through the myoinositol oxidation pathway into cell wall polysaccharides (Loewus, 1983). Phytin is first broken down to the myoinositol monophosphate, probably the 2-monophosphate, which is then hydrolysed to myoinositol and inorganic phosphate, by a specific phosphatase. Biswas *et al.* (1984) suggest a specific role for inositol metabolism.

They claim that inositol monophosphate can be converted to ribulose-5-phosphate, to which they assign a special function in nucleic acid metabolism. This idea requires more study before being accepted.

Phytin breakdown releases calcium and magnesium and phytin might therefore be a store not only of inorganic phosphates, but also of these ions.

In addition to phytase, seeds contain many phosphatases, and their activity rises during germination. These phosphatases can account for the turnover of all the phosphate esters present in the seed. An example of the changes in phosphatase and ATPase activity in germinating lettuce is shown in Fig. 5.12. In lettuce at least eight distinct phosphatases were demonstrated using electrophoretic techniques.

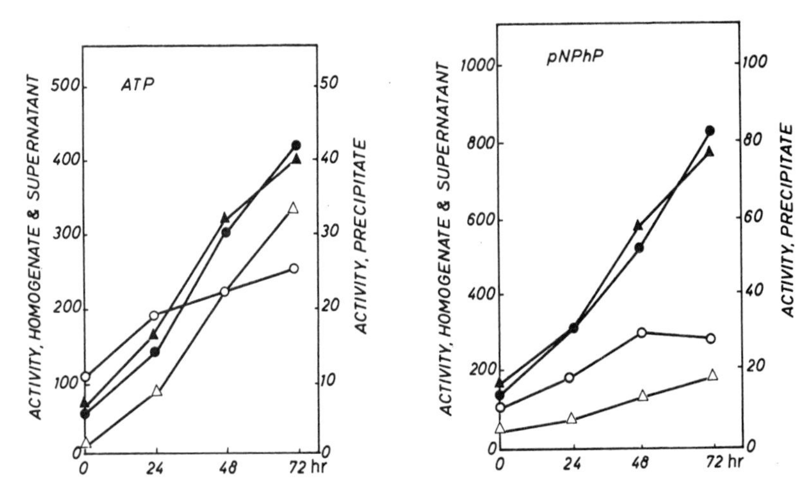

FIG. 5.12. Changes in phosphatase activity in various cell fractions during germination of lettuce. Activity as PO_4 released μmol/g dry seeds/hr for ATP and pNPhP. Homogenate (▲) ; 2000 g precipitate (○); 20000 g precipitate (△) ; 20000 g supernatant (●) (from Meyer and Mayer, 1971).

In many seeds a glycerophosphatase has been shown to be present which differs from phytase both in specificity, pH optimum and heat resistance.

Far fewer data are available about phosphate-containing compounds, other than phytin, in germinating seeds. It is clear from Table 5.6 that the total changes in lipid phosphate and ester phosphate are not very large in cotton. Nevertheless, it is clear today that the phosphate-containing compounds present in membranes such as phosphatidic acid, phosphatidyl inositol, phosphatidyl ethanolamine and phosphatidyl choline are all synthesized very early during seed germination. Evidence of this is provided by the rapid incorporation of $^{32}PO_4$ into these fractions (Katayama and Funahashi, 1969). In pea cotyledons, after only 5 hr of imbibition, the metabolism of the membranes of the endoplasmic reticulum could be detected by the incorporation of radioactive precursors (Jarden and Mayer, 1981).

Choline was incorporated very rapidly into both the ER and the plasmamembrane of cotyledons. The phospholipids of the ER and the plasmamembrane of the cells of the embryonic axis were metabolized much more rapidly than those of the cotyledons

(Di Nola and Mayer, 1985; 1986a,b; 1987). As seeds imbibe water, the integrity of their membranes is restored, as indicated by decreased leakage (described earlier). The actual amount of phospholipids in the seed membranes changes very rapidly during germination as might be expected due to the general increase in metabolic, and especially synthetic activity associated with membrane systems. As yet very little is known about the enzymes concerned with membrane metabolism and turnover, and the changes in them during germination. They appear to be of the same nature as those described in other plant tissues (Mazliak, 1980; Mayer and Marbach, 1981; Jelsema *et al.*, 1982).

The rapid transformation of the phosphate-containing components of the membrane is paralleled by the rather early onset of the metabolism of other membrane components. Both proteins and phospholipids are turned over in imbibed seeds of *Avena fatua* (Cuming and Osborne, 1978a,b). There can be little doubt that turnover, synthesis and breakdown of membrane components are among the earliest events occurring during germination. This would be fully in accord with the changes in the subcellular organelles such as the mitochondria, ribosomes and polysomes which may be observed during germination (Lott and Castelfranco, 1972; Treffry *et al.*, 1967). The appearance of a better defined endoplasmic reticulum is also characteristic of seed germination and has been described by many workers, using the electron microscope (Srivasta and Paulson, 1968; Swift and O'Brien, 1972a,b).

In general a large variety of sugar phosphate esters is present in seeds, including the intermediaries of glycolysis and triose phosphates, as well as nucleotides.

The nucleotides are a group of phosphorus-containing compounds which have a key role in the energy metabolism of the seed. Some, such as ITP and GTP, also play a vital role in various metabolic pathways, while UTP is particularly important in sucrose metabolism and UDP in cell wall synthesis. The ATP content of seeds of a number of species (*Trifolium incarnatum, Brassica napus* and *Lolium multiflorum*) was correlated with seed vigour (Ching, 1973), however it did not appear to be a clear-cut indicator.

A detailed analysis of nucleosides and nucleotides during the first 40 hr of germination of peas showed that the AMP level fell during this period (Brown, 1965). The ADP level also showed a small initial drop followed by a much bigger one between 16 and 40 hr, while the ATP level first rose for 16 hr and then fell again. The content of free adenosine fell markedly during germination, while xanthosine doubled during the first 40 hr of germination. As most of the ATP in germinating seeds is formed by oxidative phosphorylation it is not surprising that the ATP level rises as aerobic respiration increases during imbibition and germination. A very rapid rise in ATP content of wheat embryos has been observed, its level increasing 10-fold during the first hour of imbibition (Obendorf and Marcus, 1974; Pradet *et al.*, 1968). In oats the ATP content as well as the sum of all the adenine nucleotides (AMP + ADP + ATP) increased with increasing respiration, under conditions which allow normal germination to occur (Lecat, 1987). Under anaerobic conditions or at an unfavourable temperature, when no germination occurs, the rise in the nucleotides is rather small and is probably due to fermentative processes taking place under such conditions. Not only the ATP content of seeds rises rapidly during germination, but other nucleotides and nucleosides also increase in amount. The amount appearing in the growing embryonic axis is usually greater than that disappearing from the storage tissues in the same period (see literature cited by Bewley and Black, 1982). Apparently these compounds are rapidly synthesized during germination. Conversion from the nucleotide diphosphate to the triphosphate also takes place. In most cases the changes in the adenine derivatives have been measured. For example in radish the ATP level rises during germination, while the ADP level falls (Moreland *et al.*, 1974). Some decrease in AMP was also recorded although the AMP level is usually low and changes very little. This problem will be discussed further, later on in this chapter. ATPase is the enzyme which is involved in energy release reactions, which involve the hydrolysis of the phosphate ester bond, coupled with reactions which utilize energy, such as transport mechanisms. ATPase has been shown to be located on membranes, for example in the mitochondria or tonoplast and plasmalemma. Other nucleotides, NAD and NADP, have a central role in the oxido-reductive processes in the germinating seed.

The changes in the total NAD and NADP content of wheat, oats and peas was studied by Bevilacqua and Scotti (1953) and Bevilacqua (1955). They showed that the nucleotide content of the seeds and seedlings rises in all cases during germination. The rise in peas was much greater than in wheat or oats. In both wheat and oats the initial rise and fall occur at different times. In wheat endosperm an initial rise for up to 5 days was followed by a plateau, while in peas at this time a marked drop in nucleotide content in the cotyledons was observed. Rises have been reported for germinating rice seeds by Mukherji *et al.* (1968). A massive conversion of NAD to NADP has been reported at the very onset of the germination of peanuts, *Arachis* (Reed, 1970). The ratio between the oxidized and reduced forms of these nucleotides depends on respiratory processes and the size of the sinks for reducing power.

Of the enzymes concerned with the metabolism of nucleotides and sugar phosphates, only a few have been investigated in detail in seeds. In general most attention has been paid to general phosphatase activity on the one hand and to the phosphokinases responsible for phosphate transfer on the other.

III. Metabolism of Nucleic Acids and Proteins

During germination both cell elongation and cell division occur. It is therefore obvious that at some time both protein synthesis and synthesis of nucleic acids must also take place. Much attention has been given to the question — when do the various events in the normally accepted pathways of nucleic acid metabolism and protein synthesis occur in germinating seeds? There is very little evidence to indicate that the actual metabolic pathways are different in germinating seeds than in other tissues. Much of the work has been done using incorporation of radioactive precursors into the nucleic acid fractions and studies with metabolic inhibitors. When, in such studies, intact seeds are used, many difficulties are met with, because of diffusion barriers in seed coats, uneven distribution of precursors in the seed and even differential penetration of both precursors and inhibitors into different parts of the seed. Many researchers have used isolated seed parts for such studies, but the results are often difficult to interpret and care must be taken not to generalize from such investigations about the events in the whole seed.

Some of the overall changes in nucleic acid content of seeds are shown in Figs 5.1 and 5.2, and in Table 5.1. The data in Fig. 5.2 appear to indicate that in the maize seed the nucleic acid content in the storage tissue hardly changes during germination, but rises markedly in the whole seedling, due chiefly to its increase in the growing embryo axis. As can be seen from Table 5.6, in germinating cotton seeds both DNA and RNA increase in amount from the 2nd day onwards, but not at the same rate. DNA is mainly nuclear, and the amount per nucleus is fairly constant. An increase in DNA is therefore indicative of cell division.

1. Nucleic Acid Metabolism

Although DNA synthesis is detected rather late during seed germination (Table 5.6), appreciable DNA synthesis was observed prior to protein synthesis (Mary *et al.*, 1972). DNA synthesis may occur before or after cell division (see later). Cell division is often but not always a prerequisite of radicle protrusion.

DNA is apparently modified during germination. Changes can be observed in the nucleus during the transition from the quiescent to the active state of seeds. Such changes include a dispersion of the chromatin and vacuolization of the nucleus. During the normal cell cycle, mitosis is followed by growth during which there is no DNA synthesis, although there is much other synthetic activity (G1 stage). In this stage most of the nuclei contain double stranded DNA (2C). A period of DNA synthesis follows (S stage) and the DNA in most of the nuclei doubles and their DNA becomes 4 stranded (4C). The second growth period (G2) again leads to mitosis. During quiescence nuclear development in seeds is usually arrested during G1 and most of the DNA is 2C. This condition is typical of seed such as lettuce and onion. In other seeds such as maize and wheat the cell cycle may not be as synchronous and development may be arrested. The onset of quiescence can occur when some of the cells are in the G1, with 2C DNA, while others are already in the G2 and have 4C DNA. With the onset of imbibition and of active metabolism, the cell cycle of all the cells in the seeds proceeds towards mitosis, cell division and growth. However, replication of DNA will not begin immediately in all cases, as in some nuclei the 4C state is already reached at the onset of quiescence. A detailed description of some of these changes appears in a review by Deltour (1985).

The properties of DNA isolated from dry seeds differ from those of the DNA extracted from germinated embryos (Chen and Osborne, 1970). The DNA of the dry seeds is very resistant to adverse conditions such as heat, or dehydration. Little is as yet known about its properties. The ratio of histone/DNA has been shown to change during germination in pea embryos (Grellet *et al.*, 1977). Such changes are consistent with the possible changes in the transcriptive activity of the chromatin. The structure of chromatin available for transcription changes during germination. In wheat, template activity increased 10-fold and endogenous RNA polymerase activity 40-fold during a 72 hr period of germination (Sugita and Sasaki, 1982). Histone synthesis in maize embryos started well before cell division. Cell division showed a definite sequence being initiated first in the mesocotyl (Baiza *et al.*, 1986). Non-histone proteins in the chromatin also change during germination and increases in DNA polymerase activity have been reported during germination. At least in some cases the DNA content of the endosperm increases in the absence of cell division. The function of such DNA formation remains unclear. A similar increase in DNA was observed in castor bean where DNA synthesis occurred both in the endosperm and cotyledons following imbibition. The DNA synthesis in the cotyledons appears to be normal replication, while in the endosperm some of the DNA synthesis was independent of mitosis (Galli *et al.*, 1986) and may have been endoreplication involving only part of the genome (for reviews see Osborne, 1971, 1980; Mayer and Marbach, 1981).

A new interesting observation relates to changes in the amount of phosphorylated protein present in the nucleus. Some such proteins disappear and others are synthesized. Protein kinases are involved in this process and it seems that they play a regulatory role at some phase of the cell cycle.

The synthesis of DNA is often studied by following the incorporation of radioactive thymidine. This method is not always reliable. Anomalous incorporation of thymidine into cytoplasmic fractions has been reported during germination. This may well be due to the breakdown of the thymidine followed by the subsequent metabolism of the breakdown products. In rye embryos incorporation of radioactive precursors has been noted prior to the normal replication of DNA. This early incorporation of precursors into the DNA has been shown to be due to repair of damaged strands of DNA and is referred to as unscheduled DNA synthesis (Osborne *et al.*, 1984; Elder *et al.*, 1987). In seeds with longevity a loss of DNA integrity during the long-term storage of seeds is to be expected (Osborne, 1980). For this reason repair of DNA, following imbibition, may be of great importance. Such repair has now been shown to involve a DNA ligase, the enzyme forming phosphodiester bonds in DNA. Activity of this ligase rises very rapidly in fresh embryos of rye. In embryos of aged seeds this ligase activity is much lower and did not increase with germination. Seeds which are aged artificially by irradiation also lack such ligase activity, or its activity is much reduced. The same ligase, which is active in DNA repair, is also responsible at a later stage for DNA replication (Elder *et al.*, 1987).

The enzyme poly-c (adenosine diphosphate ribose) polymerase increases rapidly during germination. The product of the reaction catalysed by this enzyme associates with proteins linked to DNA and it might have a role in regulating ligase activity (Grey and Bryant, 1984).

The amount of polysomes is usually very low in dry seeds or they are entirely absent. During germination of wheat, polysomes rapidly appear in the embryos (Table 5.7). Polysome formation has been demonstrated both in electron micrographs of seeds at different stages of germination and by direct isolation of the ribosome and polysome fractions, using normal fractionation and ultracentrifugation techniques.

TABLE 5.7. *Ribosomes and polysomes in wheat embryo during germination (from Marcus, 1969). Ribosome activity determined by leucine incorporation into isolated ribosomes. Polysome content determined from adsorbance in polysome region of sucrose gradient*

Length of imbibition	Ribosome activity cpm/mg RNA	Polysome content O.D. units
0	268	0·01
15 min	6680	0·16
30 min	23200	1·61
1·5 hr	31900	2·42
6 hr	56300	3·66

Ribosome formation in cotton seed and RNA synthesis occurs during the early stages of germination (Waters and Dure, 1966). The cotton seeds also contain long-lived mRNA (messenger RNA). The *de novo* synthesis of a protease which is dependent on the presence of a pre-existing mRNA was demonstrated (Ihle and Dure, 1969). It was further shown that the mRNA was formed during embryogenesis, but translation of the mRNA was blocked at this stage, probably through hormonal control (Ihle and Dure, 1972). Thus in the cotton seed RNA synthesis and ribosome formation occurs, and at the same time utilization of pre-existing mRNA for polysome formation. In some cases ribosomes must be synthesized immediately at the onset of germination, while in other cases sufficient ribosomes are present in the embryo for polysome formation. In castor beans the dry seed already contains some ribosomal RNA and the heavy ribosomal fraction increases very rapidly during germination (Table 5.8).

TABLE 5.8. *Changes in RNA content of castor bean endosperm during germination (from Marre, 1967). Results as μg/endosperm*

	Dry seed	Seed germinated 24 hr	Seed germinated 48 hr
Total RNA	150	345	950
Heavy ribosomal RNA	73	175	550
Light ribosomal RNA	35	75	190
Soluble RNA	42	95	210

The existence of long-lived mRNA is no longer in doubt (Payne, 1976) although it may not be present in all seeds (Marre, 1967). In addition to cotton seeds, such mRNA has been shown to occur in the ribosomal fractions of wheat, radish, *Vigna* and other seeds (Shultz *et al.*, 1972; Suzuki and Minamikawa, 1983; Aspart *et al.*, 1984). In entire wheat seeds, an mRNA fraction was shown to be synthesized very early, after a few hours of germination (Rejman and Bucowicz, 1973). The precise nature and function of this mRNA is still in dispute.

A part of the long-lived mRNA seems to lack poly A and during germination a great deal of new polyA-rich RNA is formed (Delseny *et al.*, 1980/81). Thus it seems possible that the long-lived mRNA must be adenylated before it can function in the germinating seed (Mayer and Marbach, 1981). In addition to long-lived mRNA, new mRNA is formed very rapidly after the onset of germination. In cotton seeds the population of different mRNAs, and especially the relative amounts of the different RNA species, changes enormously during germination (Galau and Dure, 1980/81). A study of the population of RNA in the embryonic axes of *Phaseolus* showed that certain messages were present only in the developing embryo, while others were characteristic of the germinating axes (Misra and Bewley, 1985). Some mRNA was stored in the dry seed which was degraded as the axes were rehydrated. In isolated embryonic axes of maize new mRNA was formed rapidly following imbibition and this messenger was translated within 24 hr. Apparently new mRNA was formed even before water uptake by the embryo had been completed. This mRNA moved quite rapidly into the cytoplasm before the rRNA moved there (Dommes and Van de Walle, 1983). The stored mRNA present in radish seed embryos was formed late during embryogenesis and much of it apparently was coding for proteins involved in embryogenesis (Aspart *et al.*, 1984).

The overall conclusion from the studies on stored mRNA is that most of it is degraded very early during germination, but this does not exclude the possibility that before being degraded it may code for some proteins which are functional in the germination process. However, it is clear that new messenger is made very quickly following germination and this new messenger is the mRNA species of greatest importance during germination.

c-DNA clones are being used to study the mRNA populations in seeds, for example soybean and *Vigna*. Their use permits the differentiation between various mRNAs and consequently their assignment to definite developmental stages. A great deal of specificity in the mRNAs formed at various stages of germination and even in different parts of the germinating axes has been detected. For example, some species of mRNA accumulated in the hypocotyl of soybean between 12 and 24 hr of germination, when this part of the axis is hardly growing, yet were absent in the actively growing basal region of the axis, where they appear later towards the end of elongation. These results may be interpreted as showing that the formation of certain mRNAs are not only involved in the general maintenance of cell proliferation, but that they may be related to differentiation during early seedling formation (Datta *et al.*, 1987). In *Vigna* cotyledons, most of the mRNA carried over in the dry seed had disappeared after one day of germination and was replaced by new mRNAs synthesized after imbibition. Translational activity of poly(A)$^+$RNA increased rapidly in the first day and then remained constant (Koshiba *et al.*, 1986).

Other forms of RNA essential for protein synthesis namely rRNA and tRNA are also formed very rapidly with the onset of germination. These species of RNA are also present in dry seeds.

In addition to the presence of long-lived mRNA in the apparently free state, dry seeds such as rye, wheat and peas contain what appear to be cytoplasmic messenger ribonucleoproteins. These ribonucleoproteins are present in well defined particles (Peumans and Carlier, 1977) and represent an inactive storage form of mRNA. They could be shown to have messenger activity and to support protein synthesis when other factors were added to them (Peumans *et al.*, 1979).

Only a few of the enzymes involved in RNA metabolism have been studied in detail. One of them, RNA polymerase, whose activity increases prior to RNA synthesis, seems to be present at least in the mature wheat seed.

2. Protein Synthesis and its Dependence on Nucleic Acids

Since DNA synthesis apparently occurs fairly late during germination, while some RNA species are synthesized quite early, the question arises: at what stage of germination does protein synthesis begin, what are the limitations to protein synthesis and what is known about the various requirements for protein synthesis and the enzymes involved in it? The onset of protein synthesis during germination has been deduced from three main kinds of evidence: (1) the appearance of an enzyme activity, or its increase during germination; (2) the failure of an activity to appear in the presence of an inhibitor of protein synthesis and (3) from studies based on the incorporation of radioactive precursors into new or existing protein using two-dimensional electrophoresis combined with autoradiography. In the most convincing experiments these techniques have been combined. Invariably some doubt can be expressed about some of the evidence regarding protein synthesis in germinating seeds. Many of the investigations have been made, for experimental convenience, with isolated parts of seeds rather than with intact seeds. Nevertheless, it is quite clear today from the various studies that protein synthesis begins in seeds quite soon after imbibition. Thus in isolated embryos from wheat $^{3}H^{-2}$ leucine and ^{14}C-valine were incorporated into protein fractions after as little as 40 min of imbibition. The rate of incorporation increased sharply with length of imbibition and was about doubled after 3 hr (Marcus *et al.*, 1966). Rapid protein synthesis as measured by amino acid incorporation has been demonstrated in a variety of cases, usually in isolated embryos. This capacity exists even in dormant embryos and has been shown for example in *Agrostemma githago* (Hecker and Bernhardt, 1976). These embryos took up and incorporated ^{14}C leucine 2 hr after the onset of imbibition. The exact time of the beginning of synthesis seems to be rather variable. Almost always there is some delay in the onset of protein synthesis in the intact seed. This late onset may be caused by a number of factors, such as delay in formation of functional ribosomes, delay in monosome – polysome transition or of formation or availability of mRNA, or the lack of building blocks or of energy for protein synthesis. The slow entry, by diffusion of the labelled precursors may also contribute to the apparent late onset of synthesis.

The dependence of *in vitro* protein synthesis by ribosomes from dry wheat embryo or peanut cotyledons on a factor from imbibed seeds was first demonstrated by Marcus and Feeley (1964). This requirement could be replaced by polyuridylic acid (Table 5.9).

TABLE 5.9. *^{14}C Phenyl alanine incorporation in the presence and absence of Poly U by ribosome preparations from peanut cotyledons (after Marcus and Feeley, 1964)*

		mg ribosomal RNA added	Incorporation of amino acid cpm into protein by preparation from		
			Dry seeds	Imbibed 1 day	Imbibed 4 days
No poly U	Unwashed microsomes	0·4	3	57	78
		0·8	6	90	109
	Washed microsomes	0·2	2	75	84
		0·4	1	123	—
Poly U added	Unwashed microsomes	0·05	576	478	—
		0·10	876	523	543

Although first interpreted as demonstrating a requirement for mRNA synthesis, it is now supposed to be due to activation of pre-existing masked long-lived mRNA (Weeks and Marcus, 1971). The *in vitro* system using ribosomes isolated from dry seeds had almost no capacity for protein synthesis. As imbibition of the seeds continued, protein synthesis by the isolated ribosomes *in vitro* increased. Addition of poly-U increased protein synthesis in all the treatments, not only in the system from dry seeds. Similar observations have been made on other tissues, such as isolated *Phaseolus vulgaris* axes, castor bean cotyledons, cotton, lettuce and other seeds (Mayer and Shain, 1974). Generally the technique used has been incorporation of radioactive amino acids into protein. The blocks to protein synthesis which exist in the dry seed are removed extremely rapidly, and may not involve only the functionality, or otherwise, of the mRNA. It is becoming clear that in the isolated embryos, shortly after the onset of imbibition, polysomes are formed and that part of the preformed mRNA becomes attached to polysomes (Brooker *et al.*, 1977). Dure *et al.* (1980/81) who studied protein synthesis in cotton embryos characterized the newly synthesized protein formed *in vivo* by separation into sub-fractions, followed by two-dimensional electrophoresis. These proteins were compared with those formed *in vitro* by mRNA isolated from intact cotyledons, in the presence of a protein synthesizing system. The two ways of following protein synthesis showed a great deal of similarity and also indicated the great variety of different proteins formed during germination.

Sanchez-Martinez *et al.* (1986) followed the regulation of gene expression in developing maize embryos from 20 days after pollination up to full maturation and during the early stages of germination. They identified three groups of expressed polypeptides. (1) Polypeptides present in the developing embryo, some of which could not be detected in the mature dry seeds, while others persisted even during the first 2 hr of imbibition, but were absent after 8 hr. (2) Polypeptides absent from very young embryos, but formed during its maturation. Some of these are short lived and are absent in the dry seed, while others persist for varying lengths of time during the initial stages of germination. (3) The group of polypeptides which are absent in the developing embryo and the dry seed and are present only during germination. These data show very clearly that the mature seed contains messages for synthesis of polypeptides and that some of these messages were carried over into the dry seeds from the early periods of its maturation and development. Other evidence, showing that during germination pre-existing mRNA is translated, is based on the course of accumulation of lectins in the germinating seed (Peumans *et al.*, 1982).

In the analysis of protein synthesis during germination a crucial question arises — what are the signals determining whether pre-existing messages or newly synthesized mRNA is translated and when do the protein synthesizing systems specific to germination become active? There is some evidence to indicate that the actual desiccation step during seed ripening plays an important role in diverting protein synthesis from the synthesis of storage proteins to that of proteins characteristic of germination (Kermode *et al.*, 1986; Bewley, 1982).

Isolated aleurone tissue from *Avena fatua* was shown to incorporate ^{3}H-leucine within 10 min of the onset of imbibition into 50% of the cells of the aleurone segments (Maherchandani and Naylor, 1972). In these experiments autoradiography was used to identify the location of the proteins into which the label had been incorporated. The conclusion reached was that the dry mature aleurone tissue had an immediate capacity for protein synthesis, and that all the requirements for this synthesis were present in its cells. The protein synthesis during hormone treatment of aleurone layers has been the focus of much attention.

The *de novo* synthesis of protein in embryo-less half seeds was conclusively proved using a density gradient labelling technique in the presence of D_2O (Filner and Varner, 1967). The presence in the germinating seed of all the components necessary for protein synthesis was therefore firmly established. Also the synthesis during germination of a specific protein, fully characterized by its enzymic properties, has been demonstrated in a few cases. In lettuce seeds invertase activity is absent in the dry seeds; when imbibed in the presence of radioactive sulphate, it was possible to follow the formation of radioactive protein. Formation of radioactive invertase was demonstrated in the period between 10 and 24 hr of imbibition (Fig. 5.13).

Surprisingly little is known about the individual enzymes involved in protein synthesis and their changes during germination, although their activity clearly increases during germination. Wheat germ contains all the components required for the *in vitro* synthesis of protein, provided mRNA is present and indeed this system is used for the *in vitro* synthesis of various proteins, depending on the nature of the mRNA added. This system has been resolved into at least 12 components, and has been fully described (Seal *et al.*, 1986). However, the change in the activity and amount of these components during germination in this system, or in other seeds, has not yet been described.

Aminoacyl – tRNA synthetases which are necessary components for protein synthesis have been demonstrated in the cotyledons of *Aesculus* species and are evidently functional in the dry seeds (Anderson and Fowden, 1969, 1970). In *Phaseolus vulgaris* their activity increased during germination, the increase being equal in the plumule and radicle of the embryo (Fig. 5.14).

In addition to formation of active enzymes by *de novo* protein synthesis, there is good evidence for activation of pre-existing inactive enzymic proteins during the early stages of germination (Marre, 1967; Mayer, 1973). Since during seed formation many enzymic systems are extremely active, it was logical to suppose that some of these became inactivated during the dehydration of the maturing seeds. Many alternative ways can be suggested by which enzymes can be reversibly inactivated, such as folding, enclosure in or attachment to membranes, association or dissociation of subunits or by the addition of a section to the polypeptide chain of the active protein. In such cases during germination a reversal of these processes could occur. Thus reversible activation – inactivation would constitute a convenient and energetically economical way of storing enzymes. By such mechanisms the need for *de novo* synthesis of some proteins would be obviated. Evidence for such activation processes is based on the fact that the increase of enzyme activity during germination is not affected by inhibitors of protein synthesis, the demonstration of absence of incorporation of radioactive markers into the purified active enzyme, and the demonstration of the activation process in an *in vitro* system. Such complete evidence has been produced in only a few cases. Probably the most complete evidence is for the appearance of an amylolytic enzyme, probably amylopectin-1,6-glucosidase, in peas (Shain and Mayer, 1968a,b). It was shown that while an isolated soluble or a particulate cell fraction prepared from pea cotyledons, when incubated alone with substrate, showed only low enzyme activity, enzyme activity doubled when these fractions were incubated together (Table 5.10.). It was also shown that the soluble cell fraction could be replaced by trypsin in causing activation.

The changes in the enzyme activity, protein synthesis and synthesis of RNA and

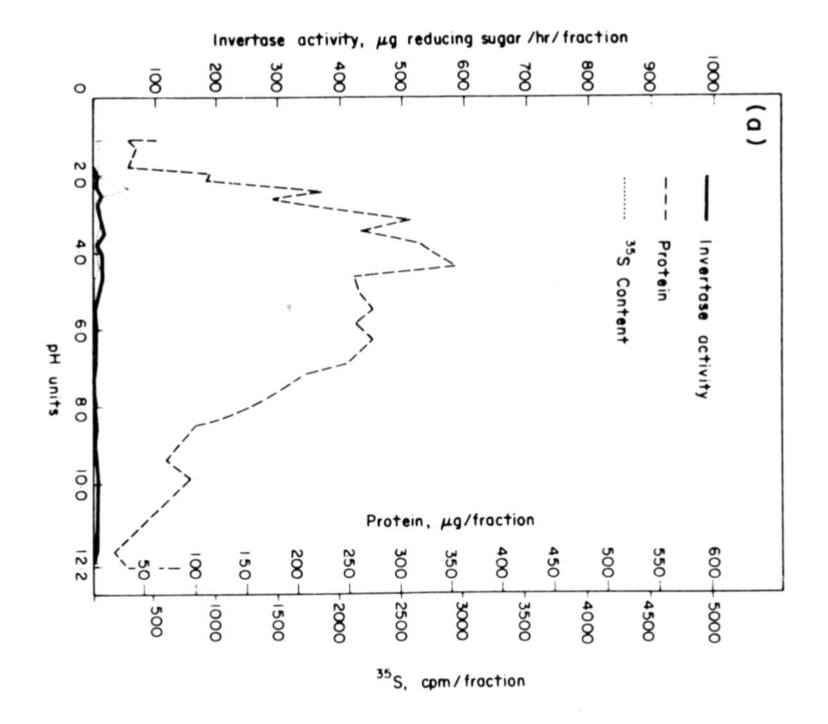

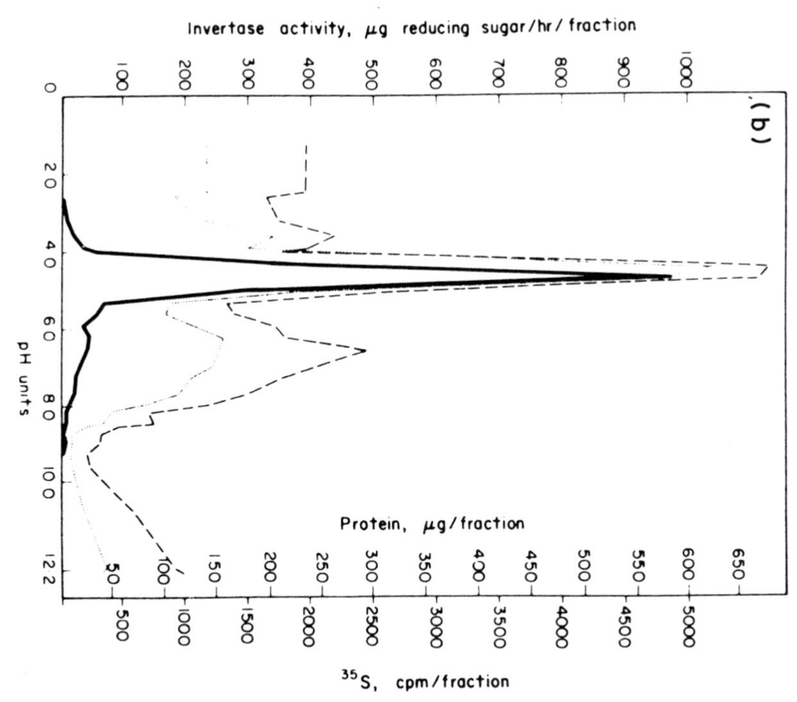

FIG. 5.13. Acid invertase activity, protein and ^{35}S content of fractions prepared from extracts of lettuce seeds germinated for 10 hr (a) or 24 hr (b) in a solution containing $^{35}SO_4^{2-}$ (Eldan and Mayer, 1974).

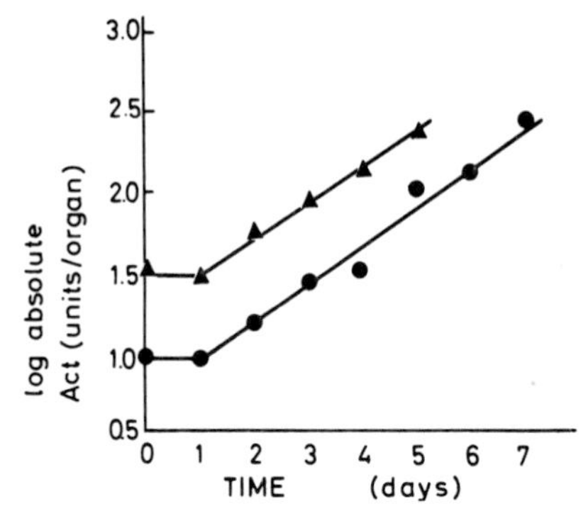

FIG. 5.14. Absolute aminoacyl-tRNA synthetase activity of plumules (●) and radicles (▲) of French bean during germination (from Anderson and Fowden, 1969).

TABLE 5.10. *Development of amylopectin-1,6-glucosidase activity when incubating particulate and soluble cell fraction, from homogenates prepared from dry pea seeds, separately or together (after Shain and Mayer, 1968a)*

	Enzyme activity (units)	
	Incubation time (hours)	
Fraction incubated	0	4
Particulate	0·300	0·110
Supernatant (soluble)	0·990	1·060
Particulate + supernatant	1·290	2·910

DNA during germination are shown schematically in Fig. 5.15. This schematic representation takes into account pre-existing enzymes in active and inactive form, pre-existing mRNA and the delayed onset of DNA synthesis.

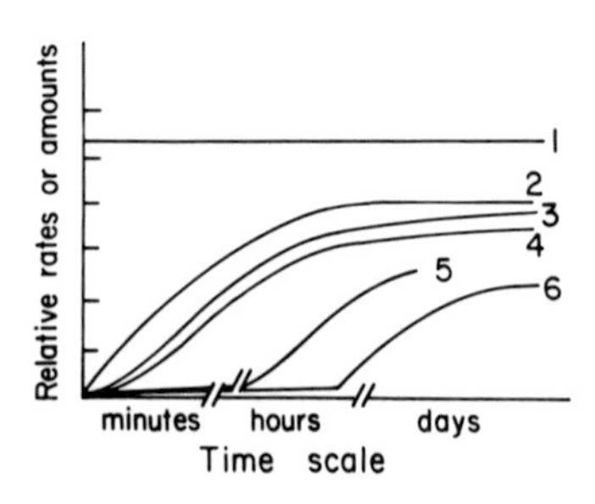

FIG. 5.15. Schematic presentation of the changes of enzyme activity and of components of the protein synthesizing systems during germination. (1) Enzymes instantly active with no increase in activity. (2) Enzymes activated during germination. (3) Enzymes expressed due to pre-existing mRNA. (4) All classes of RNA. (5) Unscheduled DNA synthesis (repair). (6) DNA replication.

It seems likely that enzyme activation may be rather widespread in seed germination. Probably a number of diverse mechanisms exist for activation. The evidence for such activation processes accumulated in recent years was reviewed by Mayer and Marbach (1981). However, it is still uncertain whether activation plays a critical role during the very early stages of germination. The relative importance of activation as opposed to *de novo* protein synthesis during the early stages of germination requires much more detailed evaluation. There can, however, be no doubt that *de novo* synthesis starts with great rapidity, and that proteins specific for the germination process are produced in this way. The emphasis therefore has shifted to some extent to the crucial problem, how and when the genome, coding for the germination specific enzymes, is switched on? Alternatively one may pose the question: how and when are the processes specific for seed formation switched off?

IV. Respiration

Germination is an energy-requiring process and is therefore dependent on the respiration of the seed. In the following account we will discuss respiration both from the point of view of overall gas exchange, and from the point of view of the biochemical mechanism.

1. Gaseous Exchange

In dry seeds it is almost impossible to measure either oxygen uptake or carbon dioxide output as gas exchange is extremely low. The problem of measuring the gas exchange of dry seeds is further complicated by the fact that most seeds are to some extent contaminated with bacteria and fungi, both on the seed coat and frequently also between the seed coat and the seed. These microorganisms also respire and it is more than likely that some, if not all, the gas exchange measured in dry seeds is in fact due to the contaminating microorganism. For example, Rose (1915) found that out of a hundred species examined, more than half were infected by fungi. When large bulks of seed are kept, as in grain silos, there is an appreciable heat production, resulting in a rise in temperature. The heat produced is presumably also due, in part, to the respiration of microorganisms. It is difficult to sterilize seeds effectively and surface sterilization may not be sufficient. Despite these reservations there is a certain amount of information which shows the existence of gas exchange in dry seeds. The level of gas exchange of dry seeds depends on the moisture content and rises as the latter rises. Bailey (1921) showed that in *Zea mays* seeds the output of carbon dioxide rose from 0.7 mg per hundred gram dry weight during 24 hr, when the seeds had a moisture content of 11%, to about 60 mg when the moisture content was 18%. Similar increases in carbon dioxide output with increasing moisture content have been shown for *Sorghum*, wheat and rice. The steepness of the rise in carbon dioxide output with increasing moisture content differs in different seeds. In the few established cases of very long-lived seeds, i.e. *Nelumbo*, it is difficult to understand how any form of respiration could have been maintained in the seeds for such long periods of time without entirely depleting the storage materials of the seeds, thus impairing their viability.

On moistening seeds with water there is an immediate gas release, which seems to be a purely physical process not peculiar to seeds and involving the liberation of gas which is supposed to be colloidally absorbed within the seeds (Haber and Brassington, 1959). This observation obviously complicates any interpretation of the gas exchange of seeds immediately after they are placed in water.

Another complicating factor in measuring respiration is the presence of the seed coat or other enveloping structures such as the glumes in the Graminae. The presence of the seed coat may reduce permeability to oxygen (see Chapters 3 and 4) and therefore the respiration of seeds without the coat may be higher than that of intact seeds (Fig. 5.16a).

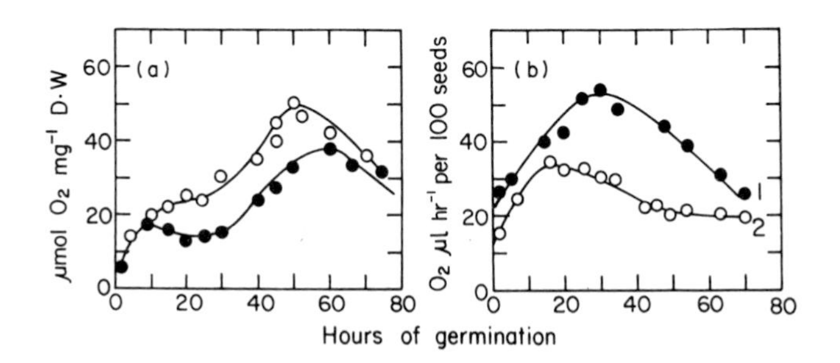

FIG. 5.16. Oxygen uptake of seeds with or without seed coats. (a) Oxygen uptake of cotyledons of pea seeds germinated with (●) or without (○) seed coats (compiled from data of Kolloffel, 1967). (b) Oxygen uptake of seeds of *Avena sativa* germinated with (●) or without (○) glumules. Germination was at 30°C, which does not allow embryo growth (compiled from data of Lecat, 1987).

At the same time glumes, which originate from the mother plant, may actively contribute to oxygen uptake (Fig. 5.16b) although this oxygen uptake is not necessarily respiratory in nature as was shown for example for oats and for barley (Lecat, 1987; Lenoir *et al.*, 1986). Respiration rises as the water content of the seed increases (Fig. 5.17 and 5.18).

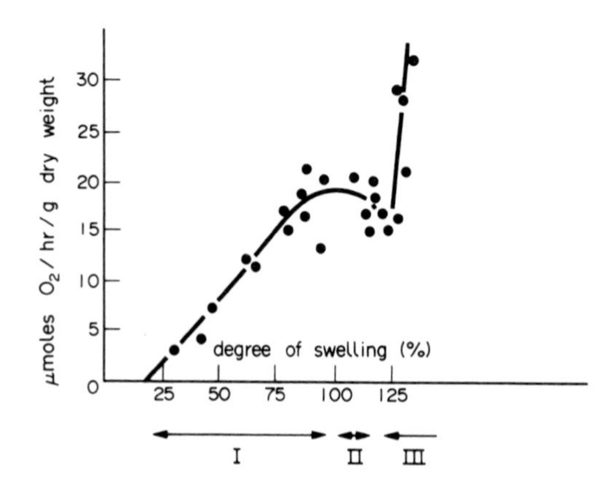

FIG. 5.17. The relation between the degree of swelling and the rate of respiration of cotyledons of intact pea seeds (from Kolloffel, 1967).

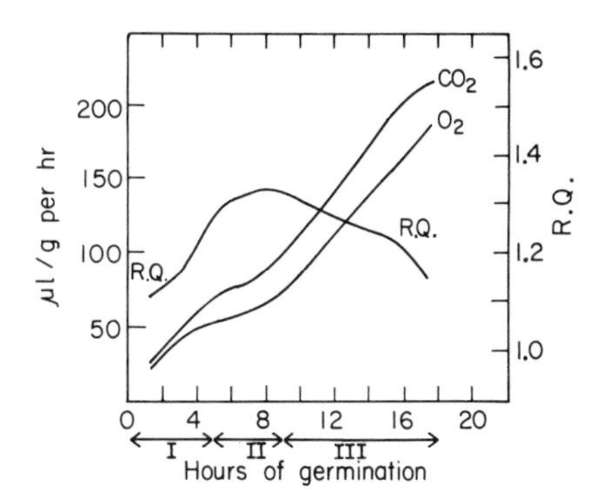

FIG. 5.18. The rate of respiration and the RQ of germinating wheat seeds at 20°C. I, II, III — Different stages of respiration, which parallel stages of hydration (Levari, 1960).

The course of increase in oxygen uptake can be divided into several phases, which appear to parallel the phases of seed hydration (see Chapter 1); an initial rapid increase, a plateau when swelling has been more or less completed and a second increase, probably accompanying embryo growth. The plateau in oxygen uptake (QO_2) appears to end at about the time when the seed envelope is broken and free gas exchange, without limitation by membranes, becomes possible (Spragg and Yemm, 1959). Eventually oxygen uptake of the cotyledons decreases again, probably due to senescence. In peas these different phases are observed in cotyledons of intact seeds (Fig. 5.17) and also in cotyledons of seeds germinated without the seed coat. In the latter, oxygen uptake is a little higher (see Fig. 5.16a) (Kolloffel, 1967). However, in *Lathyrus* seeds the different phases were observed during germination of the intact seed, but were virtually absent if the seed coat is removed before germination (Stiles, 1935).

The changes in oxygen uptake (QO_2), carbon dioxide output (QCO_2) and the respiratory quotient RQ (= QCO_2/QO_2) are illustrated in Fig. 5.18. From this figure the general increase in both QO_2 and QCO_2 with time is evident. In wheat QO_2 and QCO_2 rise almost continuously through germination. However, the plateaus for QO_2 and QCO_2 end at different times. This is due to the fact that although both oxygen uptake and carbon dioxide output rise with time, they rise at quite different rates. As a result the RQ during the early stages of germination shows very large variations. Such changes in the RQ are observed very often in seeds. Such variations in RQ point to very marked changes in the substrates used for respiration. The RQ is dependent on the state of oxidation of the substrate oxidized, the extent to which there is genuine respiration and to what extent fermentive processes occur. In seeds with very compact tissues initially fermentation occurs, and only when oxygen penetrates into the tissues respiration proper begins as indicated by the drop in RQ. In these cases there will initially be a marked carbon dioxide output due to fermentation and only a slight oxygen uptake resulting in a very high RQ although the substrate broken down may be a carbohydrate. In flax, the initial RQ is close to one, and then drops during germination as utilization of fats begins (Halvorson, 1956). Apparently the small reserve of carbohydrates is the first which is utilized for respiration.

An indication of the difference in QO_2 and RQ in different organs of the same seed is shown in Fig. 5.19.

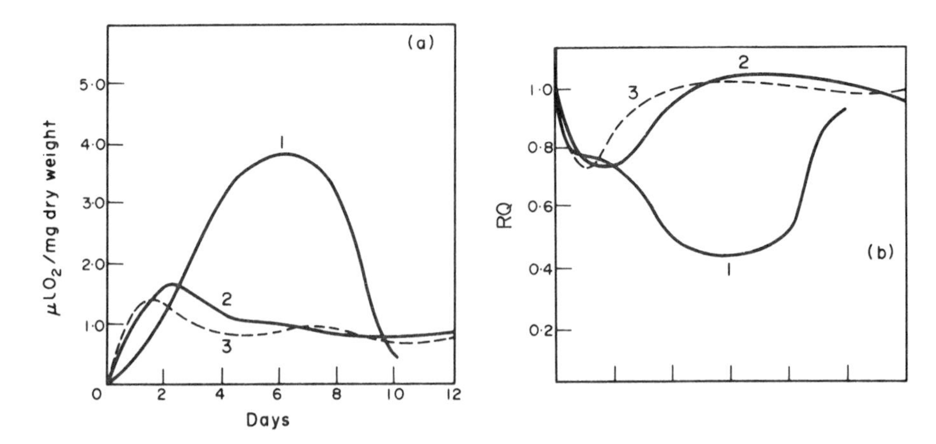

FIG. 5.19. Oxygen uptake and RQ values of various parts of castor bean seeds during germination. 1 — endosperm; 2 — cotyledons; 3 — embryonic axis (or hypocotyl). (a) oxygen uptake; (b) RQ values (Yamada, 1955).

This shows an initial rise in QO_2 in the embryo, cotyledons and endosperm. In the cotyledons and embryo oxygen uptake, expressed on a dry weight basis, rapidly attains a steady state. The rate of oxygen uptake of the endosperm continues to rise up to about 6 days and then falls again as its reserve materials finally become exhausted. These experiments were made using tissue slices of the endosperm and the hypocotyl, a method which involves wounding and will presumably alter the oxygen uptake of the organs as compared to their normal behaviour, both quantitatively and possibly also qualitatively. The RQ of the endosperm also shows very marked changes, while in the embryo and cotyledons only relatively small changes occur.

External factors which influence respiration are temperature, the oxygen and carbon dioxide content of the atmosphere and light. Each of these is liable to affect respiration. These external factors also interact in their effect on respiration.

Generally, a rise in temperature causes an increase in the rate of respiration of the seeds. However, the early experiments of Fernandes (1923) showed that the oxygen uptake at different temperatures depends not only on the actual temperature, but also on the length of time the seeds are exposed to this temperature. In other words, in studying the effect of temperature, the time factor must also be taken into account. If raised temperatures induce dormancy, no rise in the rate of respiration is observed. The respiration of flax seeds germinated at different temperatures is shown in Fig. 5.20a. As the temperature increases the rate of oxygen uptake rises and the maximal oxygen uptake is reached earlier. This last phenomenon is probably due to the more rapid germination at the higher temperatures, but the

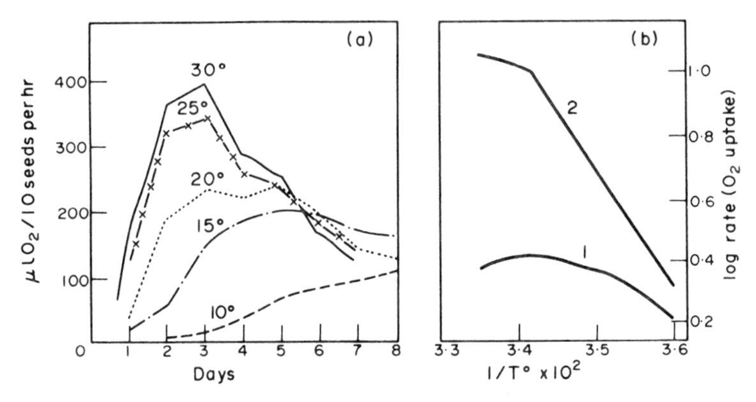

FIG. 5.20. Effect of temperature on respiration of germinating seeds. (a) Oxygen absorption of flax seeds during germination at different temperatures (Halvorson, 1956). (b) The respiration of pea seeds with or without seed coat. Seeds were germinated for 24 hr. 1 — whole seeds; 2 — seeds with seed coat removed (Spragg and Yemm, 1959).

data on the rate of germination are not available. It has also been shown that the effect of temperature depends on the presence or absence of the seed coat. For example, in peas whose seed coat had been removed, an increase in temperature raised the oxygen uptake much more than in intact pea seeds (Fig. 5.20b) (Spragg and Yemm, 1959). Temperature can only affect respiration provided oxygen can freely diffuse to the respiring tissue. If oxygen diffusion is limited, an increase in temperature will have relatively little effect.

An increase in the partial pressure of oxygen can also increase the rate of respiration of seeds. Many examples of this are available. In many cases, however, this only applies to oxygen concentrations below 20%, maximum rates being reached at this value. In many cases steady states of oxygen uptake were reached at 20% oxygen in the later stages of germination (when the roots were 1 – 3 cm long). This was shown to be the case for *Triticum spelta* and *Brassica rapa* (Reuhl, 1939) as well as for cabbage, sunflower and various cereals. However, in *Linum usitatissimum*, in the early stages of germination when the root had just emerged, oxygen uptake had not yet reached a steady state at 20% oxygen. Moreover, in other seeds such as peas maximal rates of respiration were reached at oxygen concentrations higher than in air (Al-Ani *et al.*, 1985). For peas it was shown that the effect of raised oxygen concentration depended on the stage of germination chosen for study. In seeds germinated for 36 hr, respiration in pure oxygen was appreciably higher than in air. Up to this time there was a steady rise in the percentage increase of respiration in oxygen as compared to air. After 36 hr the percentage increase dropped again and at 48 hr the rates of respiration in air and in oxygen were about equal.

Al-Ani *et al.* (1985) recently re-examined the oxygen requirement of 12 species and found that respiration was maximal at pO_2 near that in the atmosphere. Two groups of seeds, fatty and starchy ones, could be discerned (Al-Ani *et al.*, 1985). In the fatty seeds, such as soybean, lettuce, radish, turnip, sunflower, cabbage and flax, no germination occurred at pO_2 of about 2KPa (2%) while in pea, sorgum, wheat, rice and maize it proceeded slowly even at 0.1KPa. As can be seen from Fig. 5.21 rice seeds germinated 12% even at 0.003 KPa after 13 days of incubation, and after 20 days their germination was 80%.

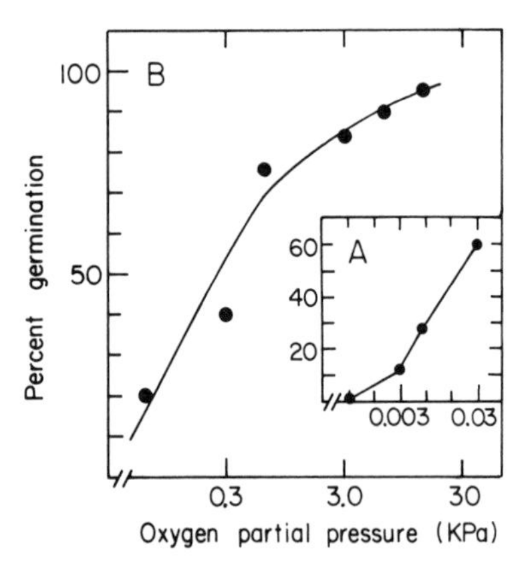

FIG. 5.21. Germination of rice seeds at different partial pressures of oxygen. (A) (Insert) Oxygen concentration in the range 0.001 – 0.03 KPa. (B) Oxygen concentration in the range 0.1 – 21 KPa. Length of germination 13 days (compiled from data of Al-Ani *et al.*, 1985).

The dependence of respiration on the external O_2 concentration was studied in detail in *Sinapis arvense* (Edwards, 1969). As can be seen from Fig. 5.22 oxygen uptake of the excised embryos was saturated at about 0.2 atmospheres early on in the imbibition of the embryo.

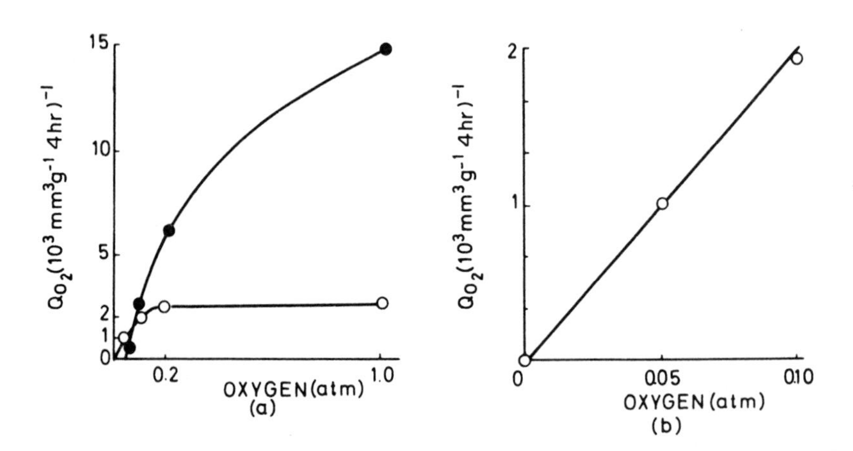

FIG. 5.22. Relation of oxygen uptake (QO_2) and external oxygen concentration (C_o) of excised embryos of *Sinapis arvensis* from 0 to 4 hr (O) and from 20 to 24 hr (●) at 25°C; (a) From 0 to 1.0 atmosphere oxygen. (b) Enlarged scale of (a) 0 to 4 hr, at 0 to 0.1 atm. 0_2 (Edwards, 1969).

However, as imbibition proceeded oxygen uptake rose as the oxygen tension increased to 1 atmosphere. From these data it appears that the seed tissues present considerable barriers to oxygen diffusion. The significance of this barrier clearly changes during germination, probably due to changes in oxygen requirements and to actual changes in the magnitude of the diffusion barrier. Such barriers may be responsible for the existence in many seeds of processes of fermentation during the early stages of germination. Although it has been shown that fermentation contributes very little to the energy balance of most germinating seeds, correlations were found between the energy charge measured during anoxia and the ability of the seeds to germinate slowly under such conditions (Raymond *et al.*, 1985; Al-Ani *et al.*, 1985). The seeds investigated could be divided into the two groups as described above. Group 1 included "fatty" seeds which did not germinate at very low oxygen pressures and whose energy charge under these conditions was 0.25 – 0.5. Group 2 included starchy seeds which did germinate under severe restrictions of the oxygen supply and whose energy charge was 0.6 or above.

Although it has repeatedly been attempted to show distinct effects of light, particularly red and far-red, on respiration the results remain unclear. It is not certain whether the respiratory metabolism was directly affected (Leopold and Guernsey, 1954; Evenari, Neumann and Klein, 1955; Gordon and Surrey, 1960).

The respiration of entire seeds is sensitive to externally applied respiratory inhibitors. Such inhibitors, provided they penetrate into the seeds and are not metabolized in it, act in much the same way as they act in other tissues. The relation between the action of respiratory inhibitors and the process of germination as a whole will be discussed in Chapter 6 as will the effect of other germination inhibitors as well as stimulators.

2. *Biochemical Aspects of Respiration*

Respiration is the process in which the substrates are oxidized through a series of steps, with the participation of oxygen as the final electron acceptor. Normally this process is coupled to ATP production. All other oxidative processes in which oxygen does not participate are not strictly respiration. They are frequently termed "anaerobic" respiration, but the term "fermentation" is a more appropriate one to use.

The most common substrates which are utilized in respiration and fermentation are the carbohydrates. Other substrates such as proteins or lipids used in respiration are first converted to three carbon units, or to acetyl CoA, which can then be utilized. Carbohydrates are generally channelled through glycolysis to form pyruvate. In respiration this is followed by the oxidation of the pyruvate to carbon dioxide and water, in the tricarboxylic acid, or Krebs, cycle. An alternative mechanism of oxidation is the direct oxidation of glucose phosphate leading to metabolism of the pentose cycle. An additional bypass of the oxidative process is the glyoxylic acid cycle.

In the main fermentative processes, using carbohydrate as the substrate, the pyruvate formed is either decarboxylated, leading to carbon dioxide formation, and the resultant acetyl derivative is reduced to alcohol, or the pyruvate is directly reduced, leading to the formation of lactic acid. Although many other fermentation processes are known in microorganisms, no information is available about their existence in plant tissues.

The main processes known with certainty to yield energy available to the organism are oxidative phosphorylation coupled to the electron transport linked to the Krebs cycle and phosphorylation at the substrate level during glycolysis. The electron transport associated with the Krebs cycle is via a chain of enzymes consisting usually of dehydrogenases, flavoproteins and cytochromes, ending in cytochrome oxidase which transfers electrons directly to oxygen. Various parts of this electron-transport system are coupled to phosphorylation and the average ratio of phosphate esterified to oxygen taken up is P/O = 3, now often referred to as ADP/O ratio.

Alternative electron-transport systems having different intermediaries and a different terminal oxidase are known. Among those that may function in plant tissues are the glutathione – ascorbic acid – ascorbic acid oxidase and the glycolic acid – glycolic acid oxidase systems. In none of the alternative electron transport systems mentioned is there any evidence to show coupling to phosphorylation and their precise function is unclear. An electron transport shunt, which is cyanide resistant, is known in plant tissues. In this shunt there is only one site for phosphorylation. Since in this pathway oxygen uptake continues and is not inhibited by cyanide, it is often referred to as alternative or cyanide insensitive respiration.

We will attempt in the following to bring such evidence as is available to show the existence of these various pathways in germinating seeds and try and evaluate their relative importance during germination.

The existence of glycolysis in many plant tissues has been shown (Stumpf, 1952; Turner and Turner, 1975). In pea seeds glycolysis has been clearly shown to exist (Hatch and Turner, 1958; Givan, 1972) and the end products of fermentation, alcohol or lactic acid, are known to accumulate during imbibition. The data in Table 5.11 show good correspondence between alcohol formation and carbon dioxide formation, as expected if only glycolysis occurs.

TABLE 5.11. *Glycolytic conversion of various substrates by pea extracts (after Hatch and Turner, 1958)*

Substrates	CO_2 production moles Calculated	CO_2 production moles Found	Alcohol production moles Calculated	Alcohol production moles Found
Fructose 21 μmoles + Fructose diphosphate 2·5 μmoles	47	40	47	41
Glucose phosphate 25·2 μmoles	50·4	46	50·4	49
Fructose diphosphate 25 μmoles	50	41	50	44

Further evidence for the existence of glycolysis has been brought by showing that mitochondria-free extracts of both peas and pea seedlings take up inorganic phosphate from the medium by glycolytic phosphorylation when fructose-1,6-disphosphate is provided as the substrate and ADP is the phosphate acceptor (Mayer, 1959a,b; Mayer and Mapson, 1962). It has also been shown that alcohol accumulates in imbibing oat caryopses and leaks out from them. Accumulation continued until germination occurred. Protrusion of the radicle apparently facilitated a better oxygen supply to the seeds and resulted in a repression of anaerobic processes (Lecat, 1987). Alcohol will accumulate in seeds if they are germinated under certain conditions, such as poor aeration for example in lettuce (Leggatt, 1948), or high temperatures as in the case of bean cotyledons (Oota *et al.*, 1956). Instead of alcohol, lactic acid may accumulate under conditions of limited oxygen supply. Lactate dehydrogenase as well as alcohol dehydrogenase has been detected in germinating seeds (Leblova, 1978). In *Zea mays* seedlings the formation of alcohol dehydrogenase was found to be increased by anaerobic conditions, both in the scutellum and embryonic axis. Apparently the substance inducing indirect alcohol dehydrogenase formation was acetaldehyde (Hageman and Flesher, 1960). The same conditions which caused an increase in dehydrogenase were accompanied by a drop in cytochrome oxidase activity in the seedlings.

The presence of enzymes of the glycolytic system have been shown in peas, for example phosphoglucomutase and phosphofructokinase (Turner and Turner, 1960, 1975) and triose isomerase (Turner *et al.*, 1965). Peas also contain two distinct hexokinases (Turner and Copeland, 1981). High activity of a number of glycolytic enzymes have also been shown to be present in extracts of dry castor bean endosperm, e.g. hexose phosphate isomerase, aldolase and fructose disphosphatase and phosphogluconate dehydrogenase (Nishimura and Beevers, 1981). The activity of these enzymes, all of which are cytoplasmic, increases during germination (Kruger *et al.*, 1983). The glycolytic system has been shown to operate at all stages of germination of *Cucurbita pepo* (Thomas and ApRees, 1972).

Comparison of the rates of glycolysis and mitochondrial activities (tricarboxylic acid cycle) in the early stages of germination of *Phaseolus mungo* showed that the activity of glycolytic enzymes was initially much higher than that of the mitochondrial activities (Morohashi and Shimokoriyama, 1975). During the first hour of imbibition mitochondrial activity increased considerably while the activity of the glycolytic enzymes showed much smaller changes.

Raymond *et al.* (1985) compared respiration and fermentation of germinating seeds of 12 species of cultivated plants. They showed that under anaerobic conditions the contribution of fermentation to energy production was negligible. Only in peas and in maize did fermentation contribute a significant, albeit small, amount to energy production.

Regulation of glycolysis is important in germination. Among the substances which appear to control glycolysis in castor bean endosperm is the level of fructose-6-phosphate which is formed from fructose-1,6-biphosphate. In addition fructose-2, 6-biphosphate has a regulatory role. It accumulated in germinating oat seeds in considerable amounts (10×10^{-3} nmol/seed), but its concentration steeply decreased after radicle emergence (Larondelle *et al.*, 1987). Recently it has been demonstrated that fructose-2,6-biphosphate regulates fructose-6-phosphate formation. Its level in turn is regulated by two enzymes, fructose-6-phosphate,2-kinase and fructose-2, 6-biphosphatase. The 2,6-biphosphate inhibits the fructose-1,6-biphosphatase and through this apparently regulates the flux between fructose 1,6-biphosphate and fructose-6-phosphate. The same compound also activates phosphoglucomutase from seeds (Kruger and Beevers, 1985). An additional enzyme involved in regulation may be pyrophosphate: D-fructose-6-phosphate,1-phosphotransferase, which is stimulated by the biphosphate (Sabularse and Anderson, 1981). Fructokinase also has a regulatory role in glycolysis, via its activation by inorganic phosphate (Givan, 1972).

Evidence for the operation of the pentose phosphate cycle in germinating seeds is of two kinds, demonstration of the enzymes taking part in it and analysis of the breakdown of glucose. Most of the enzymes participating in the pathway have been shown to be present in germinating seeds. The first two enzymes in the pathway are glucose-6-phosphate and 6-phosphogluconate dehydrogenase. Both of these enzymes increase in activity in the early stages of germination of for example oats, but decline again after 24 – 30 hr (Come *et al.*, 1988). Glucose-6-phosphate and 6-phosphogluconate dehydrogenases have also been shown in a number of other seeds, including wheat, pea and lettuce. Other enzymes of the pentose phosphate pathway are present in various plant tissues. Convincing evidence for the existence of the pentose phosphate pathway has been brought for mung beans (Table 5.12) (Chakravorty and Burma, 1959).

TABLE 5.12. *Changes in the activity of glucose-6-phosphate and phosphogluconate dehydrogenase during germination of mung beans (Phaseolus radiatus). Results are given as units/ml crude extract for total activity and units/mg protein for specific activity (after Chakravorty and Burma, 1959)*

Age of seedling in hours	Activity of glucose-6-phosphate dehydrogenase		Activity of phosphogluconate dehydrogenase	
	Total	Specific	Total	Specific
24	0·50	0·05	0·50	0·05
48	0·50	0·06	0·40	0·05
72	0·27	0·05	0·25	0·05
96	0·08	0·02	0·11	0·03

The presence of all the enzymes necessary for the oxidation of glucose-6-phosphate to ribulose phosphate and further conversion of the latter was demonstrated. The existence of enzymes carrying out a process or reaction does not prove the occurrence of the reaction *in vivo*, but it seems very likely from other information on the metabolism of seeds that the pentose phosphate cycle may be functioning. By feeding seeds during germination with glucose labelled in either position 1 or 6 and measuring the release of radioactive carbon dioxide, it is possible to estimate the relative contribution of the glycolytic and pentose phosphate pathways in the breakdown of glucose. When degradation of the glucose occurs via glycolysis and the tricarboxylic acid cycle, C1 and C6 appear in approximately equal amounts in the carbon dioxide released. When the pentose phosphate pathway is operating C1 is released preferentially. A very detailed study of this kind has been conducted in *Phaseolus mungo* (Ashihara and Matsumura, 1977). Determination of C6/C1 ratios has shown that the pentose phosphate cycle was operating both in the cotyledons and embryonic axis. However, the relative importance of the pathway changed with time and was different in various parts of the seed. In the cotyledons the activity of some of the glycolytic enzymes fell during germination, while those of the pentose phosphate pathway rose. In the embryonic axis the activity of enzymes of both the pathways increased as germination proceeded.

The pentose pathway may be particularly important during the early stages of germination, since it can provide NADPH for various synthetic processes. In addition it provides the pentoses required, for example, for nucleic acid synthesis. Apparently the operation of the pathway is needed for various anabolic processes occurring in the germinating seed, although it is not associated directly with energy production. In many cases, conditions of stress increase the percentage of the pentose phosphate cycle in the general metabolism.

It has been suggested that the operation of the pentose phosphate pathway is linked with dormancy breaking (Roberts and Smith, 1977). More recent studies failed to find differences between dormant and non-dormant lines of *Avena fatua*. In fact the C6/C1 ratio increased as dormancy was lost and dormancy breaking with GA_3 also tended to increase the ratio, while it could have been expected to decrease (Fuerst *et al.*, 1983). Present information therefore does not seem to support the idea that the pentose phosphate pathway is directly related to dormancy breaking.

Three criteria have been proposed as evidence for the operation of the tricarboxylic acid cycle. These criteria demand (a) proof of the existence of the complete cycle, (b) that the cycle can be entered at any point, and (c) that any substance which is an intermediary of the cycle can be used as a substrate. There are many methods for studying this cycle such as oxidative phosphorylation and oxygen uptake associated with specific substrates.

During the functioning of the tricarboxylic acid cycle decarboxylation and dehydrogenation of some of the intermediaries occur. Normally the cycle is closely coupled to an electron transport chain, the final hydrogen acceptor being oxygen.

The ability of mitochondria, prepared from seeds, to oxidize tricarboxylic acid cycle intermediaries has been shown for peas, peanuts, mung beans, lettuce, castor beans and other seeds. The changes in the oxidative ability of mitochondria isolated from lupin seeds and seedlings at different ages are shown in Table 5.13.

TABLE 5.13. *Ability of mitochondria isolated from lupin seeds to oxidize various substrates. Results as μl O_2/hr/mg N (Conn and Young, 1957)*

Substrate / Age of seedling	Oxygen uptake 12 hours	2 days	4 days	10 days
Succinate	19·0	70·0	167·0	208·0
α-Ketoglutarate	49·0	100·0	122·0	118·0
Malate	42·0	46·0	72·0	72·0
Citrate	30·0	25·0	74·0	90·0
Glutamate	3·0	65·0	68·0	91·0
Endogenous	4·0	4·0	9·0	5·0

It can be seen that the oxidative ability of the mitochondria generally rises as the age of the seedling increases but not at equal rates for different substrates. Proof that the tricarboxylic acid cycle is indeed a cyclic process has been brought for castor beans (Neal and Beevers, 1960). When slices of castor bean endosperm were fed with labelled pyruvate, it was possible to follow the fate of the various carbon atoms of the pyruvate. Pyruvate labelled with ^{14}C in the C1 gave rise to labelled carbon dioxide rapidly and quantitatively. When, however, the label was in the C2 or C3 positions the ^{14}C label spread slowly through various cell constituents and the labelled carbon was liberated only very slowly as carbon dioxide. This indicated that the C1 carbon is removed by immediate decarboxylation while the C2 and C3 carbons are being cycled and are only slowly decarboxylated. Some intermediaries may "leave" the cycle and enter other metabolic pathways. This may be the case in corn mesocotyls in which the C2 and C3 carbons of pyruvate were not released at all as carbon dioxide. Instead they appeared in various amino acids and in protein.

The behaviour of isolated mitochondria does not necessarily reflect their behaviour in the intact cell. It is well known today that the ability of isolated mitochondria to oxidize a certain substrate depends to a large extent on the method of isolation such as the tonicity of the medium, the presence of certain ions in the isolation medium and other preparative details. All these factors play a part in determining the activity of mitochondria, as does the exact composition of the reaction mixture in which their oxidative capacity is being tested.

During normal respiratory processes many variations and possible bypasses of certain stages in the tricarboxylic acid cycle may take place. Evidence for such bypasses usually comes from tracer studies. Such studies can demonstrate at which stage substrates can bypass known steps in the cycle. Other evidence comes from the use of inhibitors and the chromatographic identification of the intermediaries which occur during the cyclic process.

Despite the reservations which must be kept in mind with regard to work with isolated mitochondria, significant progress has recently been made in the combined study of mitochondrial ultrastructure and of mitochondrial function. Studies on the ultrastructure of lettuce and rice seeds and of cotyledons of peas, *Cucurbita maxima* and soybeans all show a quite consistent picture. As germination proceeds, the mitochondria become better defined, their structure becomes clearer and the number of cristae in them increases (Paulson and Srivasta, 1968; Srivasta and Paulson, 1968; Ueda and Tsuji, 1971; Solomos *et al.*, 1972). The development of mitochondrial structure during imbibition and growth of the tip of the embryonic axis of peas was also very clearly shown (Hodson *et al.*, 1987). This phenomenon appears to be quite general and accompanies the increasing respiratory activity of the seed. The protein content of the mitochondria isolated from pea cotyledons at various stages of germination increased between 3 and 18 hr of imbibition and their buoyant density changed, i.e. the position of the mitochondrial fraction in density gradient centrifugation shifted, and they became lighter (Nawa and Ashahi, 1971). These changes indicate that the structural changes observed in the mitochondria are accompanied by compositional changes. Such compositional changes appear also to be a kind of "maturation" during germination.

The properties of mitochondria can be altered by changes in the lipid composition of their membranes, which affect their permeability. Such changes have been observed in animal mitochondria by Seligman *et al.* (1967). Changes in structure and function of mitochondria have also been described in seeds, for example in peanuts. The mitochondria prepared from the dry embryos were deficient in cytochrome c and did not show respiratory control. As imbibition proceeds, respiratory control and cytochrome c content become more normal (Wilson and Bonner, 1971). During germination, therefore, mitochondria change from an aberrant form to a normal one, which can fulfil the normal respiratory functions. In addition there is little doubt that new mitochondria are formed during germination, as are other subcellular organelles.

As described earlier, mitochondria of cotyledons of dry peas have poorly developed cristae. The subunits of cytochrome oxidase are apparently present in the dry seed and undergo self-assembly during seed hydration (Matsuoka and Asahi, 1983). Nevertheless, mitochondria appear to be capable of protein synthesis as indicated for example by the presence in them of the enzyme methionyl-tRNA transformylase, which is known to be required for the initiation of organelle protein synthesis (Coffin and Cossins, 1986).

Similar changes, induced by dehydration, were observed in the mitochondria of maize root tips. Following a water loss of up to 60% of the initial hydrated weight, the mitochondria appeared without cristae. However, cytochemical staining for cytochrome oxidase activity clearly showed the outline of cristae. Moreover mitochondria from such dehydrated roots, which apparently rehydrated during isolation, showed dehydrogenase and cytochrome oxidase activities which were greater than in the controls (Nir *et al.*, 1970). There is therefore no simple answer to the question of what happens to mitochondrial function during seed maturation, which involves dehydration and rehydration during the early stages of germination. Functionality can be achieved in a number of ways and the process need not be the same in all seeds. The ways in which mitochondrial development in cotyledons and axes of seeds are interrelated has been studied repeatedly and although there are indications that the axis exercises some control over the cotyledons, the question remains essentially unresolved, and much of the evidence remains contradictory. This topic has recently been discussed by Murray (1984).

The increase in the functionality of entire mitochondria during germination is also indicated by increases in individual enzymes of the Krebs cycle and of the electron transport chain constituents. In peas succinate dehydrogenase, fumarate dehydratase and aconitate hydratase all increase during germination (Kolloffel and Sluys, 1970). Cytochrome oxidase and cytochrome c-NADH reductase increase quite rapidly in lettuce seeds during germination (Fig. 5.23). The rate of change in enzyme activity during the early stages does not correspond to the rate of change in water uptake by the seeds and it appears that the reductase undergoes an activation only partially dependent on water uptake (Fig. 5.24).

Respiratory control changes markedly in peas in the first 18 hr of germination. The respiratory control increases from 1.3 to 2.1 when malate was the substrate and from 1.3 to 3.3 with α-ketoglutarate as the substrate (Kolloffel and Sluys, 1970). Oxidative phosphorylation has been demonstrated in mitochondria from lupin seedlings with P/O ratios of over 3 when α-ketoglutarate was oxidized (Conn and Young, 1957) and for peas with ADP/O of about 2.0 (Kolloffel and Sluys, 1970). The exact period at which oxidative phosphorylation begins during germination is in doubt. Generally in mitochondria isolated from dry seeds oxidative phosphorylation is low or too low to be measured. In lettuce, attempts to show oxidative phosphorylation in dry seeds of their mitochondria were unsuccessful (Gesundheit and Poljakoff-Mayber, 1962). Nevertheless, Wilson and Bonner (1971) were able to

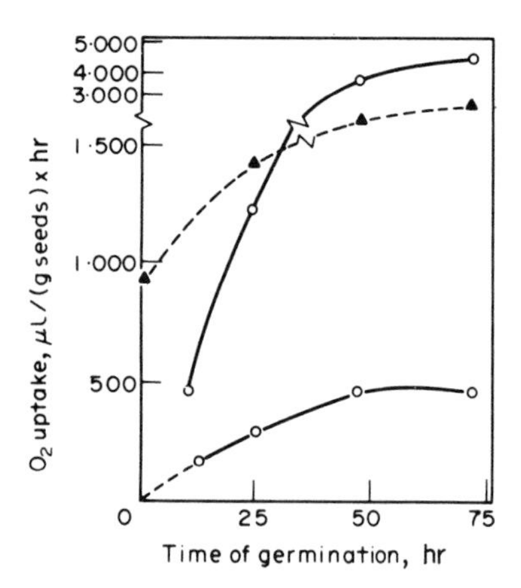

FIG. 5.23. Comparison of measured oxygen uptake of lettuce seeds during germination and potential oxygen uptake which could be mediated by activities of cytochrome c oxidase and cytochrome c reductase. Results expressed as 0_2 μl per hour related to 1 g seeds. (1) ○——○ Upper curve. Measured oxygen uptake. (2) ▲——▲ Potential oxygen uptake as calculated from cytochrome c oxidase activity. (3) ○——○ Lower curve. Potential oxygen uptake as calculated from NADH-cytochrome c reductase activity (Eldan and Mayer, 1972).

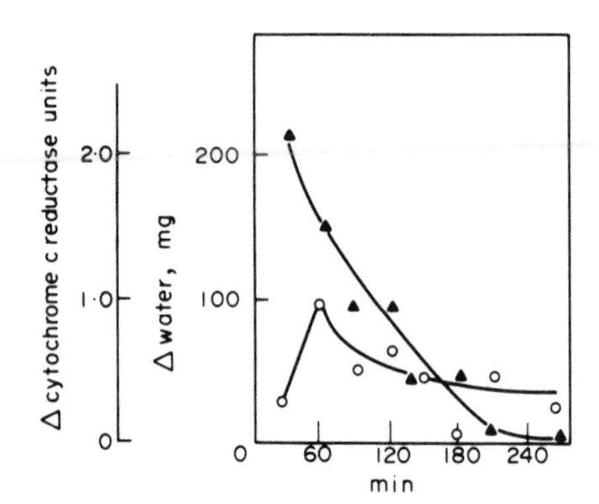

FIG. 5.24. Comparison of the rate of water uptake and the rate of development of NADH-cytochrome c reductase activity during imbibition of lettuce. Rate of change: ○——○ NADH-cytochrome c reductase activity. ▲——▲ Water uptake (Eldan and Mayer, 1972).

show some phosphorylation even in mitochondria from dry peanut embryos. In germinating wild peas, *P. elatius*, ADP/O ratios of 2.0 were found after 4.5 hr of germination when succinate was the substrate (Table 5.14). Thus in this seed phosphorylation began quite soon after the onset of imbibition.

Significant increases in ATP content and rapid conversion of AMP and ADP to ATP indicate the onset of oxidative phosphorylation. Such changes have been demonstrated in lettuce seeds and wheat embryos (Pradet *et al.*, 1968; Obendorf and Marcus, 1974). The rate of phosphorylation seems to reach a steady state very

TABLE 5.14. *Respiratory control and ADP/O ratio in mitochondria isolated from cotyledons of P. elatius after varying times of germination. Succinate as substrate (from Marbach and Mayer, 1976).*

Germination time	RCR	ADP/O
0	1·24	—
4·5	1·73	2·05
16	2·03	1·96
24	2·47	1·82
48	3·40	2·20

quickly as the ATP level and the energy charge reached a plateau after 6 hr of imbibition in the wheat embryos (Fig. 5.25). Rapid onset of oxidative phosphorylation is also indicated by the rapid rise of the energy charge of lettuce seeds when they are transferred from anoxic to hypoxic conditions (Raymond and Pradet, 1980). There is little doubt left that seeds can produce ATP readily by the normal cytochrome pathway in the presence of oxygen and that this production is cyanide sensitive and that it can occur within minutes of imbibition.

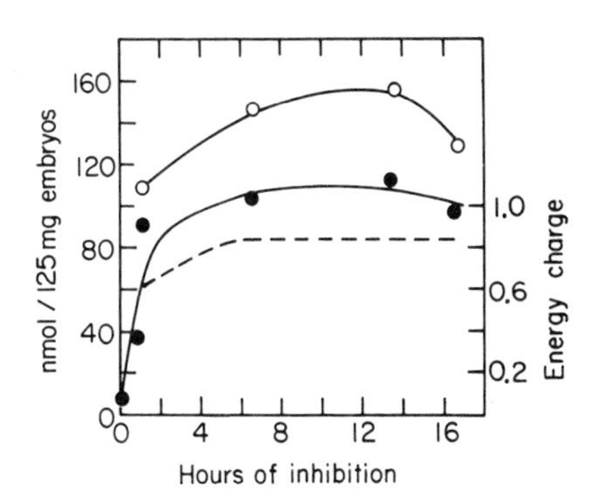

FIG. 5.25. Changes in ATP ●, total adenylates ○ and energy charge ----- in wheat embryos, excised from imbibing wheat seeds (compiled from data of Obendorf and Marcus, 1974).

This leaves unresolved the question why it is difficult to isolate mitochondria from dry seeds which are coupled and able to carry out phosphorylation? Perl (1980/81) has proposed a different mechanism for ATP formation. He suggests that ATP arises by the action of pyruvate kinase and adenylate kinase from AMP and phosphoenol pyruvate. This mechanism, which has been shown to operate *in vitro*, could also account for the increased energy charge observed in imbibing seeds. The contribution of this to energy metabolism of the seed has never been proved convincingly.

Changes in the relative levels of AMP, ADP and ATP during germination might function as a control mechanism according to the concept of energy charge (Atkinson, 1968) as suggested by Pradet *et al.* (1968). An example of such regulation of the activity of a key enzyme in plant respiration by ATP and AMP is pyruvate kinase from cotton seeds. Its activity is inhibited by malate citrate and ATP but stimulated by AMP and fumarate (Duggleby and Dennis, 1973). This enzyme shows features different from those of pyruvate kinase isolated from other tissues.

The control mechanism of respiration as an overall process in the germinating seeds is still rather poorly understood (Mayer, 1977). Except for the existence in mitochondria of some of the components of the electron transport chain already mentioned, cytochrome c, cytochrome oxidase and cytochrome c-NADH reductase, information is scant. It seems probable that the other components are present, since the mitochondria can usually carry out normal oxidative and phosphorylative activity. However, direct evidence for most other components of the electron transport chain in seeds is still missing. A cyanide insensitive electron transport pathway is known in plant tissues which is associated with the mitochondria and is probably a branching off the normal electron transport chain (see earlier). The existence of the cyanide resistant pathway is usually established by two criteria, persistence of some oxygen uptake in the presence of cyanide, and inhibition of the residual oxygen uptake by salicyl hydroxamic acid (SHAM) or propylgallate. However, it must be remembered that experiments using cyanide as an inhibitor are liable to artefacts, because of the extreme volatility of cyanide at the usual pHs used in germination experiments. In fact it has been suggested that in many cases the apparent resistance of germination to cyanide is an experimental artefact (Yu *et al.*, 1981). The use of unsealed containers, together with the release of HCN, led to a loss of cyanide, so that the concentrations tested were much lower than expected. In sealed containers 1 mM cyanide inhibited germination of most species and 10 mM cyanide totally inhibited germination even of peas, mung beans and cucumber seeds, insensitive to the lower concentration. For this reason failure to observe inhibition of oxygen uptake by cyanide must always be very carefully examined (Pradet, 1964).

The cyanide insensitive pathway has been detected in germinating seeds. However, the period at which it begins to operate seems to be extremely variable. In soy beans it apparently operates quite early (Yentur and Leopold, 1976), while in seeds of *Cicer* cyanide resistant respiration reaches a peak quite late in the process of germination (Burgillio and Nicolas, 1977).

Efforts to show the importance of the alternative, or cyanide resistant, pathway of electron transport in germination have been made in recent years. Generally the alternative pathway mediates only a small part of total oxygen uptake. In most cases its contribution to oxygen uptake did not change very radically during germination. The only really significant exception reported was in seeds of *Vigna mungo* and *V. radiata* in which the alternate pathway did appear to increase transiently in the cotyledon. In the axis of a number of species the contribution of the alternative pathway was significant, but did not change during the early stages of germination (Morohashi, 1986).

Thus there is little doubt that the pathway operates in germinating seeds, both in cotyledons and embryonic axes, but there is very little to suggest that it has any special role in germination. A special function of the pathway has been proposed in the case of *Xanthium*, in which it is suggested that a certain part of the overall oxygen uptake must be mediated by it to permit germination (Esashi *et al.*, 1987). How this is supposed to function is by no means clear.

In addition to the conventional electron-transport system associated with the mitochondria there is scattered evidence for alternative electron-transport systems. In wheat germination it has been shown that reduced NADP can be oxidized by molecular oxygen in the presence of two enzymes, one of which is peroxidase (Conn *et al.*, 1952). This is of interest in view of the evidence previously recalled for the formation of NADPH during the pentose phosphate shunt. The mechanism of oxidation of NADPH and the possible energy release during its oxidation are at present in dispute. A soluble NADPH-oxidase has been shown to be present in dry lettuce seeds as well as in lettuce seedlings. The latter enzyme may be a phenolase (Mayer, 1959a). A complete alternative electron transport system has been shown to be present in pea seedlings. The system consists of a dehydrogenase, NADP, glutathione, and ascorbic acid in the presence of ascorbic acid oxidase. This system was able to mediate 20 – 25% of the total respiration of the young seedling and apparently was functioning in this way in the intact seedlings after about 3 days of germination. In dry pea seeds the system did not seem to function because ascorbic acid oxidase was absent (Mapson and Moustafa, 1957).

Many other oxidative enzymes are known to occur in plant tissues. Among those which have been shown to be present in seeds are catalase, peroxidase, lipoxidase as well as phenolase. The changes in these enzyme systems with germination have been followed in many cases. Unfortunately nothing is known as to how these enzymes are integrated into multi-enzyme systems and how, if at all, they are related to respiration. Phenolase has been frequently supposed to function as a terminal oxidase which can oxidize reduced coenzymes in the presence of a phenol, but direct evidence is lacking. Peroxidases and possibly also catalases may be connected in some way to direct oxidation of flavoproteins by molecular oxygen, but again convincing evidence is lacking. Perhaps these are involved in scavenging the free radicals which can form in the seeds as a result of metabolic processes.

In summary, in seeds the main respiratory pathways are: (1) glycolysis, linked to the tricarboxylic acid cycle and the electron transport chain, and (2) the pentose phosphate pathway. The former process is responsible for ATP production in germinating seeds, while the latter produces carbon skeletons and reducing power for anabolism. The mitochondria are the site of the tricarboxylic acid cycle and the electron transport chain which becomes active during germination.

In mitochondria the increase in activity is accompanied by structural changes, including cristae formation. The increasing functionality is caused by an increase of all the enzyme systems required for operations of the Krebs cycle. The mitochondria also develop a normally functioning electron transport chain. In the dry seed various parts of either the Krebs cycle or the electron transport chain may be deficient. The deficiency need not necessarily be the same in every tissue, nor are the exact time sequences of development of enzyme activity and functionality of the mitochondria the same in all species. Many questions are still left unanswered, particularly whether and to what extent alternate respiratory pathways exist and play a role in seed germination and dormancy? It may also be asked to what extent do respiratory pathways respond to changing environmental factors?

Bibliography

Akazawa T. and Miyata S. (1982) *Essays Biochem.* **18**, 41.
Al-Ani A., Bruzau F., Raymond P., Saint-Ges V., Leblanc J-M. and Pradet A. (1985) *Plant Physiol.* **79**, 885.
Albaum H. G. and Cohen P. P. (1943) *J. Biol. Chem.* **149**, 19.
Albaum H. G. and Umbreit W. W. (1943) *Am. J. Bot.* **30**, 533.
Anderson J. W. and Fowden L. (1969) *Plant Physiol.* **44**, 60.
Anderson J. W. and Fowden L. (1970) *Biochem. J.* **119**, 677.
Ashihara H. and Matsumura H. (1977) *Phytochemistry* **8**, 461.
Ashton F. M. (1976) *Ann. Rev. Plant Physiol.* **27**, 95.
Ashton W. M. and Williams P. C. (1958) *J. Sci. Food Agr.* **9**, 505.
Aspart L., Meyer Y., Laroche M. and Penon P. (1984) *Plant Physiol.* **76**, 664.
Atkinson D. E. (1968) *Biochemistry* **7**, 403.
Bailey C. F. (1921) *Univ. of Minnesota Agr. Expt. Stat. Tech. Bull.* **3**.
Baiza A., Aguilar R. and Sanchez-de-Jiminez E. (1986) *Physiol. Plant.* **68**, 259.
Baumgartner B., Tokuyasu K. T. and Chrispeels M. J. (1978). *J. Cell Biol.* **79**, 10.
Beevers H. (1979) *Ann. Rev. Plant Physiol.* **30**, 159.
Bevilacqua L. R. (1955) *Rend. Acad. Naz. Lincei ser. VII*, **18**, 214.
Bevilacqua L. R. and Scotti R. (1953) *Acad. Ligure de Sci. et Let.* **X, 1.**
Bewley J. D. (1982) in *Encyclopedia of Plant Physiology*, vol. 14a, p. 559 (Eds D. Boulter and B. Parthier).
Bewley J. D. and Black M. (1982) *Physiology and Biochemistry of Seeds*, vol. 2, Springer-Verlag, Berlin.
Bewley J. D. and Halmer P. (1980/81) *Is. J. Bot.* **29**, 118.
Biswas B. B., Ghosh B. and Majunder A. L. (1984) in *Subcellular Biochemistry* vol. 10, p. 237 (Ed. D. R. Roodyn), Plenum Press, New York.
Boatman S. G. and Crombie W. M. (1958) *J. Exp. Bot.* **9**, 52.
Breidenbach R. W. and Beevers H. (1967) *Biochem. Biophys. Res. Comm.* **27**, 462.
Brooker J. D., Cheung C. P. and Marcus A. (1977) in *The Physiology and Biochemistry of Seed Dormancy and Germination*, p. 347 (Ed. A. A. Kahn), North-Holland, Amsterdam.
Brown G. E. (1965) *Biochem. J.* **95**, 509.
Burgillio F. R. and Nicolas G. (1977) *Plant Physiol.* **60**, 524.
Chakravorty M. and Burma D. P. (1959) *Biochem. J.* **73**, 48.
Chapman C. W. (1987) *Phytochemistry* **36**, 3127.
Chen D. and Osborne D. J. (1970) *Nature* **225**, 336.
Chibnall A. C. (1939) *Protein Metabolism in the Plant*, Yale Univ. Press.
Ching T. M. (1963) *Plant Physiol.* **38**, 722.
Ching T. M. (1966) *Plant Physiol.* **41**, 1313.
Ching T. M. (1968) *Lipids* **3**, 482.
Ching T. M. (1973) *Plant Physiol.* **51**, 400.
Coffin J. W. and Cossins E. A. (1986) *Phytochemistry* **25**, 2481.
Come D., Corbineau F. and Lecat S. (1988) *Seed Sci. Technol.* (In press).
Conn E. E., Kraemer L. M., Pei-Nan Lin and Vennesland B. (1952) *J. Biol. Chem.* **194**, 143.
Conn E. E. and Young L. C. T. (1957) *J. Biol. Chem.* **226**, 23.
Crombie W. M. and Comber R. (1956) *J. Exp. Bot.* **7**, 166.
Cruickshank D. H. and Isherwood F. A. (1958) *Biochem. J.* **69**, 189.
Cuming A. C. and Osborne D. J. (1978a) *Planta* **139**, 209.
Cuming A. C. and Osborne D. J. (1978b) *Planta* **139**, 219.
Damodoran M., Ramashwamy R., Venkatesan T. R., Mahadevan S. and Randas K. (1946) *Proc. Ind. Acad. Sci.* **B23**, 86.
Datta K., Parker H., Averyhart-Fullard V., Schmidt A. and Marcus A. (1987) *Planta* **170**, 209.
Davies H. V. and Chapman J. M. (1979) *Planta* **146**, 579.
Davies H. V. and Chapman J. M. (1979) *Planta* **146**, 585.
Delseny M., Aspart L. and Cooke R. (1980/81) *Is. J. Bot.* **29**, 246.
Deltour R. (1985) *J. Cell Sci.* **75**, 43.
De Mason D. (1983) *Ann. Bot.* **52**, 71.
De Mason D. (1985) *Protoplasma* **126**, 159.
De Mason D., Sexton R., Gorman M. and Reid J. S. G. (1985) *Protoplasma* **126**, 168.
Di Nola L. and Mayer A. M. (1985) *Phytochemistry* **24**, 2549.

Di Nola L. and Mayer A. M. (1986a) *Phytochemistry* **25**, 2255.
Di Nola L. and Mayer A. M. (1986b) *Phytochemistry* **25**, 2725.
Di Nola L. and Mayer A. M. (1987) *Phytochemistry* **26** , 1591.
Dommes J. and Van de Walle C. (1983) *Plant Physiol.* **73**, 484.
Duggleby R. G. and Dennis D. T. (1973) *Arch. Biochem. Biophys.* **155**, 270.
Duperon E. (1958) *C. R. Acad. Sci. Paris* **246**, 298.
Duperon R. (1960) *C. R. Acad. Sci. Paris* **251**, 260.
Dure L. S. (1960) *Plant Physiol.* **35**, 925.
Dure L. S., Galau G. A. and Greenway S. (1980/81) *Is. J. Bot.* **29**, 293.
Edelman J., Shibko S. I. and Keys A. J. (1959) *J. Exp. Bot.* **10**, 178.
Edwards M. (1969) *J. Exp. Bot.* **20**, 876.
Egami F., Ohmachi K., Iida K. and Taniguchi S. (1957) *Biochimia* **22**, 122.
Eldan M. and Mayer A. M. (1972) *Physiol. Plant.* **26**, 67.
Eldan M. and Mayer A. M. (1974) *Phytochemistry* **13**, 389.
Elder R. H., Dell 'Aquila A., Mezzina M., Sarasin A. and Osborne D. J. (1987) *Mutation Res.* **181**, 61.
Ergle D. R. and Guinn G. (1959) *Plant Physiol.* **34**, 476.
Esashi Y., Kirota A., Abe M. and Fuwa N. (1987) *Plant Cell Physiol.* **28**, 151.
Evenari M., Neumann G. and Klein S. (1955) *Physiol. Plant.* **8**, 33.
Fernandes D. S. (1923) *Rec. Trav. Bot. Neerl.* **20**, 107.
Filner P. and Varner J. E. (1967) *Proc. Nat. Acad. Sci. U.S.* **58**, 1520.
Fine J. M. and Barton L. V. (1958) *Contr. Boyce Thomson Inst.* **19**, 483.
Forest J. C. and Wightman F. (1971) *Can. J. Bot.* **49**, 709.
Forest J. C. and Wightman F. (1972) *Can. J. Bot.* **50**, 538.
Fowden L. (1965) in *Plant Biochemistry*, p. 361 (Eds J. Bonner and J. E. Varner), Academic Press, New York.
Fuerst E. P., Upadhyaya M. K., Simpson G. M., Naylor J. M. and Adkins S. W. (1983) *Can. J. Bot.* **61**, 667.
Fukui T. and Nikuni Z. (1956) *J. Biochem. Japan* **43**, 33.
Galau G. A. and Dure L. S. (1980/81) *Is. J. Bot.* **29**, 81.
Galli M. G., Balzaretti R. and Sgorbati S. (1986) *J. Exp. Bot.* **37**, 1716.
Gesundheit Z. and Poljakoff-Mayber A. (1962) *Bull. Res. Counc. Isr.* **D11**, 25.
Gibbons G. C. (1979) *Carlsberg Res. Comm.* **44**, 353.
Gifford D. J., Imeson H. C., Thakore E. and Bewley J. D. (1986) *J. Exp. Bot.* **37**, 1879.
Givan C. V. (1972) *Planta* **108**, 24.
Gordon S. A. and Surrey K. (1960) *Radiation Res.* **12**, 325.
Graf E., Empson K. L. and Eaton J. (1987) *J. Biol. Chem.* **262**, 11647.
Gram N. H. (1982a) *Carlsberg Res. Commun.* **47**, 143.
Gram N. H. (1982b) *Carlsberg Res. Commun.* **47**, 173.
Grellet F., Delseny M. and Guitton Y. (1977) *Nature* **267**, 724.
Grey J. E. and Bryant J. A. (1984) *Phytochemistry* **23**, 477.
Haber A. H. and Brassington N. (1959) *Nature, Lond.* **183**, 619.
Hageman R. H. and Flesher D. (1960) *Arch. Biochem. Biophys.* **87**, 203.
Halmer P. (1985) *Physiol. Veg.* **23**, 107.
Halvorson H. (1956) *Physiol. Plant.* **9**, 412.
Harvey B. M. R. and Oaks A. (1974) *Plant Physiol.* **53**, 453.
Hatch M. D. and Turner J. F. (1958) *Biochem. J.* **69**, 495.
Hecker M. and Bernhardt D. (1976) *Phytochemistry* **15**, 1105.
Hodson M. J., Di Nola L. and Mayer A. M. (1987) *J. Exp. Bot.* **38**, 525.
Huang A. H. C. and Moreau R. A. (1978) *Planta* **141**, 111.
Huang A. H. C., Moreau R. A. and Lin K. D. F. (1978) *Plant Physiol.* **61**, 339.
Huang A. H. C., Trelease R. N. and Moore T. S. (1983) *Plant Peroxisomes*, Academic Press, New York.
Ihle J. N. and Dure L. S. (1969) *Biochem. Biophys. Res. Comm.* **36**, 705.
Ihle J. N. and Dure L. S. (1972) *J. Biol. Chem.* **247**, 5048.
Ingle J., Beevers L. and Hageman R. H. (1964) *Plant Physiol.* **39**, 705.
Ingle J. and Hageman R. H. (1965) *Plant Physiol.* **40**, 48.
Jarden R. and Mayer A. M. (1981) *Phytochemistry* **20**, 2669.
Jelsema C. L., Morre D. J. and Ruddat M. (1982) *Bot. Gaz.* **143**, 26.

Joshi A. C. and Doctor V. M. (1975) *Lipids* **10**, 191.
Katayama M. and Funahashi S. (1969) *J. Biochem. Tokyo* **66**, 479.
Kermode A. R., Bewley J. D., Dasgupta J. and Misra S. (1986) *Hort. Sci.* **21**, 1113.
Kirsop B. H. and Pollock J. R. A. (1957) *European Brewery Convention*, p. 84.
Klein S. (1955) Ph.D. Thesis, Jerusalem (in Hebrew).
Kolloffel C. (1967) *Acta Bot. Neerl.* **16**, 111.
Kolloffel C. and Sluys J. V. (1970) *Acta Bot. Neerl.* **19**, 503.
Kornberg H. L. and Beevers H. (1957) *Biochim. Biophys. Acta* **26**, 531.
Koshiba T., Tomura H. and Miura M. (1986) *Plant Cell Physiol.* **27**, 1069.
Kruger N. J. and Beevers H. (1985) *Plant Physiol.* **77**, 358.
Kruger N. J., Kombrink E. and Beevers H. (1983) *FEBS letters* **153**, 409.
Larondelle Y., Corbineau F., Dethier M., Come D. and Hers H. G. (1987) *Eur. J. Biochem.* **166**, 605.
Lea P. J.and Joy K. W. (1983) in *Mobilisation of Reserves in Germination*, p. 77 (Eds C. Nozzolillo, P. J. Lea and F. A. Loewus), *Recent Advances in Phytochemistry*, vol. 17, Plenum Press.
Leblova S. (1978) in *Plant Life in Anaerobic Environments*, p. 155 (Eds D. O. Hook and R. M. M. Crawford), Ann Arbor Science.
Lecat S. (1987) Ph.D. Thesis, Pierre and Marie Curie University, Paris.
Lechevallier D. (1960) *C. R. Acad. Sci. Paris* **250**, 2825.
Leggatt C. W. (1948) *Can. J. Res.* **C26**, 194.
Lenoir C., Corbineau F. and Come D. (1986) *Physiol. Plant.* **68**, 301.
Leopold A. C. and Guernsey F. (1954) *Physiol. Plant.* **7**, 30.
Levari R. (1960) Ph.D. Thesis, Jerusalem (in Hebrew).
Lin Y. H. and Huang A. H. C. (1983) *Arch. Biochem. Biophys.* **225**, 360.
Lin Y. H., Moreau R. A. and Huang A. H. C. (1982) *Plant Physiol.* **70**, 100.
Lin Y. H., Winner L. H. and Huang A. H. C. (1983) *Plant Physiol.* **73**, 460.
Loewus F. A. (1983) *Rec. Adv. Phytochemistry* **17**, 173.
Loewus M. W. and Loewus F. A. (1982) *Plant Physiol.* **70**, 765.
Lott J. N. A. and Castelfranco P. (1972) *Can. J. Bot.* **48**, 2233.
McCleary B. V. (1983) *Phytochemistry* **22**, 649.
McLeod A. M. (1957) *New Phytologist* **56**, 210.
McLeod A. M., Travis D. C. and Wreay D. G. (1953) *J. Inst. Brew.* **59**, 154.
Maherchandani N. and Naylor J. M. (1972) *Can. J. Bot.* **50**, 305.
Majunder A. N. L., Mandal N. C. and Biswas B. B. (1972) *Phytochemistry* **11**, 503.
Mapson L. W. and Moustafa E. M. (1957) *Biochem. J.* **62**, 248.
Marbach I. and Mayer A. M. (1976) *Physiol. Plant.* **38**, 126.
Marcus A. (1969) *Sym. Soc. Exp. Biol.* **23**, 143.
Marcus A. and Feeley J. (1964) *Proc. Nat. Acad. Sci. U.S.* **51**, 1075.
Marcus A. and Velasco J. (1960) *J. Biol. Chem.* **235**, 563.
Marcus A., Feeley J. and Volcani T. (1966) *Plant Physiol.* **41**, 1167.
Marre E. (1967) *Curr. Topics Develop. Biol.* **2**, 75.
Marshall J. J. (1972) *Wallerstein Lab. Comm.* **35**, 49.
Mary Y. Y., Chen D. and Sarid S. (1972) *Plant Physiol.* **49**, 20.
Matheson N. K. and Richardson R. H. (1976) *Phytochemistry* **15**, 887.
Matsuoka M. and Asahi T. (1983) *Eur. J. Biochem.* **134**, 223.
Mayer A. M. (1958) *Enzymologia* **19**, 1.
Mayer A. M. 1959a) *Enzymologia* **20**, 13.
Mayer A. M. (1959b) *Proc. Int. Bot. Cong. vol.* **2**, p. 256.
Mayer A. M. (1973) *Seed Sci. Technol.* **1**, 51.
Mayer A. M. (1977) in *The Physiology and Biochemistry of Seed Dormancy and Germination*, p. 357 (Ed. A. A. Khan), North-Holland, Amsterdam.
Mayer A. M. and Mapson L. W. (1962) *J. Exp. Bot.* **13**, 201.
Mayer A. M. and Marbach I. (1981) *Progress Phytochem.* **7**, 95.
Mayer A. M. and Shain Y. (1974) *Ann. Rev. Plant Physiol.* **25**, 167.
Mazliak P. (1980) *Progress Phytochem.* **6**, 49.
Mettler I. J. and Beevers H. (1980) *Plant Physiol.* **66**, 55.

Meyer H. and Mayer A. M. (1971) *Physiol. Plant.* **24**, 95.
Miflin B. J. and Lea P. J. (1977) *Progress Phytochem.* **4**, 1.
Miflin B. J. and Lea P. J. (1982) in *Encyclopedia of Plant Physiology*, New Series, vol. 14A, p. 5. (Ed. B. Boulter), Springer Verlag, Berlin.
Mikkonen A. (1986) *Physiol. Plant.* **68**, 282.
Misra, S. and Bewley, J. D. (1985) *J. Exp. Bot.* **36**, 1644.
Moreland D. E., Hussey Y., Shriner C. R. and Farmer F. S. (1974) *Plant Physiol.* **54**, 56.
Morohashi Y. (1986) *J. Exp. Bot.* **37**, 262.
Morohashi Y. and Shimokoriyama M. (1975) *J. Exp. Bot.* **26**, 932.
Mukherji S., Dey B. and Sircar S. M. (1968) *Physiol. Plant.* **21**, 360.
Muntz K., Bassuner R., Lichtenfeld C., Scholz G. and Webber E. (1985) *Physiol. Veg.* **23**, 75.
Murray D. R. (1984) in *Seed Physiology*, vol. 2, p. 247 (Ed. D. R. Murray), Academic Press, New York.
Muto S. and Beevers H. (1974) *Plant Physiol.* **54**, 23.
Nawa Y. and Ashahi T. (1971) *Plant Physiol.* **48**, 671.
Neal G. E. and Beevers H. (1960) *Biochem. J.* **74**, 409.
Negbi M. and Sargent J. A. (1986) *Bot. J. Linnean Soc.* **93**, 247.
Nir I., Pojakoff-Mayber A. and Klein S. (1970) *Plant Physiol.* **45**, 173.
Nishimura M. and Beevers H. (1981) *Plant Physiol.* **67**, 1255.
Oaks A. (1983) in *Mobilisation of Reserves in Germination*, p. 53 (Eds C. Nozzolillo, P. J. Lea and F. A. Loewus), *Recent Advances in Phytochemistry*, vol. 17, Plenum Press.
Oaks A. and Bidwell R. G. S. (1970) *Ann. Rev. Plant Physiol.* **21**, 43.
Obendorf R. L. and Marcus A. (1974) *Plant Physiol.* **53**, 779.
Okamoto K. and Akazawa T. (1980) *Plant Physiol.* **65**, 81.
Oo C. and Stumpf P. K. (1983a) *Plant Physiol.* **73**, 1028.
Oo C. and Stumpf P. K. (1983b) *Plant Physiol.* **73**, 1033.
Oota T., Fujii R. and Osawa S. (1953) *J. Biochem. Tokyo* **40**, 649.
Oota Y., Fujii R. and Sunobe Y. (1956) *Physiol. Plant.* **9**, 38.
Osborne D. J. (1977) in *The Physiology and Biochemistry of Seed Dormancy and Germination*, p. 319 (Ed. A. A. Khan), North-Holland, Amsterdam.
Osborne D. J. (1980) in *Senescence in Plants*, p. 13 (Ed. K. V. Thimann), CRC Press, Boca Raton.
Osborne, D. J., Dell' Aquila, A. and Elder, R. H. (1984) *Folia Biologica* (Praha) Special Publication, 155.
Ouelette B. F. F. and Bewley J. D. (1986) *Planta* **169**, 333.
Paech K. (1935) *Planta* **24**, 78.
Palmiano E. P. and Juliano B. O. (1972) *Plant Physiol.* **49**, 751.
Paulson R. E. and Srivasta L. M. (1968) *Can. J. Bot.* **46**, 1437.
Payne P. I. (1976) *Biol. Rev.* **51**, 329.
Peers F. G. (1953) *Biochem. J.* **53**, 102.
Perl M. (1980/81) *Is. J. Bot.* **29**, 307.
Pernollet J. -C. (1978) *Phytochemistry* **17**, 1473.
Pernollet J. -C. (1982) *Physiol. Veg.*, **20**, 250.
Peumans W. J. and Carlier A. R. (1977) *Planta* **136**, 195.
Peumans W. J., Caers L. I. and Carlier A. R. (1979) *Planta* **144**, 485.
Peumans W. J., Stinissen H. M. and Carlier A. R. (1982) *Planta* **156**, 41.
Pradet A. (1964) *Compt. Rend. Acad. Sci. Paris* **258**, 1610.
Pradet A., Narayanan S. and Vermeersch J. (1968) *Bull. Soc. Franc. Physiol. Veg.* **14**, 107.
Prentice N. (1972) *Agr. Food Chem.* **20**, 764.
Prianishnikov D. N. (1951) *Nitrogen in the Life of Plants*, Kramer Business Service.
Pridham J. B., Walter M. W. and Worth H. G. (1969) *J. Exp. Bot.* **20**, 317.
Psaras G. (1984) *Ann. Bot.* **54**, 187.
Radley M. (1968) *Soc. Chem. Ind. London Monograph* **31**, 53.
Raymond P. and Pradet A. (1980) *Biochem. J.* **190**, 39.
Raymond P., Al-Ani A. and Pradet A. (1985) *Plant Physiol.* **79**, 879.
Rebeiz C. A., Breidenbach R. W. and Castelfranco P. (1965) *Plant Physiol.* **40**, 286.
Reddy N. R., Sathe S. K. and Salunkle D. P. (1982) *Adv. Food Res.* **28**, 1.
Reed J. (1970) *Dissert. Abstr. Int.* **31**, 6.
Reid J. S. G. (1971) *Planta* **160**, 131.
Rejman E. and Buchowicz J. (1973) *Phytochemistry* **12**, 271.

Reuhl E. (1936) *Rec. Trav. Bot. Neerl.* **33**, 1.
Roberts E. H. (1969) *Symp. Soc. Exp. Biol.* **23**, 161.
Roberts E. H. and Smith R. D. (1977) in *The Physiology and Biochemistry of Seed Dormancy and Germination*, p. 385 (Ed. A. A. Khan), North-Holland, Amsterdam.
Rose D. H. (1915) *Bot. Gaz.* **49**, 425.
Ryan C. A. (1973) *Ann. Rev. Plant Physiol.* **24**, 173.
Sabularse D. C. and Anderson R. L. (1981) *Biochem. Biophys. Res. Commun.* **103**, 848.
Sanchez-Martinez D., Puigdomenech P. and Pages M. (1986) *Plant Physiol.* **82**, 543.
Schultz G. A., Chen D. and Katchalski E. (1972) *J. Mol. Biol.* **66**, 379.
Seal S. N., Schmidt A. and Marcus A. (1986) *Methods in Enzymology* **118**, 128.
Seligman A. M., Ueno H., Morizono Y., Wasserkrug H. L., Katzoff L. and Honker J. G. (1967) *J. Histochem. Cytochem.* **15**, 1.
Semenko G. I. (1957) *Fiziol. Rasteny* **4**, 332.
Shain Y. and Mayer A. M. (1965) *Physiol. Plant.* **18**, 853.
Shain Y. and Mayer A. M. (1968a) *Physiol. Plant.* **21**, 765.
Shain Y. and Mayer A. M. (1968b) *Science* **162**, 1283.
Shain Y. and Mayer A. M. (1968c) *Phytochemistry* **7**, 1491.
Shargool P. D. and Cossins E. A. (1968) *Can. J. Biochem.* **46**, 393.
Smith B. P. and Williams H. H. (1951) *Arch. Biochem. Biophys.* **31**, 366.
Smith S. M. and Leaver C. J. (1986) *Plant Physiol.* **81**, 762.
Sodek L., Lea P. J. and Miflin B. J. (1980) *Plant Physiol.* **65**, 22.
Solomos T., Malhotra S. S., Prasad S., Malhotra S. K. and Spencer M. (1972) *Can. J. Biochem.* **50**, 725.
Sopanen T. (1980) Publication No. 6, Dept. of Botany, Univ. of Helsinki (Ph.D. thesis).
Spragg S. P. and Yemm E. W. (1959) *J. Exp. Bot.* **10**, 409.
Srivasta L. M. and Paulson R. E. (1968) *Can. J. Bot.* **46**, 1447.
Stiles W. (1935) *Bot. Rev.* **1**, 249.
Stumpf P. K. (1952) *Ann. Rev. Plant Physiol.* **3**, 17.
Stumpf P. K. and Bradbeer C. (1959) *Ann. Rev. Plant Physiol.* **10**, 197.
Sugita M. and Sasaki K. (1982) *Physiol. Plant.* **54**, 41.
Suzuki Y. and Minamikawa T. (1983) *Plant Cell Physiol.* **24**, 1371.
Swain R. R. and Dekker E. E. (1966) *Biochim. Biophys. Acta* **122**, 87.
Swift J.G. and O'Brien T. P. (1972a) *Aust. J. Biol. Sci.* **25**, 9.
Swift J. G. and O'Brien T. P. (1972b) *Aust. J. Biol. Sci.* **25**, 469.
Taylor J., Novellie L. and Liebenberg N. (1985) *J. Exp. Bot.* **36**, 1287.
Theimer R. R. and Rosnitscek I. (1978) *Planta* **139**, 249.
Thomas S. M. and ApRees T. (1972) *Phytochemistry* **11**, 2177.
Tomlinson R. V. and Ballou C. E. (1962) *Biochemistry* **1**, 166.
Trelease R. N. and Doman D. C. (1984) in *Seed Physiology*, vol. 2, p. 201, Academic Press, New York.
Treffry T., Klein S. and Abrahamsen M. (1967) *Austr. J. Biol. Sci.* **20**, 859.
Turner D. H. and Turner J. F. (1960) *Biochem. J.* **74**, 486.
Turner D. H., Blanch E. S., Gibbs M. and Turner S. F. (1965) *Plant Physiol.* **40**, 1146.
Turner J. F. and Turner D. H. (1975) *Ann. Rev. Plant Physiol.* **26**, 159.
Turner J. F. and Copeland L. (1981) *Plant Physiol.* **68**, 1123.
Ueda K. and Tsuji H. (1971) *Protoplasma* **73**, 203.
Varner J. E., Ram Chandra G. and Chrispeels M. J. (1965) *J. Cell. Comp. Physiol.* **66** (Suppl. 1), 55.
Virtanen A. I., Berg A. M. and Kari S. (1953) *Acta Chem. Scand.* **7**, 1423.
Vogel R., Trautschold I. and Werle E. (1968) *Proteinase Inhibitors,* Academic Press, New York.
Waters L. C. and Dure L. S. (1966) *J. Mol. Biol.* **19**, 1.
Weeks D. P. and Marcus A. (1971) *Biochim. Biophys. Acta* **232**, 67, 1.
Wetter L. R. (1957) *J. Am. Oil Chemists* **34**, 66.
Wilson K. A., Rightmire B. R., Chen J. C. and Tan-Wilson A. L. (1986) *Plant Physiol.* **82**, 71.
Wilson S. B. and Bonner W. D. (1971) *Plant Physiol.* **48**, 340.
Yamada M. (1955a) *Sc. Papers Coll. Gen. Ed. Univer. Tokyo* **5**, 149.
Yamada M. (1955b) *Sc. Papers Coll. Gen. Ed. Univer. Tokyo* **5**, 161.
Yamada M. (1957) *Sc. Papers Coll. Gen. Ed. Univer. Tokyo* **7**, 97.

Yamaguchi J., Mori H. and Nishimura M. (1987) *FEBS letters* **213**, 329.
Yamamoto Y. (1955) *J. Biochem. Tokyo* **42**, 763.
Yamamoto Y. and Beevers H. (1960) *Plant Physiol.* **35**, 102.
Yentur S. and Leopold A. C. (1976) *Plant Physiol.* **57**, 274.
Yocum L. E. (1925) *J. Agron. Res.* **31**, 727.
Yu K. S., Mitchell C. A. and Robitaille H. A. (1981) *Plant Physiol.* **68**, 509.

6

Germination Stimulators and Inhibitors

Their Effects and their Possible Regulatory Role

Whenever a seed becomes dormant or ceases to be dormant, something in the seed must change. Although this is an obvious conclusion to draw, very little is known about the nature of the changes which occur. Experimentally it is inconvenient to have to wait till seeds become dormant or emerge from dormancy, as this involves time and does not permit an easy comparison of the different stages of development simultaneously. An alternative approach to the problem is to induce dormancy or to break it, either naturally or by artificial means.

Plant hormones, or growth regulators, are substances known to affect plants and regulate their growth and development. They also play an important role in fruit and seed development. We will use the terms hormones, growth substances and growth regulators interchangeably. There is a great deal of evidence that plant hormones are very important in controlling dormancy and germination (see Chapter 4). Therefore the effects of gibberellic acid (GA), cytokinins, abscisic acid (ABA) and ethylene on dormancy and on metabolism during germination have been studied. In addition attempts have been made to measure the changes in the endogenous level of the different hormones during dormancy and dormancy-breaking treatments. It was hoped that this would throw some light not only on the effect of these substances, but will also elucidate to some extent the metabolic pathways affected by the hormones in bringing about germination, but success has been meagre.

Other substances which can induce or break dormancy, such as coumarin and thiourea have also been used in attempts to identify the events leading to germination. No really comprehensive view of what is happening during germination and dormancy breaking has yet been formulated.

In the following we will first consider the effects of naturally occurring plant growth substances such as gibberellins, cytokinins, ethylene and abscisic acid, and will then briefly refer to other compounds including coumarin and thiourea. The latter are normally exogenously applied compounds and are used as a tool to investigate metabolic changes during dormancy and germination.

I. Effect of Natural Growth Substances

1. Gibberellic Acid

As already mentioned (Chapter 4) gibberellins are able to break dormancy in some seeds. There are some 72 known gibberellins, 61 of which have been identified in higher plants. Many of them have been found in seeds. For example 16 gibberellins have been found in *Phaseolus* species and 20 were identified in *Sechium edule* (Albone *et al.*, 1984). The most active gibberellin in higher plants is probably GA_1, which is known to promote extension growth. When applied externally to dormant seeds the most effective gibberellins are GA_3 and GA_{4+7}. Most of the other gibberellins appear to be inactive. Treatment of cereal seeds with gibberellin induces rapid hydrolysis of starch in the endosperm of cereals, hydrolysis of proteins in the aleurone layer and the appearance in the seed of free phosphate. The data in Fig. 6.1 show some of the changes which are observed in barley seeds in response to exogenously applied GA_3. The observed changes were dependent on the GA_3 concentration.

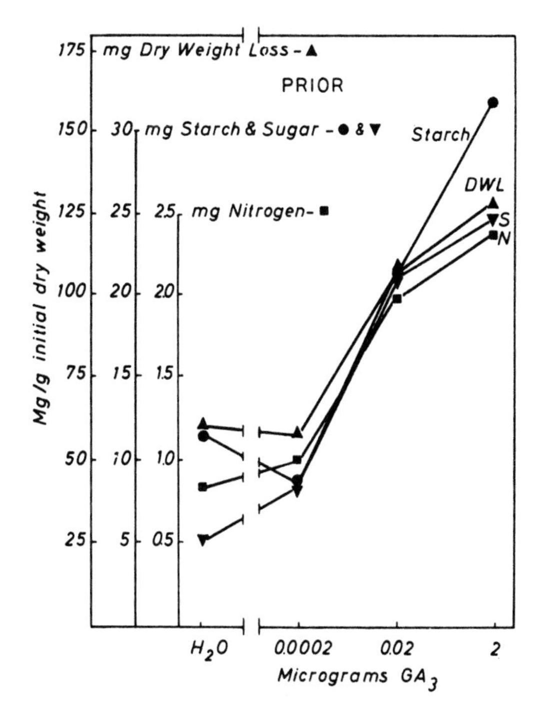

FIG. 6.1. GA_3 concentration – response curve for "Prior" barley treated at 30°C for 22 hr (Paleg, 1962).

Kirsop and Pollock (1958) were the first to show that the hydrolysis of the starch in the endosperm of cereals depended on a hormone-like substance produced by the embryo. The hormone was identified as GA_3, independently by Yomo (1960) and by Paleg (1960a,b).

Most of the metabolic studies with gibberellins have been made using GA_3, which is effective in cereal seeds. The GA produced by barley embryos is probably GA_1, but its identity is still in dispute. In many, if not all, seeds GA_{4+7} is far more effective than GA_3. Since GA_3 cannot be converted to GA_4 or GA_7 one must assume that it is acting at the same site, but less effectively and therefore much higher concentrations are required.

Intensive work has been carried out on the effect of GA_3 on the metabolism of cereal seeds during germination. This work was facilitated by the early observation that the effect of GA_3 on the endosperm could be studied in half seeds of cereals, from which the embryo had been removed. Such half seeds respond to GA_3 applied exogenously in the same way as the entire seed responds to the hormone from the embryo. The starchy endosperm is almost completely dissolved. A rapid hydrolysis of starch occurs and reducing sugars are formed and released into the external medium. This series of events formed the basis for a bioassay of gibberellic acid.

Attention was next focused on the mechanism by which GA_3 brings about its effect. The work of Varner *et al.* (1965) led to the important discovery that gibberellic acid controlled the synthesis of α-amylase in the aleurone layer (Fig. 6.2). The newly formed α-amylase then hydrolyses the starch in the endosperm. The appearance of α-amylase in the aleurone depended on *de novo* synthesis of protein which in turn depended on the synthesis of new mRNA. It could be prevented by inhibitors of DNA-dependent RNA synthesis.

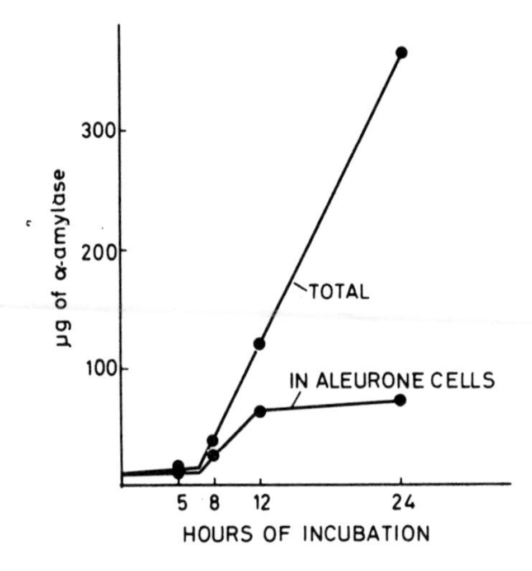

FIG. 6.2. Time course of α-amylase synthesis by 10 aleurone layers incubated with 1 μM GA_3. Enzyme activity was measured in the medium surrounding the aleurone layers and in the supernatant of a 0.2 M NaCl extract of the aleurone layers. The term total refers to the sum of the two activities (Chrispeels and Varner, 1967).

α-Amylase is secreted from the epithelial layer of the scutellum. This was first reported for rice but is probably much more general in nature (Akazawa and Hara-Nishimura, 1985; Okamoto *et al.*, 1980). The secretory step involves glycosylation of a protein produced in the epithelial cells. The release of sucrose from the aleurone layer was also shown to be GA_3 dependent and to require protein synthesis (Table 6.1).

The results of new research, obtained using the recently developed highly sensitive radio-immunoassay for hormones have thrown some doubt on the idea that transport of gibberellin occurs from the embryo to the aleurone (Weiler and Wieczorek, 1981).

TABLE 6.1. *Dependence of release of sucrose from barley aleurone layer on GA_3 (from Chrispeels et al., 1973)*

Treatment	Duration of incubation (hr)	Sucrose released (μg/aleurone layer)
Control	10	120
+ cycloheximide	10	71
+ GA_3	10	251
+ GA_3 + cycloheximide	10	70

cycloheximide 10μg/ml
GA_3 2 μM

These more recent results indicate that in barley GA_4 synthesis may occur directly in the aleurone and control the synthesis of α-amylase in it. This GA synthesis appears to depend on some factor from the embryo. Nevertheless this aleurone localized synthesis of GA has not been proved beyond doubt and the classical concept that GA is released from the embryo and transported, perhaps by diffusion, to the aleurone may still be valid (Weiler and Wieczorek, 1981; Atzorn and Weiler, 1983). Clarification of this controversy requires further, detailed analysis. The situation may differ in different cereals, e.g. barley or rice.

The sensitivity of the aleurone to the effect of the hormones is a developmentally determined step and apparently is induced by seed dehydration during maturation. The induction of sensitivity may perhaps be due to membrane changes (Armstrong *et al.*, 1982).

The role of gibberellins and their effect on α-amylase synthesis has been reviewed by Jacobsen and Chandler (1987). It must be pointed out that, although α-amylase formation is a convenient marker to identify GA effect and has been used to clarify the nature of the effects induced by this hormone, α-amylase formation occurs after germination. It is very doubtful, therefore, whether induction of α-amylase plays an important role in breaking dormancy or in facilitating germination (Jones and Stoddart, 1977). Nevertheless, gibberellins, either endogenous or exogenous, are considered to be necessary factors in inducing germination (Groot and Karssen, 1987), although their exact role is not yet understood.

Much effort has been devoted to link the effect of gibberellins to membrane metabolism. One of the hypotheses suggested that GA_3 was responsible for an increase in the amount of endoplasmic reticulum (Jones, 1969). A careful study has shown that in the aleurone system GA_3 does not increase the ER nor does it markedly change the amount of enzymes in the ER (Jones and Jacobsen, 1978; Chrispeels and Jones, 1980/81). It did, however, change the properties of the ER and especially its buoyant density and slightly changed the time of appearance of some marker enzymes of the ER.

Sing and Paleg (1968a,b) have suggested that phospholipids in cellular membranes serve as receptors for the gibberellins. They studied the interaction between GA_3 and cold treatment in dwarf and tall varieties of wheat, and showed that cold treatment increased the sensitivity to GA_3 in the varieties which contained one of the dwarfing alleles. The increased sensitivity appeared to be due to an increase in the number of receptor sites in the membrane (Sing and Paleg, 1985a,b,c; Sing and Paleg, 1986b,c). Twenty hr at 5°C significantly increased the synthesis of phosphatidyl inositol, phosphatidyl choline and phosphatidyl ethanolamine in cells of the aleurone layer of the dwarf varieties, but not in the tall ones. The treatment with the low temperature could be replaced by treatment with IAA (Sing and Paleg, 1986b). These data suggest that at least some of the effects of GA may be at the level of membranes. It appears therefore that GA_3 may affect the metabolic processes concerned with membrane formation in general (Mayer and Shain, 1974) and it may be involved in polysome formation.

Another phenomenon which has been observed to occur before germination is the weakening of the resistance of the endosperm to rootlet protrusion in gibberellin deficient mutants of tomato (Groot and Karssen, 1987). These mutants were dwarf mutants and only germinated if provided with GA. GA_{4+7} was 1000 times more effective in inducing germination than commercial GA_3. The requirement for GA was absolute. The resistance of the endosperm was measured in the area through which the radicle emerges (Fig. 6.3). Structural changes in the endosperm cells of lettuce facing the tip of the radicle have also been observed by Georghiou *et al.* (1983), although they did not measure the actual change in resistance in their experiments.

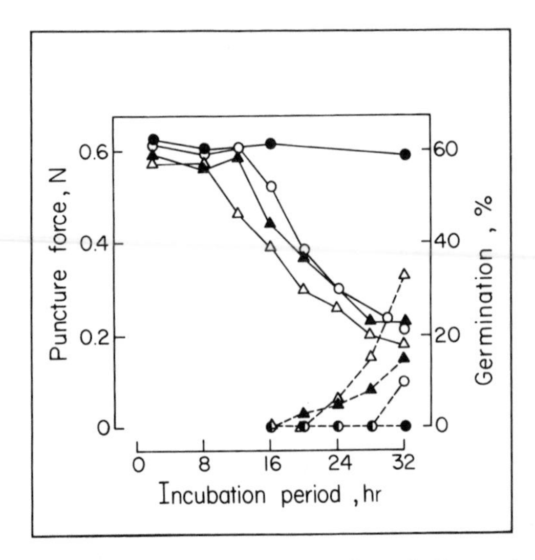

FIG. 6.3. Changes with time of the median force (in Newtons) required to puncture the layers opposing the radicle tip (solid lines) and of germination (broken lines) of wild type (▲,△) and mutant (●,○) tomato seeds incubated in water (closed symbols) or 10μM GA_{4+7} (open symbols) (Groot and Karssen, 1987)

From the above discussion it is clear that there is as yet no clear idea about the role of GA in breaking dormancy and inducing germination. It is possible that the ratio or interrelationship between GA and ABA is a key factor which regulates germination. The effect of GA in inducing the formation of hydrolytic enzymes may be a factor which regulates the mobilization of reserves. Biochemical investigations using GA deficient mutants may provide the much needed insight into the dormancy-breaking mechanisms of gibberellic acid.

2. Cytokinins

The dormancy-breaking action of cytokinins has already been described in Chapter 4. Cytokinins are converted from inactive to active forms by mechanisms which are unknown (Van Staden, 1973).

There is no clear-cut correlation between the level of endogenous cytokinins and dormancy breaking. In embryos of *Acer pseudoplatanus*, from fruits undergoing stratification at 5°C, the level of free cytokinins increased during the first 20 days of stratification, but then decreased again to almost their initial level during the rest of the treatment. Bound cytokinins decreased continuously during the entire period of stratification (Julin-Tegelman and Pinfield, 1982). When the seeds were germinated following stratification the same again occurred, an increase in cytokinins followed by a decrease.

In apple seeds an increase in cytokinins has also been observed in the 3rd to 5th week of stratification, followed by a decrease. But in the case of the apple seeds the increase in cytokinin level coincided with the beginning of the release from dormancy (Kopecky *et al.*, 1975).

A detailed investigation of cytokinin metabolism during the germination of dormant embryos of *Fraxinus americana* has been carried out using ^{14}C-labelled zeatin (Tzou *et al.*, 1973). When $8-{}^{14}$C zeatin was fed to the embryos it was apparently metabolized to 9-zeatin-β-ribonucleotide and riboysyl mono, di- and triphosphates accumulated. Zeatin ribonucleoside was as effective as zeatin in breaking dormancy, but zeatin accumulated to a much lower level than the ribonucleotide. The metabolism of zeatin took place in embryos of both dormant and non-dormant seeds and was not inhibited by ABA. However, ABA did antagonize the dormancy-breaking effect of zeatin.

Metabolism of exogenously supplied cytokinins is not necessarily the same as that of endogenous ones (McGaw *et al.*, 1983).

Lewak and Bryzek (1974) compared the effects of kinetin and benzyl adenine on the photosensitivity of apple embryos and on the induction of acid phosphatase during stratification. The difference between the effect of these two compounds led them to conclude that there is more than one cytokinin in the seeds and the equilibrium between them differs at different developmental stages. It should be pointed out that neither of these compounds is a natural hormone.

The idea that regulation may be due to more than one cytokinin is supported by the work of Hewett and Wareing (1973) who separated five cytokinins from buds of *Populus*, without identifying them. The maximal cytokinin content coincided with the beginning of sprouting.

It is well established that cytokinins become incorporated into RNA, but the significance of this in hormone action remains to be established.

It is known that cytokinins are involved in embryo growth and in expansion of the cotyledons during stimulation of germination in lettuce (Ikuma and Thimann, 1963). Treatment of *Acer pseudoplatanus* seeds with cytokinin resulted in increased elongation of the radicle (Pinfield and Stobart, 1972). Cytokinin secretion by the embryonic axis and its controlling effect of events in the cotyledons seems to be a general phenomenon in dicotyledonous seeds and has been shown in a number of species (Ilan and Gepstain, 1980/81). The response of the cotyledons is probably due to the endogenous cytokinins present, or to those transported to them from the embryonic axis, or applied exogenously.

The work of Ashton *et al.* shows very clearly that the embryonic axis of squash seeds secretes cytokinins, which induce formation of isocitric lyase and proteolytic enzymes in the cotyledons (Penner and Ashton, 1967). Cytokinins released by the endosperm appear to reduce the ion release from the aleurone in wheat in some way and to affect triglyceride metabolism in the aleurone layer (Eastwood and Laidman, 1971). Thus cytokinins may have a role similar to GA_3. A number of possible modes of action of cytokinins have been suggested. On the basis of the presence of cytokinins in tRNA it has been suggested that they act by controlling protein metabolism at the translational level, but the evidence is far from convincing (Thomas, 1977). Other suggestions invoke an effect on cyclic AMP or on membrane permeability (Thomas, 1977). Unfortunately, none of these proposals answers the questions — what are the mechanisms by which cytokinins act and what are the sites of their action? All reviews on the mode of action of the cytokinins (Kende, 1971; Hall, 1973; Thomas, 1977; Van Staden, Davey and Brown, 1982) show that most of the crucial questions remain unanswered. It is fairly obvious that cytokinins interact with other growth regulators such as gibberellins and ethylene. Unfortunately, this does not really help in understanding how it is acting.

At present there is no reason to assume that at the metabolic level cytokinins in germination act differently than in other developmental stages in the plant. However, this assumption remains to be proven experimentally.

The observed lack of correlation between the effect of cytokinin and its internal concentration could perhaps be explained by the hypothesis of Trewavas (1982), who proposed that it is not the level or concentration of a hormone which determines its activity, but the sensitivity to the hormone. If the action of the hormone is due to hormone – receptor complexes then the number of receptor sites rather than the hormone concentration will determine physiological response. Although this supposition is attractive, in the absence of any proof of the existence of receptors or of changes in their concentration within the cell, this idea remains speculative.

3. *Abscisic Acid*

The function of abscisic acid in imposing dormancy on many seeds has already been described (see Chapter 4). ABA can also prevent precocious germination. For example soybean embryos cannot germinate until the midstage of their development, about 20 days post-anthesis, when their ABA content has decreased markedly. If the ABA content was reduced artificially, by washing, the embryos are able to germinate. The percentage germination was correlated with the length of the washing period, which in turn was related to the ABA content (Ackerson, 1984a,b). The ABA content of two cultivars of wheat, which differed in their ability to germinate, increased at about the same rate until the embryos reached their maximal fresh weight, about 40 days post-anthesis. From this stage on, the ABA content of the embryos capable of germinating was lower than that of the non-germinating, sprouting-resistant variety (Fig. 6.4) (Walker-Simmons, 1987).

The ability to synthesize ABA, as indicated by the ability to transform mevalonic acid and isopentosyl pyrophosphate to ABA, was demonstrated in immature seeds

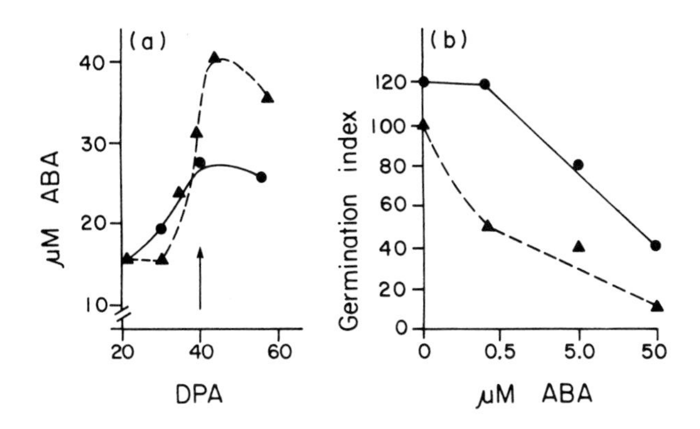

FIG. 6.4. (a) ABA accumulation in embryos of two varieties of wheat differing in their ability for precocious germination. (b) Germination of embryos, 40 days after anthesis, of the two varieties as affected by external concentrations of ABA. Germination index is an arbitrary expression taking into account the rate of germination, the length of the germination experiment and the final percentage of the embryos germinating. (●) sprouting non-resistant variety; (▲) sprouting resistant variety (compiled from data of Walker-Simmons, 1987).

of wheat (Milborrow and Nodle, 1970) and for barley embryos (Cowan and Railton, 1987).

The degree of dormancy also correlated well with the ABA content in *Fraxinus americana* (Sondheimer *et al.*, 1968) and in various other species (Milborrow, 1974). ABA is not evenly distributed in the seed. In dormant apple seeds the ABA content (ng/g dry wt) in the endosperm membrane and in the seed coat was approximately three times as high as in the embryonic axis or the cotyledons (Subbaiah and Powell, 1987). The differences in rate of accumulation of ABA in the various parts of the seed may account for the fact that when ABA is assayed in entire seeds, as a function of time from pollination, no linear relationship is observed. The curve showing the change in total ABA content of pea seeds is biphasic. The first phase coincides with the development of the seed coat and the endosperm; a maximum of ABA is attained when the development of the seed coat reaches its maximum. The second phase coincides with growth of the cotyledons and the embryonic axis. ABA accumulation in these organs is a function of growth (Wang *et al.*, 1987). These seeds were also shown to be able to convert ^{2}H-1′-deoxy ABA and ^{3}H-1′-deoxy ABA to ABA, as determined by the incorporation of the label into metabolites, including ABA, which appeared mainly in the seed coat. Although 1′-deoxy ABA has not been shown to be a metabolite of ABA in plant tissues, its conversion to ABA in pea seeds is indicative of their ability to synthesize and metabolize ABA.

It is characteristic of exogenously applied abscisic acid that it must be continuously present. As soon as ABA is removed germination can take place (Milborow, 1974). It seems likely that this can be accounted for, at least in part, by the inactivation and metabolism of the residual ABA in the seed. Direct evidence for inactivation of exogenously applied ABA has been provided in the case of axes of *Phaseolus vulgaris* and *Fraxinus americana*. The metabolism of ABA is complex and as in the case of other growth substances includes conjugation in various forms as well as degradation (Milborrow, 1983; Walton, 1987) (see also Chapter 4). Some of the degradation products are dehydrophaseic acid and hydroxymethyl ABA.

When following the fate of labelled ABA during metabolism care must be taken to distinguish between those products which occur naturally and those which do not. This is important because in most experiments the ^{14}C – ABA used is a racemic mixture and the two enantiomers (S and R) are metabolized at different rates and probably also by different routes (Walton, 1987).

ABA interacts with other growth regulators, particularly with cytokinins and with gibberellic acid (Ketring, 1973). Zeatin can partially reverse the effect of ABA, but does not do so by changing the rate of ABA inactivation. Although ABA alone inhibited cotyledon expansion and radicle elongation it did not reverse the stimulatory effect of cytokinins on radicle growth (Pinfield and Stobart, 1972).

The effects of ABA on metabolism are varied. There is at least some evidence that ABA interferes in nucleic acid metabolism. Using techniques of autoradiography it has been shown that ABA inhibits the incorporation of ^{3}H-uridine and ^{3}H-thymidine into the embryos of *Fraxinus excelsior* (Villiers, 1968). Protein synthesis itself was not directly inhibited by ABA. Direct studies of DNA and RNA metabolism in pea embryos also point to an effect of ABA at the level of the nucleic acids and especially RNA (Khan and Heit, 1969). This work showed that ABA depressed incorporation of ^{32}P into various nucleic acid fractions of the embryo.

Direct evidence for ABA effects on RNA and protein metabolism has been provided for excised cotton embryos. ABA prevented translation of pre-existing mRNA which coded for known enzymes (Table 6.2). Again the effect of ABA was dependent on its continued presence. The inhibition caused by ABA was prevented

TABLE 6.2. *Effect of ABA on formation of enzyme activity in immature excised embryos of cotton germinated in petri dishes (from data of Ihle and Dure, 1972) (Embryo age is expressed in mg. The younger the excised embryo the smaller its weight)*

Embryo age	Days germinated	Enzyme units/cotyledon pair Carboxypeptidase	Isocitratase
110 mg	3	12·5	20·5
+ Actinomycin D	3	12·5	21·1
+ ABA	3	0	0
+ Actinomycin + ABA	3	12·0	22
90 mg	4	7·0	12·6
+ Actinomycin D	4	7·1	13·6
+ ABA	4	0	0
+ ABA + Actinomycin D	4	7·5	12·5
+ GA_3	4	7·2	12·6
+ ABA + GA_3	4	0	0

(ABA concentration 1×10^{-6}M. Actinomycin D concentration 20 mg/ml. GA_3 concentration 5×10^{-5} M).

when Actinomycin D, a known inhibitor of RNA synthesis, was added simultaneously (Table 6.2). Actinomycin D alone did not affect the synthesis of the enzymes studied. It appeared that the inhibitory activity of ABA on translation required the continued production of some RNA, which appeared to be non-ribosomal RNA (Ihle and Dure, 1972). The action of ABA is exceedingly complex.

In the cotton embryos, GA_3 did not reverse the inhibition of enzyme synthesis induced by ABA (Table 6.2). However, in peas the effects of ABA were reversed both by kinetin and by GA_3 (Khan and Heit, 1969).

The effect of ABA in reversing the GA_3 induced formation of hydrolytic enzymes in cereal seeds can also be accounted for by its effect on nucleic acid metabolism. However, the effects observed were not simple and there was no simple stoichometry between the effect of ABA and GA_3 (Jacobsen, 1973). Moreover, GA_3 does not reverse the ABA-induced dormancy in all cases.

A number of observations relate to the interaction between GA and ABA. ABA blocks accumulation of mRNA responsible for α-amylase formation (Higgins *et al.*, 1977). The degradation and activity of mRNA coding for α-amylase is also changed by the presence of ABA (Jacobsen and Higgins, 1978). It is now clear that both GA and ABA regulate the transcription of genes coding for mRNA and rRNA, which are involved in α-amylase formation (Jacobsen and Beachy, 1985). The translation of this mRNA is, as in the case of cotton, also blocked by ABA (Ho and Varner, 1976; Ho, 1979). When barley aleurone was treated with ABA, synthesis of most of the GA-induced proteins was blocked, but formation of many new polypeptides was induced. An inhibitor of α-amylase was also found as a result of ABA treatment (Mundy, 1984). ABA also induced the formation of an enzyme which converts ABA to phaseic acid (Ho, 1983; Lin and Ho, 1986; Quatrano, 1987). Even in this relatively well-studied system the mechanism of action of ABA is complex and inadequately understood.

Ihle and Dure (1972) showed that excised cotton embryos germinate readily. They make the interesting suggestion that, in cotton plants, ABA is transported from the mother plant to the developing embryo and its function is to stop formation of hydrolytic enzymes at some stage of embryogenesis, until the seed is cut off from its supply of ABA. In their view ABA prevents vivipary of the cotton seed in the boll. This would be in accord with the general finding that the ABA content of fruits is much higher than that of seeds (Milborow, 1974) and that it prevents precocious germination.

It seems possible that ABA, dehydration and dormancy are closely interlinked. The idea that ABA is responsible for switching on or off the genes regulating seed dormancy has been suggested repeatedly. It is supported by the observation that dehydration has been shown to result in ABA accumulation.

Several attempts have been made to demonstrate direct effects of ABA on embryogenesis of seeds. Apparently ABA can modulate the formation of the subunit of β-conclycinin in soybean cotyledons and that it can either stimulate or inhibit embryo growth and formation of other proteins depending on the developmental stage (Bray and Beach, 1985; Ackerson, 1984a,b). However, this work does not establish that ABA has a direct effect on a given gene, or gene set, involved in embryogenesis. The study of the early methionine-labelled polypeptide message (Em) in immature embryos of wheat showed clearly that ABA modulates the level of Em expression.

It was also shown that this Em expression was not seen in germinating embryos, but that embryo specificity of Em was decided by some other signal than ABA (Williamson *et al.*, 1985). This critical signal might be the dehydration system. Recently it has been shown that ABA changes the level of specific mRNAs in developing embryos. This is supported by the finding that two embryo-specific RNA sequences, modulated by ABA, were identified (Quatrano, 1986; Williamson and Quatrano, 1988).

Another specific effect of ABA, which appears to demonstrate its regulatory role, was shown by Yomo and Varner (1973) in excised pea cotyledons. The formation of protease activity was repressed by ABA. This repression of activity was due to repression of enzyme formation which in turn was caused by the accumulation of amino acids. The repression was not removed by incubation with GA_3. In fact GA_3 in peas had no effect on protease activity. In contrast α- and β-amylase activity in the excised cotyledons increased much more rapidly than in the intact seed and this was prevented by ABA. The concentration of ABA which inhibited amylase formation failed to inhibit either oxygen uptake or ^{14}C-leucine incorporation (Yomo and Varner, 1973). Thus, as in cotton, the effect of ABA was specific and it appears that ABA indeed has some regulatory role.

The inhibition of the development of some of the enzyme systems described above does not appear to be sufficient to explain the germinating inhibiting effect of ABA. Apparently the synthesis of some key proteins must be suppressed. It is not unlikely that such proteins are membrane-linked.

Ion movement has been shown to respond to ABA in germinating radish seeds. Proton extrusion and potassium ion (K^+) uptake were both inhibited and membrane potentials were changed by ABA (Ballarin-Denti and Cocucci, 1979) (Fig. 6.5).

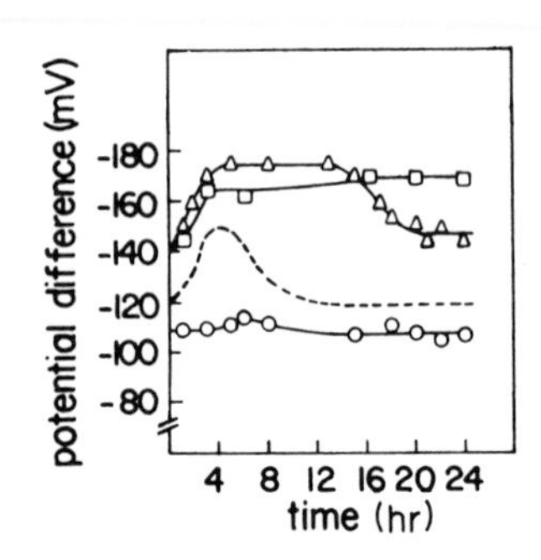

FIG. 6.5. Effect of fusicoccin (FC), ABA and FC + ABA on the potential difference levels during the early phases of germination. The PDs were measured in the cortical cells of the embryonic axes of radish seeds. Control ----- ; FC,10^{-5} M △——△ ; ABA 8 × 10^{-5} M ○——○ ; FC + ABA □——□ (Ballarin-Denti and Cocucci, 1979).

In the control seeds a transient rise in the potential difference of cortical cells in the embryonic axis of radish was observed. This was prevented by ABA quite early on in germination. Such effects of ABA suggest that its effect is due to inhibition of cell expansion. Inhibition of water uptake rather than complex effects on protein and nucleic acid metabolism have also been proposed by Schopfer *et al.* (1979) as the mechanism of germination inhibition. They have provided some evidence that ABA lowers the ability of the embryo to take up water under osmotic stress. The proposed mechanism was that ABA directly affected the extensibility coefficient of the cell wall (Schopfer and Plachy, 1985).

It is interesting to note that the fungal metabolite fusicoccin, which mimics the action of GA, also reverses the effect of ABA. Fusicoccin induces hydrogen ion (H^+) extrusion and potassium ion (K^+) uptake and these are among the earliest results of its physiological action (Marre, 1979). These observations would suggest that membrane potential and ion uptake processes play a role in germinating seeds as they do in many other physiological processes (Van Steveninck and Van Steveninck, 1983).

From the above discussion it can be seen that, at least in some cases, ABA affects transcription of nucleic acid. However, it remains to be shown that this action of ABA or its effect on cell wall extensibility or membrane permeability is in fact the mechanism by which germination is inhibited. It seems even more doubtful that dormancy induction, as opposed to germination inhibition, is in fact due to ABA alone. A clear-cut causal relationship still remains to be established, a point which has been considered in detail by Black (1983). It seems fairly clear that the effect of ABA in blocking α-amylase production is not the mechanism of germination inhibition. Probably other systems exist which are relevant to the still problematic mechanism of dormancy induction.

In the last few years a great deal of new information and new data on the effects of ABA on germination have accumulated. Unfortunately these have not yet produced any coherent view on the way it acts. ABA probably has more than one target for its action and its effect on germination still remains unexplained.

4. Ethylene

Considerable progress has been made in understanding ethylene synthesis, metabolism and action in plant tissue. However, the knowledge on the role of ethylene in germination of seeds remains scanty (Lieberman, 1979). Germination of some species of seeds can be stimulated by externally applied ethylene (Ketring, 1977), but this effect is apparently not related to the breaking of dormancy. Ethylene appears to affect events occurring after dormancy has been partially or fully released for example by chilling (Kepeczynski and Karssen, 1985). Germinating seeds are able to produce ethylene, but the role of this endogenous supply is still not clear. Ethylene production has been measured in lettuce seeds in relation to thermoinhibition of germination and the target for ethylene action in these seeds was located in the hypocotyl (Abeles, 1986). Direct evidence for a requirement for ethylene for the germination of *Amaranthus* has been brought by Kepeczynski and Karssen (1985).

The pathway for the biosynthesis of ethylene in seeds is the same as in other plant tissues (Yang and Hoffman, 1984).

The role of auxin (IAA) in ethylene formation in seeds has not been studied. IAA is present in bound form in seeds in large amounts and can be liberated from these bound forms during germination, but the physiological significance of this is not known (Reinecke and Bandurski, 1987).

The mode of action of ethylene in plant tissues is still unclear. It has been suggested that ethylene binding sites exist in plant cells. This suggestion was based on the demonstration of ethylene binding capacity of extracts of mungbean sprouts (Sisler, 1980, 1984) and of developing *Phaseolus vulgaris* cotyledons (Jerie *et al.*, 1979). The binding was to a protein which was membranal in nature. Whether these binding sites are in fact the actual receptors which are causally related to ethylene action is still undetermined. It is also unclear whether ethylene must be metabolized in order that it can act as a hormone, although the fact that it is metabolized is now well established (Beyer, 1985; Hall, 1986; Sanders *et al.*, 1986). In other systems, such as fruit and storage tissue, it has been shown that one of the effects of ethylene is gene activation and formation of specific mRNA. Other effects of ethylene probably also exist, for example on membrane integrity.

In addition, numerous effects of ethylene on metabolism have been reported, such as changes in the level of hydrolytic enzymes, endogenous auxins and perhaps polyribosome formation (Ketring, 1977).

Exposure of pea cotyledons to fairly high concentrations (130 μl/l) of ethylene induced a five-fold increase in the cyanide insensitive, SHAM sensitive, pathways in mitochondria isolated from these cotyledons (Duncan and Spencer, 1987). Whether this is related to the effect of ethylene on germination remains to be determined. Most of these effects are probably secondary in nature.

From the above discussion it is clear that as far as germination is concerned, the mechanism of action of ethylene is not clear and insufficient information is available about its site of action or how it acts.

5. The regulatory role of hormones

As outlined in this chapter plant hormones play an important role in regulation of germination and dormancy. Most of the substances included in the term plant hormones either stimulate or inhibit germination. Interactions exist between their effects and those of environmental factors such as light or temperature. Externally applied hormones affect various metabolic pathways, as well as developmental stages such as regulation of precocious germination. It is generally assumed that endogenous hormones and those applied exogenously act in a similar fashion, although this assumption is by no means always justified. The effect of gibberellins and ABA on induction of protein synthesis has been studied extensively. It is now certain that the induction of α-amylase synthesis and that of other hydrolytic enzymes in cereals seeds is dependent on gibberellins. As mentioned earlier this effect of GA cannot be the key process which induces germination or breaks dormancy.

The post-germinative induction of enzyme synthesis has been used as a model for studying plant hormone action. The primary result of GA action in cereal seeds is the increase in the level of mRNA in the aleurone layer, i.e. the induction of transcription. The possibility that the effect of GA is not directly on transcription itself, but on the stability of the transcripts, seems to be rather unlikely, but cannot be completely discounted. α-Amylase in barley exists as different isozyme forms and expression of the genes responsible for their formation is apparently differentially regulated (Chandler *et al.*, 1987). Other observations are also suggestive of effects of GA additional to those on the transcription of mRNA. Pollard (1969) studied the sequential appearance of various enzymes following GA treatment. Enzymes such as β-1, 3-glucanase, phospho-monoesterase, ATPase and others became active before the onset of α-amylase synthesis. The appearance of some of them was not dependent on *de novo* protein synthesis and enzyme activation was probably involved. In these cases the effect of GA_3 could be mimicked by cyclic AMP (Pollard, 1971; Newton and Brown, 1986). GA can also induce the activation of the transfer of lipase activity from protein bodies to lipid bodies (Fernandes and Staehelin, 1987).

It appears that GA_3 has two distinct target sites, one in the embryo and the other in the aleurone layer of the endosperm. The latter is not directly concerned with germination while the former may be. Chen and Park (1973, 1977) regard the embryo as the direct site of gibberellin action and showed that the concentration of GA which induces α-amylase formation failed to break dormancy in *Avena fatua*. Although the nature of its action is unclear, GA might be triggering the formation of some proteins, which can induce cell division and growth, which lead to germination. Briggs (1973) suggested that the embryo is the source of gibberellic acid, which is transported from it to other parts of the seed. This is unlikely to be the event which triggers germination. It may, however, affect the mobilization of reserve materials and their transfer to the embryo, which is acting as a metabolic sink.

The effect of GA can be reversed in the cereals by the application of ABA. However, ABA is not a competitive inhibitor of GA. ABA apparently blocks the transcription of some genes, which are induced by GA. At the same time ABA induces the transcription of other genes. ABA has been reported to induce synthesis of proteins typical for maturing seeds and blocks the synthesis of proteins characteristic of germinating seeds and young seedlings (Quatrano, 1987b). The interaction of the two hormones GA and ABA and the right balance between them may play an important part in the control of dormancy and germination. Other hormones such as cytokinins and even IAA and ethylene may be part of the hormonal balance which regulates the sequence of events occurring first in the developing seeds and then during its maturation, dormancy and germination.

A great deal of information exists on the effect of exogenously applied compounds, but knowledge on the synthesis or formation of growth regulators or on the site of their action is lacking in most cases. It is clear from most of the work on germination that the levels of stimulatory and inhibitory growth regulators are changing both before and during germination. Moreover, although it is now clear that the growth regulator content responds to external conditions, such as stratification, almost nothing is known about how such changes are brought about. Are the growth regulators present and activated or deactivated, are they present in bound form or must each and everyone be synthesized? From studies of the effect of exogenously supplied hormones it has become clear that most of them can bind to proteins. Putative binding proteins for auxins, gibberellins, ethylene, ABA and cytokinins have been described. Nevertheless there is still no clear evidence that these binding sites are involved in mediating the specific effects of some of the substances we have considered. There is some indication that the binding proteins are membranal in nature. These might perhaps be the receptors to which the hormones bind, and as a result of the binding the activity or transport of the hormones may be regulated. Such interpretations are further complicated by the fact that there may be more than one binding site for each compound.

The question of the nature of the receptor site is a crucial one. The germination inhibitors and stimulators are all relatively simple molecules, particularly compared with the steroids active in animal tissues. The molecules contain relatively little intrinsic information. In order to exert the wide spectrum of effects which have been noted in germination, they presumably must interact with and modify other molecules. It is not necessary to ascribe all effects of hormones to changes in gene expression. Changes in membrane composition or properties and enzyme activation or provision of energy may also be modified by growth regulators and these must involve the same basic interaction with suitable receptor sites.

Two somewhat divergent approaches exist in the interpretation of the effect of plant hormones. One suggests that cells may, at any given time, be competent or non-competent to respond to a given growth regulator, due to the switching on or off of the relevant parts of its genome. The other approach emphasizes the concept of changing sensitivity. The sensitivity of a tissue to hormones may change during development and especially as a result of external signals such as temperature or light. The changes in sensitivity are envisaged as being due to changes in the number of receptor sites for a hormone on a cellular membrane.

In recent years, the techniques of immuno-assay and immuno-localization of plant hormones, and methods of preparing monoclonal antibodies have been greatly improved. As a result GA and ABA can be localized and quantitized much more reliably than was previously possible. These improvements will perhaps solve some of the problems raised above.

The whole problem of the synthesis, mode and site of action of plant hormones is by no means resolved, but the relevance of these problems to the process of germination is obvious.

II. The Effects of Coumarin and Thiourea

The description in this chapter of the effects of thiourea and coumarin is more detailed than is warranted by their significance in controlling germination. However, since most of the work was carried out with lettuce seeds, it also serves as the background for the analysis of the metabolism of one seed species during germination.

Generally speaking, seeds were treated with coumarin to reduce germination from the 50% level, in the dark, to zero and their behaviour compared with other seeds stimulated by thiourea to give 100% germination. By this means the correlation between germination percentage and various metabolic events was studied. The effect of these two compounds will therefore be considered jointly.

1. Effect on Storage Materials

Dry lettuce seeds contain a very large amount of fat and a very small amount of free fatty acids. After about 24 hr, as normal germination proceeds, the lipid content of the seeds begins to fall. In the presence of coumarin this fall is prevented entirely, while in the presence of thiourea it is delayed for almost 24 hr (Fig. 6.6a). Free fatty acids, on the other hand, rise continuously in the seeds treated with thiourea, while in untreated seeds the free fatty acid content first rises and then falls again (Fig. 6.6b).

In seeds treated with coumarin, germination was inhibited and accumulation of

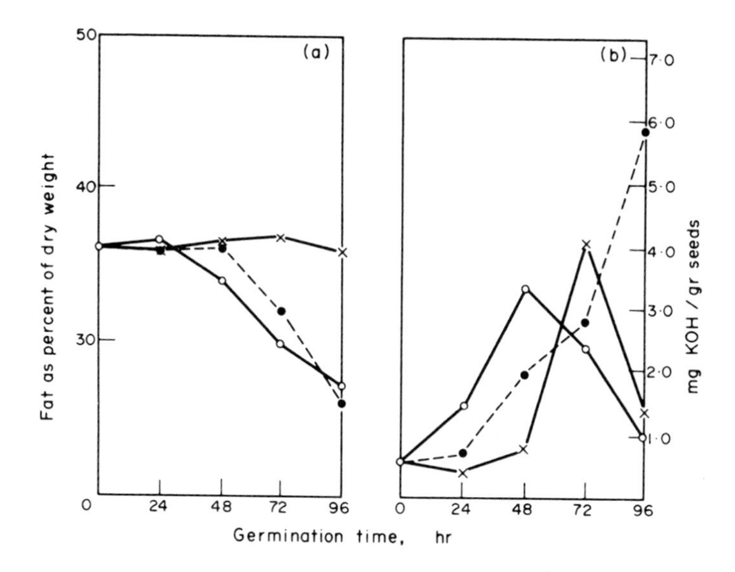

FIG. 6.6. Effect of coumarin and thiourea on fat metabolism of lettuce seeds during germination (Poljakoff-Mayber and Mayer, 1955). (a) Changes in total lipids. (b) Changes in free fatty acids. ○——○ Seeds germinated in water. ×——× Seeds germinated in coumarin (100 ppm). ●----● Seeds germinated in thiourea (1250 ppm).

free fatty acids delayed. Eventually it rose and then fell again. At least two lipases are apparently concerned with lipid metabolism, a neutral and an acid one. The acid lipase in the control seeds begins to rise only when germination is well under way, and reaches about 50%. This acid lipase is totally inhibited by thiourea, although germination is stimulated. The neutral lipase increased markedly in the two treatments which allow germination to take place. Coumarin prevents the rise in activity of both lipases (Fig. 6.7). However, *in vitro* it inhibits only the neutral lipase. It appears that storage lipid metabolism does not have any primary role in the breaking of dormancy. When germination does occur, the growing embryo acts as a sink for the products of lipid metabolism and lipid metabolism is triggered in some way.

The changes in lipid metabolism should be considered together with changes in sugars, into which the lipids may be converted. In Fig. 6.8 some of these changes are illustrated. It can be seen that during germination, in water or in thiourea, the metabolism of glucose and sucrose in the seeds is essentially the same for the first 72 hr. Sucrose content first drops during 48 hr and then rises again and glucose begins to increase after 24 hr. In seeds germinated in coumarin there is no glucose formation although sucrose is broken down.

Nitrogen metabolism in germinating lettuce seeds is characterized by an increase in the amount of soluble nitrogenous compounds, the total amount of nitrogen remaining constant for the first 3 days. Between 48 and 72 hr of germination soluble nitrogen increased in parallel with seedling growth. When germination was prevented in some way, for example by treatment with coumarin, the rise in soluble nitrogen was prevented (Table 6.3).

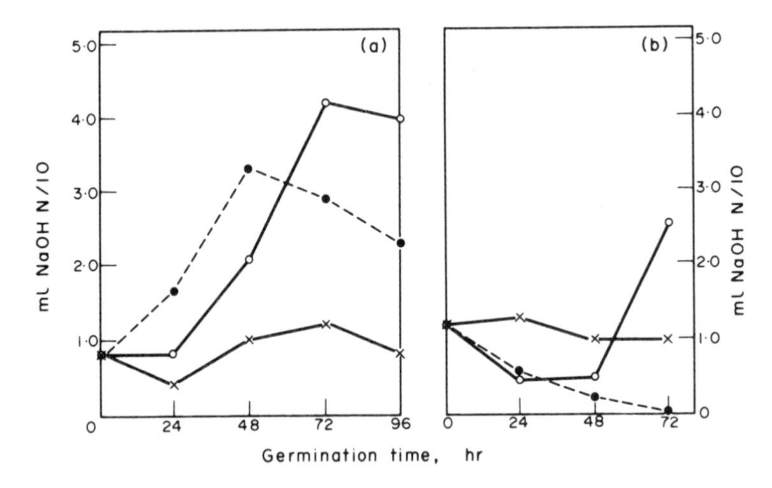

FIG. 6.7. Effect of coumarin and thiourea on lipase activity of lettuce seeds during germination (Rimon, 1957). Results as titer of alkali equivalent to fatty acids liberated by lipase. (a) Neutral lipase. (b) Acid lipase. (Symbols as in Fig. 6.6).

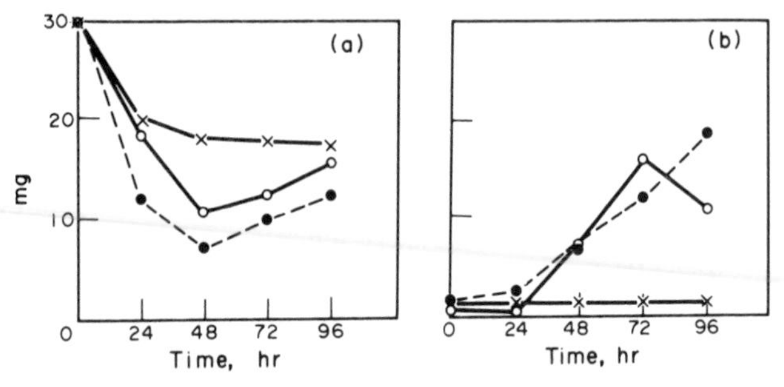

FIG. 6.8. Effect of coumarin and thiourea on sugar metabolism in lettuce seeds during germination (Poljakoff-Mayber, 1952). Results as mg equivalents of glucose. (a) Sucrose. (b) Reducing sugars. (Symbols as in Fig. 6.6).

TABLE 6.3. *The effect of coumarin on changes in soluble nitrogen during germination of lettuce seeds. Soluble N given as percentage of initial weight of seeds (Klein, 1955)*

Time of germination (hr)	Seeds germinated			
	In water		In coumarin	
	% germ.	Soluble N	% germ.	Soluble N
0		0·260		0·260
24	17	0·228	0	0·264
48	46	0·365	6·5	0·272
72	50	0·556	11·5	0·273

Storage proteins are apparently broken down during normal germination and growth, but this breakdown is prevented if germination is inhibited by coumarin. Coumarin also inhibits a proteinase present in the seeds (Poljakoff-Mayber, 1953). This proteinase has a pH optimum of 6.8 and is present in the dry seeds. It appears to be responsible for the breakdown of an endogenous trypsin inhibitor. Removal of this inhibitor seems to be involved in permitting the formation of a trypsin-like enzyme, required for breakdown of storage proteins (Shain and Mayer, 1965).

Such changes as have been observed in the storage materials all seem to be directly initiated by the germination process. In so far as coumarin depressed such changes, and light or thiourea stimulated them, the changes appear to be a direct result of the effect of the treatments during germination and not their cause. In no case did germination inhibitors or stimulators appear to affect the metabolism of such compounds as ascorbic acid or riboflavin except indirectly, through their effect on germination.

2. *Effect on Respiration*

It is very difficult to compare the respiratory activity of seeds treated with germination inhibitors or stimulators with those of the controls. Coumarin for example may inhibit germination completely but the comparison with the controls will be between seeds whose respiration rises because they germinated and seeds which do not germinate at all. Klein (1955) tried to overcome this difficulty by using light-sensitive lettuce seeds. These germinate relatively little in the dark. When treated with coumarin they do not germinate at all. However, by choosing a suitable light stimulus, the effect of coumarin can be overcome and the germination restored to almost the same level as in the dark. Under these conditions the oxygen uptake of the coumarin-treated seeds is consistently higher than that of the water controls (Fig. 6.9).

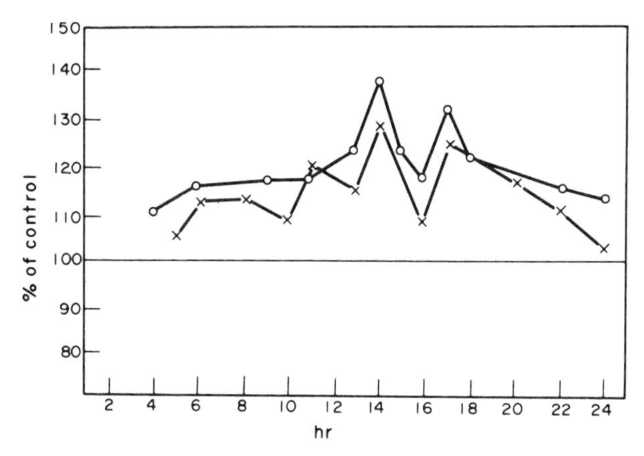

FIG. 6.9. Effect of coumarin on respiration of germinating lettuce seeds (Klein, 1955). The seeds of the light-sensitive variety Grand Rapids were germinated in 75 ppm coumarin and given a light stimulus two hours after the beginning of imbibition. Germination was 7%. Results as percentage of water controls. o——o Carbon dioxide output. x——x Oxygen uptake.

The carbon dioxide behaved similarly. However, it is not clear in this type of experiment whether the rise in respiration was due solely to the coumarin treatment or whether the light, which increased germination, also affected respiration. Similar results have been obtained by Ishikawa (1958) for pea and *Setaria* seeds. He also noted that coumarin at certain concentrations raises the oxygen uptake of the seeds, as did 2, 4-dinitrophenol (DNP). Coumarin may act by uncoupling respiration from ATP formation during germination. Coumarin also increases the oxygen uptake of mitochondria isolated from lettuce seeds (Poljakoff-Mayber, 1955). More direct evidence for the action of coumarin on oxidative phosphorylation was provided by Ulitzur and Poljakoff-Mayber (1963). *In vitro*, coumarin inhibited oxidative phosphorylation and lowered the P/O ratio. Some of the more general observations on the effect of coumarin on germination and growth (Mayer and Poljakoff-Mayber, 1961) could be explained on the basis of an uncoupling action. However, in barley seeds no evidence has been found that coumarin acts as an uncoupler (Van Sumere *et al.*, 1972).

A significant difference between lettuce seeds and barley is in their ability to metabolize coumarin. In lettuce seeds coumarin is rapidly broken down to a number of metabolic products, which have not been fully identified (Sivan *et al.*, 1965), while no evidence for metabolism of coumarin was found in barley (Van Sumere *et al.*, 1972). In lettuce seeds, the inhibition of respiration by coumarin is evident in the third stage (see Chapter 5), when the increase in oxygen uptake is closely associated with seedling growth. No effect of coumarin is observed during the first and second stages, which involve the initial rise in oxygen uptake, during imbibition, and plateau respectively.

Thiourea had virtually no effect on either QO_2, or QCO_2 of lettuce seeds, and consequently the RQ did not change appreciably (Poljakoff-Mayber and Evenari, 1958). Various attempts to study the effect of coumarin on enzymes involved in respiration did not result in any major clarification of its mechanism of action. Some effects on phosphate release and ATP levels were observed (Mayer *et al.*, 1957; Mayer *et al.*, 1966; Gesundheit and Poljakoff-Mayber, 1962; Mayer, 1958). Thiourea had some effect on the appearance of activity of oxidative enzymes such as stimulation of peroxidase and inhibition of catalase, and it induced a much earlier appearance of enzymes of the tricarboxylic acid cycle (Poljakoff-Mayber and Evenari, 1958; Poljakoff-Mayber, 1953, 1956). None of these effects, nor those of coumarin and thiourea on phenolase (Mayer, 1967), can be directly related to germination promotion or inhibition.

3. Other Effects

The effect of coumarin has been investigated in other tissues and some of the results may be relevant to germination. Coumarin has been shown to induce swelling in isolated mesophyll cells (Harada *et al.*, 1972) and to inhibit cellulose synthesis specifically (Hara *et al.*, 1973). The incorporation of a radioactive label into cellulose fractions of the cell walls of mungbean seedlings was inhibited 70% by coumarin at a concentration of 100 ppm. Cellulose synthesis in the alga *Prototheca* was also inhibited by coumarin (Hopp *et al.*, 1978). The inhibition of cellulose synthesis by coumarin seems to be rather a general phenomenon; it also occurs in cotton seed fibres and *Acetobacter xylinum* (Delmer, 1983). The precise mechanism of the inhibition is still unclear, although it appears to involve some intermediate in cellulose formation, as well as some transfer reaction. Whether the effect of coumarin on germination is mediated through the same processes seems doubtful and there is no evidence about a role of cellulose synthesis early on in germination. Perhaps there are other essential transfer reactions which are also blocked by coumarin. Two fairly recent reports show that coumarin at low concentrations can enhance RNA, DNA and protein synthesis in embryonic axes of cotton seeds, while at higher concentrations all of these processes are inhibited (Khafagy and Mousa, 1982). A correlation between inhibition of germination, inhibition of radicle growth and PAL activity has been observed in seeds of lettuce (Kudakasseril and Minocha, 1984). In both cases cause and effect cannot be clearly distinguished.

A number of attempts have been made to clarify the mechanism of action of coumarin by studying structure – activity relationships. The purpose of these studies was to determine the molecular configuration of coumarin required for its function. However, these studies have been disappointing, all the authors have reached the same conclusion — that no definite answer is provided by these studies (Mayer and Evenari, 1952; Berrie *et al.*, 1968; Harada and Koizumi, 1971).

Coumarin was shown to interfere in amino acid uptake and incorporation by barley and lettuce seeds and by yeast (Van Sumere *et al.*, 1972). The effect was more pronounced in barley embryos than in intact seeds. Since the effect on uptake occurred rather late in germination it is doubtful whether it can be the cause for the germination inhibition.

A different aspect of the changes caused by exogenously-applied germination inhibitors and stimulators is their effect on the endogenous, naturally-occurring inhibitors and stimulators. After 12 hr of germination in solutions of coumarin, the formation of an inhibitor of coleoptile growth was induced and the disappearance of natural growth inhibitors, initially present in the seeds, was prevented. Thiourea induced the formation of both an additional germination promoter and an inhibitor (Blumenthal-Goldschmidt, 1958).

An interaction exists between coumarin and gibberellic acid. The latter can reverse the inhibitory action of coumarin in germination (Mayer, 1958). Cycocel (2-chloroethyl-trimethyl ammonium chloride) can also reverse the germination inhibition induced by coumarin. This compound also reverses the inhibition of germination by IAA (Khan and Tolbert, 1966). These results are in agreement with the previously mentioned effect which coumarin may have on the balance of endogenous growth regulators in the seed.

Coumarin may act by affecting respiratory metabolism both directly, by interfering at the phosphorylation stage, and indirectly, by affecting the availability of phosphate. It may act by preventing the formation of certain enzyme systems by preventing their liberation from some bound form.

Coumarin is often supposed to function by blocking SH groups. This is primarily based on the reversal by BAL of the inhibition of growth caused by coumarin. However, Ishikawa (1958) showed that BAL does not ordinarily reverse germination inhibition caused by coumarin. Mayer and Evanari (1952), on the basis of structure – activity studies, concluded that a blocking of SH groups could not explain coumarin's effect on germination.

Thiourea has been shown to interact with ABA and to cancel some of its inhibitory effects. It has also been shown to stimulate markedly the uptake of potassium ions (K^+) by embryonic axes of *Cicer*. High concentrations, or prolonged exposure, to thiourea have cytotoxic effects, which result in inhibition of mitosis in the root tips and decreased incorporation of thymidine into DNA (Rodriguez *et al.*, 1983; Aldosoro *et al.*, 1981).

A study of the interaction between coumarin and thiourea showed that coumarin, in essence, makes seeds more sensitive to thiourea, i.e. the maximum stimulatory effect on germination becomes evident at lower thiourea concentrations. Higher thiourea concentrations, which in the absence of coumarin caused a maximum effect, are much less effective in its presence (Table 6.4). Coumarin and thiourea interact not only in their effect on germination but also in the subsequent growth of the se edlings. The similarity between thiourea and light in stimulating germination is further reflected by the fact that coumarin makes seeds more sensitive to both agents (Mayer and Poljakoff-Mayber, 1961).

TABLE 6.4. *The effect of coumarin on the germination of lettuce seeds in the presence of thiourea (Poljakoff-Mayber et al., 1958)*

	% Germination			
Thiourea concentration $\times 10^{-2}$ M	0	1·0	2·5	5·0
Coumarin concentration ppm				
0	33·0	57·8	59·1	63·4
10	2·7	15·6	34·3	14·4
20	0·8	8·5	13·8	3·6

Low concentrations of thiourea inhibit the germination of *Trifolium subterraneum*, but stimulate it at high ones (Grant-Lipp and Ballard, 1970). This indicates that thiourea acts at more than one site. Structure – activity studies on germination stimulation by thiourea have not advanced knowledge on its mode of action significantly. However, there is some indication that the tautomerism which thiourea and its derivatives undergo may determine its effectiveness as a germination stimulator (Mayer and Poljakoff-Mayber, 1958; Garrard and Biggs, 1966). Despite the fact that thiourea can complex copper and despite its effect on copper-containing enzymes, its action cannot be simply attributed to its copper-complexing action, as the effect of thiourea on germination is not simply reversed by the addition of cupric ions. The relation between thiourea and copper ions in the seeds, as reflected in the activity of copper-containing enzymes, seems to be much more complex (Mayer, 1961). Ascorbic acid, which by itself does not affect germination at all, greatly enhances the stimulatory effect of thiourea. This effect was not due to interference in enzyme systems concerned with ascorbic acid metabolism, but rather to some additional action of ascorbic acid (Poljakoff-Mayber and Mayer, 1961).

Thiourea also affects the respiratory mechanism, possibly by rapidly channelling all respiration in the direction of energy yielding processes.

It has been suggested that the effect of a number of stimulators, such as nitrate, cyanide and thiourea, may be due to their inhibition of catalase (Hendricks and Taylorson, 1975). This would lead to accumulation of hydrogen peroxide (H_2O_2) which in turn would affect the ratio between the glycolytic pathway and the pentose phosphate pathway, increasing the latter, due to the oxidation of NADPH by H_2O_2. Although this theory is attractive, it does not appear to be true, at least for *Xanthium* seeds (Esashi *et al.*, 1979).

Reynolds (1974), on the basis of structure – activity studies, suggested that the – SH of thiols is responsible for stimulation via the interaction with hydrophobic groups in macromolecules (presumably proteins) which are supposed to be essential for germination. This idea requires experimental proof.

From the results above it does not appear that any clear-cut conclusion can be drawn about the mechanism by which coumarin inhibits germination or thiourea stimulates it. No studies have been carried out to determine whether coumarin or thiourea act at a more fundamental level such as transcription or translation, or whether they could be affecting membrane properties. The emphasis has shifted to the study of the effect of plant hormones and in particular GA and ABA.

III. The Effect of Various Metabolic Inhibitors

The various metabolic inhibitors such as DNP, cyanide, azide, ethionine and various herbicides presumably evoke their effect on germination by their effect on metabolism. There is little to indicate that they affect germination differently than other stages of growth, differentiation and development. In the following we will briefly consider a few of these substances.

DNP (2,4-dinitrophenol) is an uncoupler of oxidative phosphorylation. It also inhibits germination and at low concentrations stimulates oxygen uptake, for example in pea and lettuce seeds (Ishikawa, 1958; Klein, 1955). Its uncoupling action is probably the cause of its inhibition of germination.

An unusual effect of DNP has been observed in seeds of *Trifolium subterraneum* (Ballard and Grant-Lipp, 1967). In dormant seeds of this species 0.1 mM DNP induced germination. Other uncoupling agents have been reported to have a similar effect on dormancy.

The effect of some other respiratory inhibitors on germination is shown in Fig. 6.10. Among the six compounds examined, cyanide had the most unusual effect, because at very low concentrations it stimulates germination, although it is known to inhibit respiration in seeds (see Chapter 5). The stimulating concentration of cyanide had only a marginal effect on oxygen uptake (Taylorson and Hendricks, 1973). Taylorson and Hendricks confirmed that at low concentrations, 0.1 – 1.0 mM, cyanide can promote the germination of lettuce and also of *Amaranthus* and *Lepidium*. In these species cyanide is rapidly metabolized. In the presence of cysteine it is converted to cyanoalanine, which in turn is converted to asparagine. The products of cyanide metabolism are incorporated into seed proteins during germination as was demonstrated by the use of ^{14}CN. According to Taylorson and Hendricks the incorporation of cyanide into protein via asparagine and aspartic acid is adequate to explain the promotion of germination. However, such an interpretation requires that the protein into which the ^{14}C is incorporated will be specific

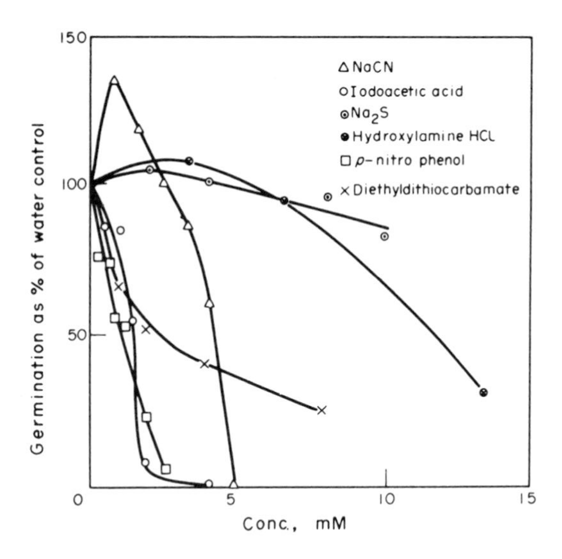

FIG. 6.10. Effect of respiratory inhibitors on germination of lettuce seeds (Mayer *et al.*, 1957). The seeds used were of the light-sensitive variety Grand Rapids. Tests carried out at 26°C in the dark.

for germination and that its amount in the seed is somehow limiting to germination. It also would require that the formation of such a protein specifically requires prior aspartic acid formation. Such a sequence seems rather unlikely. Nevertheless these results are of importance because they show that a compound which is toxic can be metabolized by germinating seeds and because they demonstrate the actual incorporation of part of such a molecule into seed proteins.

Among the other metabolic inhibitors tested on the germination of lettuce seeds, very high sensitivity to iodoacetate and *p*-nitrophenol was observed, 2 – 3 mM blocking germination completely. Diethyldithiocarbamate was less effective and Na_2S and hydroxylamine were almost ineffective (Fig. 6.10).

Ethionine at 5×10^{-3} M and azide at 10^{-3} M also inhibited the germination of lettuce. The effect of ethionine was partially reversed by light but that of azide was not (Mancinelli, 1958a,b).

Penicillin and streptomycin, which are frequently used to prevent bacterial contamination of germinating seeds, can also inhibit germination (Mancinelli, 1958a,b). These compounds were effective at 3.5×10^{-3} M and 8.5×10^{-4} M respectively. Many other fungal metabolites, such as ramulosin and aflatoxin, also inhibit germination. Their precise mode of action is unknown.

An entirely different type of compound affecting germination is that of the herbicides, including the morphactins. These do not represent any homogeneous group since many widely different substances are used as herbicides. Many of these substances when applied directly to seeds will prevent their germination. In agricultural practice herbicides are not normally used to prevent germination. They are, however, frequently used to kill the freshly-emerging weed seedling, while leaving untouched or less affected the deeper sown crop seeds. In this case their selectivity is chiefly based on depth of penetration into the soil rather than on any inherent specific physiological differences in sensitivity to the herbicide of the cultivated species and the weed species to be eradicated. The compound most widely investigated has been 2,4-dichlorophenoxy acetic acid (2,4-D). This substance has been taken to represent all auxin-type herbicides and has served as a model in many reactions in which the mode of action of indolyl acetic acid was being investigated. The literature on the action of herbicides is huge and cannot be reviewed here, particularly as the number of active compounds is constantly increasing (Audus, 1964; Moreland, 1967, 1980; Schneider, 1970). Many of these compounds are actively and rapidly metabolized (Lawrence, 1984), and this aspect must be taken into account when considering how they are acting on the germination process.

An interesting and surprising germination inhibitor is D_2O. In some species, e.g. *Vicia*, pure D_2O delays germination and shifts the temperature optimum, or increases their light requirement as is the case for *Paulownia* (Blake *et al.*, 1968; Labouriau, 1977; Grubisic and Konjevic, 1986). In other species, e.g. *Digitalis* or *Lobelia*, it is highly toxic. No explanation for these differences has yet been given.

From the preceding discussion on the effect of germination inhibitors and stimulators it appears that at present there is no reason to assume that germination can be inhibited by only one kind of mechanism. It is not evident even that any given germination inhibitor acts on a single metabolic process. On the contrary, it appears that many compounds which inhibit germination do so simply because they interfere in a general way with normal metabolism. It is, however, possible that specific germination inhibitors act on some single stage of metabolism. Too few data are at present available to reach any definite conclusion on this point.

In addition to the hormones and other substances discussed in this chapter, plants react to various secondary metabolites which are present in the tissues. Under certain ecological conditions, in defined niches, such secondary metabolites may act on seeds and may affect the germination of competing species and therefore have a competitive survival value. Some of these compounds will be discussed in the following chapter.

Bibliography

Abeles F. B. (1986) *Plant Physiol.* **81**, 780.
Ackerson R. C. (1984a) *J. Exp. Bot.* **35**, 403.
Ackerson R. C. (1984b) *J. Exp. Bot.* **35**, 411.
Akazawa T. and Hara-Nishimura J. (1985) *Ann. Rev. Plant Physiol.* **36**, 441.
Albone K. S., Gaskin P., MacMillan J. and Sponsel U. M. (1984) *Planta* **162**, 560.
Aldosoro J. J., Matilla A. and Nicolas G. (1981) *Physiol. Plant.* **53**, 139.
Armstrong C., Black M., Chapman J. M., Norman H. A. and Angold R. (1982) *Planta* **154**, 573.
Atzorn R. and Weiler E. W. (1983) *Planta* **159**, 289.
Audus L. J. (Ed.) (1964) *The Physiology and Biochemistry of Herbicides.* Academic Press, London.
Ballard L. A. T. and Grant-Lipp A. E. (1967) *Science* **156**, 398.
Ballarin-Denti A. and Cocucci M. (1979) *Planta* **146**, 19.
Berrie A. M. M., Parker W., Knights B. A. and Hendrie M. R. (1968) *Phytochemistry* **7**, 567.
Beyer E. M. (1985) in *Ethylene and Plant Development*, p. 121 (Ed. J. A. Roberts and G. A. Tucker), Butterworths, London.
Black M. (1983) in *Abscisic Acid*, p. 331 (Ed. F. T. Addicott), Praeger, New York.
Blake M. F., Crane F. A., Uphaus R. A. and Katz J. J. (1968) *Planta* **78**, 35.
Blumenthal-Goldschmidt S. (1958) Ph.D. Thesis, Jerusalem (in Hebrew).

Bray E. A. and Beachy R. N. (1985) *Plant Physiol.* **79**, 746.
Briggs D. E. (1973) in *Biosynthesis and its Control in Plants*, p. 219 (Ed. B. V. Milborrow), Academic Press, New York.
Chandler P. M., Jacobsen J. V., Zwar J. A., Ariffin Z. and Huiet L. (1987) in *Molecular Biology of Plant Growth Control*, p. 23 (Ed. J. E. Fox and M. Jacobs), Alan R. Liss, New York.
Chen S. C. and Park W-M. (1973) *Plant Physiol.* **52**, 174.
Chrispeels M. J. and Varner J. E. (1967) *Plant Physiol.* **42**, 398.
Chrispeels M. J.and Jones R. L. (1980/81) *Is. J. Bot.* **29**, 225.
Chrispeels M. J., Tenner A. J. and Johnson K. D. (1973) *Planta* **113**, 35.
Cowan A. K. and Railton I. D. (1987) *J. Plant Physiol.* **131**, 423.
Delmer D. (1983) *Adv. Carbohydrate Chem. Biochem.* **41**, 105.
Duncan I. and Spenser M. (1987) *Planta* **170**, 44.
Eastwood D. and Laidman D. L. (1971) *Phytochemistry* **10**, 1459.
Esashi Y., Ohhara Y., Okazaki M. and Hishinuma K. (1979) *Plant Cell Physiol.* **20**, 349.
Fernandez D. E. and Staehelin L. A. (1987) in *Molecular Biology of Plant Growth Control*, p. 323 (Ed. J. E. Fox and M. Jacobs), Alan R. Liss, New York.
Garrard L. A. and Biggs R. H. (1966) *Phytochemistry* **5**, 103.
Georghiou K., Psaras G. and Mitrakos K. (1983) *Bot. Gaz.* **144**, 207.
Gesundheit Z. and Poljakoff-Mayber A. (1962) *Bull. Res. Counc. Isr.* **11D**, 25.
Grant-Lipp A. E. and Ballard L. A. T. (1970) *Z. Pflanzenphysiol.* **62**, 83.
Groot S. P. C. and Karssen C. M. (1987) *Planta* **171**, 525.
Grubisic D. and Konjevic R. (1986) *Plant Sci.* **48**, 37.
Hall R. H. (1973) *Ann. Rev. Plant Physiol.* **24**, 415.
Hall M. A. (1986) in *Hormones, Receptors and Cellular Interactions in Plants*, p. 69 (Eds C. M. Chadwick and D. R. Garrod), Cambridge University Press, Cambridge.
Hara M., Umetsu N., Miyamoto C. and Tamori K. (1973) *Plant Cell Physiol.* **14**, 11.
Harada H. and Koizumi T. (1971) *Z. Pflanzenphysiol.* **64**, 350.
Harada H., Ohyama K. and Cheruel J. (1972) *Z. Pflanzenphysiol.* **66**, 307.
Hendricks S. B. and Taylorson R. B. (1975) *Proc. Natl. Acad. Sci. USA* **72**, 306.
Hewett E. W. and Wareing P. F. (1973) *Physiol. Plant.* **28**, 393.
Higgins J. T. V., Zwar J. A. and Jacobsen J. V. (1977) Acides nucleiques et synthèse des proteines chez les vegetaux. *Colloques International CNRS* No. **261**, p. 48.
Ho T-h.D. (1979) *Plant Physiol.* **63**, Suppl. p. 79, *Abst.* 444.
Ho T-h.D. (1983) in *Abscisic Acid*, p. 147 (Ed. F. T. Addicott), Praeger, New York.
Ho T-h.D. and Varner J. E. (1976) *Plant Physiol.* **57**, 175.
Hopp H. E., Romero P. A. and Pont Lezica R. (1978) *FEBS Letters* **86**, 259.
Ihle J. N. and Dure L. S. (1972) *J. Biol. Chem.* **247**, 5048.
Ikuma H. and Thimann K. V. (1973) *Plant Cell Physiol.* **4**, 113.
Ilan I. and Gepstain S. (1980/81) *Is. Bot.* **29**, 193.
Ishikawa S. (1958) *Kummamoto J. Science Ser.* **B4**, 9.
Jacobsen J. V. (1973) *Plant Physiol.* **51**, 198.
Jacobsen J. V. and Beachy L. R. (1985) *Nature* **316**, 276.
Jacobsen J. V. and Chandler P. M. (1987) in *Plant Hormones and their Role in Plant Growth and Development*, p. 164 (Ed. P. J. Davies), Nijhoff, Dordrecht.
Jacobsen J. V. and Higgins T. J. V. (1978) in *Phytohormones and Related Compounds. A Comprehensive Treatise*, Vol. 1, p. 583 (Eds D. S. Letham, P. B. Goodwin and J. T. V. Higgins), Elsevier, North-Holland, Amsterdam.
Jerrie P. H., Shaari A. R. and Hall M.A. (1979) *Planta* **144**, 503.
Jones R. L. (1969) *Planta* **87**, 119.
Jones R. L. and Jacobsen J. V. (1978) *Bot. Mag. Tokyo Special Issue* **1**, 83.
Jones R. L. and Stoddard J. L. (1977) in *The Physiology and Biochemistry in Seed Dormancy and Germination*, p. 77 (Ed. A. A. Khan), North-Holland, Amsterdam.
Julin-Tegelman A. and Pinfield N. (1982) *Physiol. Plant.* **34**, 318.
Kende H. (1971) *Itern. Rev. Cytol.* **31**, 301.
Kepczynski J. and Karssen C. M. (1985) *Physiol. Plant.* **63**, 49.
Kepczynski J., Rudnicki R. M. and Khan A. A. (1977) *Physiol. Plant.* **40**, 292.
Ketring D. L. (1973) *Seed Sci. Technol.* **1**, 305.

Ketring D. L. (1977) in *The Physiology and Biochemistry of Seed Dormancy and Germination*, p. 157 (Ed. A. A. Khan), North-Holland, Amsterdam.
Khafagy E. Z. and Mousa A. M. (1982) *Z. Pflanzenphysiol.* **107**, 321.
Khan A. A. and Heit C. E. (1969) *Biochem. J.* **113**, 707.
Khan A. A. and Tolbert N. E. (1966) *Physiol. Plant.* **19**, 76.
Kirsop B. H. and Pollock J. R. A. (1958) *J. Inst. Brew.* **64**, 22.
Kopecky F., Sebanek J. and Blazkva J. (1975) *Biol. Plant.* **17**, 81.
Kudakasseeril G. J. and Minocha S. C. (1984) *Z. Pflanzenphysiol.* **114**, 163.
Labouriau L. G. (1977) *J. Thermal Biol.* **2**, 111.
Lawrence D. K. (1984) in *The Biosynthesis and Metabolism of Plant Hormones*, p. 23 (Eds. A. Crozier and J. R. Hillman), Cambridge University Press, Cambridge.
Klein S. (1955) Ph. D. Thesis, Jerusalem (in Hebrew).
Lewak St. and Bryzek B. (1974) *Biol. Plant.* **16**, 334.
Lin L-S. and Ho T-h.D. (1986) *Plant Physiol.* **82**, 289.
Lieberman M.(1979) *Ann. Rev. Plant Physiol.* **30**, 533.
Mancinelli A. (1958a) *Ann. di Botanica* **26**, 56.
Mancinelli A. (1958b) *Ann. di Botanica* **26**, 67.
Marre E. (1979) *Ann. Rev. Plant Physiol.* **30**, 273.
Mayer A. M. (1958) *Enzymologia* **19**, 1.
Mayer A. M. (1959) *Nature* **184**, 826.
Mayer A. M. (1961) *Physiol. Plant.* **14**, 322.
Mayer A. M. and Evenari M. (1952) *J. Exp. Bot.* **3**, 246.
Mayer A. M. and Poljakoff-Mayber A. (1958) *Bull. Res. Council Israel* **6D**, 103.
Mayer A. M. and Poljakoff-Mayber A. (1961) in *Plant Growth Regulation*, p. 735, Iowa State University Press, Iowa.
Mayer A. M., Poljakoff-Mayber A. and Appelman W. (1957) *Physiol. Plant.* **10**, 1.
Mayer A. M., Poljakoff-Mayber A. and Krishmaro N. (1966) *Plant Cell Physiol.* **7**, 25.
Mayer A. M. and Shain Y. (1974) *Ann. Rev. Plant Physiol.* **25**, 167.
McGaw B. A., Scott I. M. and Horgan R. (1983) in *The Biosynthesis and Metabolism of Plant Hormones*, p. 165 (Eds A. Crozier and J. R. Hillman), Cambridge University Press, Cambridge.
Milborrow B. V. (1974) *Ann. Rev. Plant Physiol.* **25**, 259.
Milborrow B. V. (1983) in *Abscisic Acid*, p. 79 (Ed. F. T. Addicott), Praeger, New York.
Milborrow B. V. and Nodle R. C. (1970) *Biochem. J.* **119**, 727.
Moreland D. E. (1967) *Ann. Rev. Plant Physiol.* **18**, 365.
Moreland D. E. (1980) *Ann. Rev. Plant Physiol.* **31**, 597.
Mundy J. (1984) *Carlsberg Res. Comm.* **49**, 439.
Newton R. P. and Brown G. E. (1986) in *Hormones, Receptors and Cellular Interactions in Plants*, p. 115 (Eds C. M. Chadwick and D. R. Garrod), Cambridge University Press, Cambridge.
Okamoto K., Kitano N. and Akazawa T. (1980) *Plant and Cell Physiol.* **21**, 201.
Paleg L. G. (1960a) *Plant Physiol.* **35**, 293.
Paleg L. G. (1960b) *Plant Physiol.* **35**, 902.
Paleg L. G. (1962) *Plant Physiol.* **37**, 798.
Penner D. and Ashton F. M. (1967) *Plant Physiol.* **42**, 791.
Pinfield N. J. and Stobart A. K. (1972) *Planta* **104**, 134.
Poljakoff-Mayber A. (1952) *Pal. J. Bot. Jer. Ser.* **5**, 186.
Poljakoff-Mayber A. (1952) *Bull. Res. Council, Israel* **2**, 239.
Poljakoff-Mayber A. (1953a) *Enzymologia* **16**, 122.
Poljakoff-Mayber A. (1953b) *Pal. J. Bot. Jer. Ser.* **6**, 101.
Poljakoff-Mayber A. (1955) *J. Exp. Bot. Jer. Ser.* **6**, 313.
Poljakoff-Mayber A. and Evenari M. (1958) *Physiol. Plant.* **11**, 84.
Poljakoff-Mayber A. and Mayer A. M. (1955) *J. Exp. Bot.* **6**, 287.
Poljakoff-Mayber A. and Mayer A. M. (1961) *Ind. J. Plant Physiol.* **3**, 125.
Poljakoff-Mayber A., Mayer A. M. and Zacks S. (1958) *Bull. Res. Council, Israel* **6D**, 117.
Pollard C. J. (1969) *Plant Physiol.* **44**, 1227.
Pollard C. J. (1971) *Biochim. Biophys. Acta* **252**, 553.
Quatrano R. S. (1987a) in *Plant Hormones and their Role in Plant Growth and Development*, p. 494 (Ed. P. J. Davies), Nijhoff, Dordrecht.

Quatrano R. S. (1987b) in *Oxford Surveys of Plant Molecular and Cell Biology*, p. 467 (Ed. B. J. Miflin), Oxford University Press, Oxford.
Reinecke D. M. and Bandurski R. S. (1987) in *Plant Hormones and their Role in Plant Growth and Development*, p. 24 (Ed. P. J. Davies), Nijhoff, Dordrecht.
Reynolds T. (1974) *J. Exp. Bot.* **25**, 375.
Rimon D. (1957) *Bull. Res. Council, Israel* **6D**, 53.
Rodriguez D., Matilla A., Aldasoro J., Hernandez-Nistal J. and Nicolas G. (1983) *Physiol. Plant.* **57**, 267.
Sanders I. O., Smith A. R. and Hall M. A. (1986) *Physiol. Plant.* **66**, 723.
Schneider G. (1970) *Ann. Rev. Plant Physiol.* **21**, 499.
Schopfer P. and Plachy C. (1985) *Plant Physiol.* **77**, 676.
Schopfer P., Bajracharya D. and Plachy C. (1979) *Plant Physiol.* **64**, 822.
Shain Y. and Mayer A. M. (1965) *Physiol. Plant.* **18**, 853.
Singh S. P. and Paleg L. G. (1985a) *Austr. J. Plant Physiol.* **12**, 209.
Singh S. P. and Paleg L. G. (1985b) *Austr. J. Plant Physiol.* **12**, 277.
Singh S. P. and Paleg L. G. (1985c) *Austr. J. Plant Physiol.* **12**, 549.
Singh S. P. and Paleg L. G. (1986a) *Austr. J. Plant Physiol.* **13**, 409.
Singh S. P. and Paleg L. G. (1986b) *Plant Physiol.* **82**, 685.
Singh S. P. and Paleg L. G. (1986c) *Plant Physiol.* **82**, 688.
Sisler E. C. (1980) *Plant Physiol.* **66**, 44.
Sisler E. C. (1984) in *Ethylene*, p. 45 (Eds Y. Fuchs and E. Chakutz), Nijhoff-Junk, Hague.
Sivan A., Mayer A. M. and Poljakoff-Mayber A. (1965) *Israel J. Bot.* **14**, 69.
Sondheimer E., Tzou D. S. and Galson E. C. (1968) *Plant Physiol.* **43**, 1443.
Subbaiah T. and Powell L. (19897) *Physiol. Plant.* **71**, 203.
Taylorson R. B. and Hendricks S. B. (1973) *Plant Physiol.* **52**, 23.
Thomas T. H. (1977) in *The Physiology and Biochemistry of Seed Dormancy and Germination*, p. 111 (Ed. A. A. Khan), North-Holland, Amsterdam.
Trewavas A. J. (1982) *Physiol. Plant.* **55**, 60.
Tzou D. S., Galson E. C. and Sondheimer E. (1973) *Plant Physiol.* **51**, 894.
Ulitzur S. and Poljakoff-Mayber A. (1963) *J. Exp. Bot.* **14**, 95.
Varner J. E., Ram Chandra G. and Chrispeels M. J. (1965) *J. Cell. Comp. Physiol.* **66**, suppl. 1, pp. 55-68.
Van Staden J. (1973) *Physiol. Plant.* **28**, 222.
Van Staden J., Davey J. E. and Brown N. A. C. (1982) in *The Physiology and Biochemistry of Seed Development Dormancy and Germination*, p. 137 (Ed. A. A. Khan), Elsevier Biomedical Press.
Van Steveninck, R. F. M. and Van Steveninck, M. E. (1983) in *Abscisic Acid*. p. 171 (Ed. F. T. Addicott) Praeger, New York.
Van Sumere C. F., Cottenie J., de Greef J. and Kint J. (1972) *Rec. Adv. Phytochemistry* **4**, 165.
Villiers T. A. (1968) *Planta* **82**, 342.
Walker-Simmons M. (1987) *Plant Physiol.* **84**, 61.
Walton D. C. (1980/81) *Is. J. Bot.* **29**, 168.
Walton D. C. (1987) in *Plant Hormones and their Role in Plant Growth and Development*, p. 113 (Ed. P. J. Davies), Nijhoff, Dordrecht.
Wang T. L., Cook S. K., Francis R. J., Ambrose M. J. and Hedley C. L. (1987) *J. Exp. Bot.* **38**, 1921.
Weiler E. W. and Wieczorek U. (1981) *Planta* **152**, 159.
Williamson J. D., Quatrano R. S. and Cuming A. C. (1985) *Eur. J. Biochem.* **152**, 50.
Williamson J. D. and Quatrano R. S. (1988) *Plant Physiol.* **86**, 208.
Yang S. F. and Hoffman N. E. (1984) *Ann. Rev. Plant Physiol.* **35**, 155.
Yomo H. (1960) *Hakko Kyokaishi* **18**, 603.
Yomo H. and Varner J. E. (1973) *Plant Physiol.* **51**, 708.

7

The Ecology of Germination

The Seed in its Natural Environment

In previous chapters the various aspects of germination of seeds have been considered. In the following an attempt will be made to relate these observations to the behaviour of seeds in their natural habitat. Various factors regulate the germination of seeds in their natural habitat, some of which are internal, whereas others are external environmental factors. Any of these can determine whether a given seed will germinate in a certain place or not. The demonstration in the laboratory of the existence of a regulating mechanism is not proof of its operation under natural conditions. To prove this and to show that the plant derives some definite advantage from a given regulatory mechanism is much more difficult. The only advantage about which one may justifiably speak is where some special mode of germination has survival value for the species in a specific environment. This implies that the existence of the mechanism under study enables a species to exist under a given set of conditions, while in the absence of this mechanism it will not survive. Survival value may also take the form of giving a species a better chance to establish itself in competition with other species occupying the same habitat. Unfortunately the proof of survival value of germination-regulating mechanisms is not easily obtainable and relatively few detailed studies have been made.

A spread of germination over a period of time can have survival value. It can protect the species from eradication as a result of adverse conditions which may follow germination. If all seeds germinated simultaneously, under such adverse conditions, and the plants were unable to complete their life cycle, then no further renewal of the species would be possible.

The spread of germination over a period of time has been observed in many species (Koller, 1972). This occurs because not all of the yield of seeds produced during a given year will germinate in the following year. Many of the seeds remain dormant in the soil or their germination in the soil is inhibited. The situation resulting from this behaviour is very complex and involves both the behaviour of the seeds themselves and interaction with their environment. Consequently the seeds of many species accumulate in the soil and form a *seed bank*. Such seed banks supply seeds for germination in subsequent years. The seeds accumulating in the banks usually belong to species with small seeds having very precise requirements for germination. The whole problem of seed banks has been discussed by Thompson (1987).

General theoretical models for germination strategy of seeds under varying conditions of survival have been proposed by Cohen (1966, 1968) and Cohen and Levin (1987). These models imply that in any given environment, there will be species with high seed dispersal and low dormancy as well as species with low seed dispersal and high dormancy. The first group will result in spread of the seeds in space while the second will result in a spread of germination over time, a fraction of the seeds germinating each year. A comparison of different plant species in a given environment might therefore show that the extent of dispersal and of dormancy will be negatively correlated. Dispersal may be more common in environments which are unfavourable for most of the time and which become favourable infrequently and for only short periods.

A positive correlation between dispersal and dormancy may be expected when the variability of the environment increases. Such positive correlations may occur under two sets of conditions: (a) when the fraction of dispersing seeds which succeed in reaching a habitable patch and the survival of dormant non-germinating seeds increase simultaneously or (b) when they both remain constant for most species in the habitat (Cohen and Levin, 1987).

The ecological conditions prevailing in a given habitat will affect germination, the determining factor being probably the micro-climatic conditions prevailing in the immediate vicinity of the seed. Normally, seeds are shed so as to fall either on soil or leaf litter, or in some cases they may fall in a region more or less covered with water. In other words, seeds are usually situated in or on the soil, or in water. The conditions affecting germination under such circumstances will depend on the nature of the soil, its chemical composition and its physico-chemical structure and on the seed's depth in the soil or under water. Depth will influence aeration as well as penetration of light and the daily and yearly fluctuations in temperature. The chemical composition of the soil, or of the water, may affect germination in a number of ways. Possibly the soil may consist of leaf litter or partially decomposed matter which may contain substances inhibiting germination. The soil may have a salt content which will osmotically retard or prevent germination or its ionic composition may be toxic to germination. Soil structure will affect aeration and water capacity and also may determine the ability of a seedling to emerge above the soil or the ability of the roots of the seedling to establish themselves in the soil.

Although seed size and shape of a species show a remarkable constancy, and are genetically determined, nevertheless somatic polymorphism is quite common (Harper *et al.*, 1970). In polymorphic seeds, produced on the same plant, different ecological roles and differences in dormancy and in germination behaviour are often associated with the different seed forms. These differences in seed shape, size or weight may ensure differences in seed distribution in space. In other cases the polymorphism is probably due simply to the shedding of the seeds at different stages of development, which may result in differences in requirements for after-ripening and therefore of time of germination. Silvertown (1984) has suggested that a major cause of polymorphism is the rate of development of the seeds within a fruit or on a given motherplant. The effect of the motherplant is discussed in the following section.

Variability in the behaviour of different seeds in a natural population is not necessarily due to somatic polymorphism. Such variability may be due to genetic differences between the seeds, which are not always easy to determine (Quinn, 1977). The size, shape, structure and composition of seeds can determine their germination behaviour in different environments. However, one must be careful in ascribing ecological functions to all such variations, until they have been proved experimentally.

The mother plant is the first habitat in which the seed finds itself during seed development and maturation. The mother plant to some extent preconditions subsequent germination behaviour. In the following we will discuss external and biotic factors liable to affect the seed in its habitat during germination, and will try to relate the existence of certain germination-regulating mechanisms in the seed to factors in the habitat.

I. Effect of the Mother Plant

The development, maturation, and drying of seeds is a complex process, and essentially outside the scope of this chapter. However, it can be intuitively understood that some conditions which prevail during seed development could influence its subsequent germination behaviour. Among the most obvious of these factors are the photoperiod and thermoperiod to which the mother plant is exposed and the position of the developing seed on the parent plant.

TABLE 7.1. *The effect of photoperiod during growth of the mother plant on subsequent germination of Chenopodium album (after Karssen, 1970)*
LD, SD = Long day, short day

Culture condition of parent plant		LD	SD	SD with interrupted night
2 weeks after harvest	Light	10	92	24
	Dark	0	77	12
4 months after harvest	Light	24	100	100
	Dark	3	100	96

A good example of the effect of photoperiodism in the habitat of the mother plant on the subsequent behaviour of the seeds is provided by *Chenopodium album* (Table 7.1). The seeds of plants exposed to long days had lower germination than those from plants exposed to short days (Karssen, 1970). Photoperiod was also found to affect seed coat development. In *Ononis sicula*, seeds in fruits exposed to short days had permeable seed coats. However, if the fruits were exposed to long days the permeability of seed coats was much decreased (Gutterman and Heydecker, 1973). Further evidence of this kind of effect was found in *Cucumis*. In this case storage conditions of the fruits changed subsequent germination of the seeds in the dark. Exposure of fruits of *Cucumis sativum* to red light resulted in 92% dark germination, while exposure to far-red light or storage in the dark resulted in 28% and 23% dark germination respectively (Gutterman and Porath, 1975). Exposure to a different photoperiod had a similar effect. It was also observed that injection of etherel, which liberates ethylene, into the fruits changed germination behaviour depending on the day length in which the fruits were stored. Detailed experiments have been carried out on *Amaranthus retroflexus* seeds (Kigel *et al.*, 1977). The photoperiod condition under which the parent plants were grown affected post-harvest dormancy.

Seeds from parent plants grown in short days (8 hr) had a higher dark germination and responded at 30°C more rapidly to short irradiation than seeds grown in long days (16 hr) (Kigel *et al.*, 1979). Thus all these data clearly indicate that the parent plant, by some phytochrome-mediated mechanism, influences seed development in a way which affects subsequent germination behaviour. Temperature has a similar effect. Seeds of *Syringa* plants grown at 15°C were less dormant at low temperature than seeds produced on plants grown at 18 – 24°C (Juntilla, 1973).

The position of the seed on the parent plant has a marked effect on subsequent germination behaviour. The softening of the seeds of *Medicago* follows a definite sequence according to the position of the seed in the burr. Softening begins at the calyx end, and proceeds along the spiral of the burr (McComb and Andrews, 1974). In *Aegilops* the order of the caryopses in the spikelet had such an influence (Gutterman, 1980/81). Well defined positional effects have been observed in determining dormancy of celery seeds — *Apium graveolens*. The seeds derived from the primary umbel were less dormant than those from the tertiary and quaternary umbel, and somewhat heavier (Thomas *et al.*, 1979). The case of the difference in behaviour of the two seeds of *Xanthium* has been frequently discussed (see also Chapters 3 and 4). These examples, and those which follow, constitute instances of seed polymorphism which permit a spread of germination over time. The adaptive value of polymorphism in the genus *Rumex* has been related to microclimatic conditions prevailing in the soil and to differences in the survival of the heavy as compared to the lighter seeds (Cavers and Harper, 1966; 1967).

Many other polymorphic responses of seed germination have been described. In *Bidens pilosa*, long and short achenes differ in their germination requirements. The short achenes might produce a back-up population of seeds (Forsyth and Brown, 1982). The *Avena fatua* seeds found in different spikelets differ in their response to nitrate and red light (Hilton, 1985). Qualitative as well as quantitative differences have been reported in the phenolic content of polymorphic seeds of *Atriplex* (Khan and Ungar, 1986), more phenolics being present in the small seeds. It may be recalled that the germination of seeds of a number of species greatly improved by exposing the seeds to high relative humidities. In their different positions on the parent plant they might be drying at different rates. The thermoperiod to which the parent plant was exposed may also play an important role in the subsequent dormancy of the ripe seeds, as was shown for *Amaranthus* (Kigel *et al.*, 1979).

Finally, it is tempting to speculate that hormonal effects are important in seed parental effects. Under different conditions of light or temperature the balance of growth promoters and inhibitors reaching the developing seed from the parent plant might be significantly different. If it can be shown that the growth substance content of the seed is a determining factor in seed germination, this could provide a basis for explaining at least some parental effects. However, as already discussed (Chapters 4 and 6), such growth substance effects are still far from clear and much more convincing evidence is required.

The ecological function of these parental effects is still unclear. It is likely that the conditions prevailing during seed maturation are to some extent predictive of conditions which will prevail immediately after the seeds are shed. The positional effects are probably yet another way of ensuring the spread of germination over time.

II. Environmental Factors in the Habitat of the Germinating Seed

I. Water

The water content of soil in different habitats can vary from saturation as in swamps and water-logged soils, to zero for sandy soils in arid regions. The water capacity may vary within wide limits between different soil types. In a given soil it can also vary with climatic conditions, during different seasons of the year and coverage of the soil by plants. Frequently habitats are classified according to their total moisture content, as hydric, mesic or xeric. However, for germination, the availability of water at a given period of time is the determining factor. Availability will be determined by osmotic factors, binding of water by soil colloids, capillary forces and soil composition and texture. In addition availability of water will be determined by competition with other organisms in the soil requiring water.

The moisture content of the soil may show seasonal periodicity. In some regions, high moisture content is associated with high temperatures, i.e. when summer rains are followed by hot, dry periods. Sometimes, high moisture content occurs when the soil temperatures are at or near freezing so that, although moisture is present, it is in fact not available. Seasonal variability in moisture content is, however, by no means universal. In those regions where precipitation is more or less equally distributed throughout the year, no water shortage exists in the macrohabitat, but in certain microhabitat niches temporary water shortage may occur. On the other hand, where rainfall is seasonal, where only summer or winter rains occur, the moisture content of the soil varies considerably, from very high immediately after heavy precipitation, to near dryness before the next rain, even within the rainy period.

A special condition exists in habitats of high salinity. Soils may be coastal, where the source of salinity is chiefly sea spray which, by causing accumulation of salt in the soil, also changes water availability. In areas with a high water table, seepage may occur and may cause swamp formation or salinization. Inland, saline sodic soils are frequently marked not only by high salt content but also by being relatively impervious, having poor drainage and poor aeration and by often being flooded for part of the year. A very special case is constituted by the tidal areas in which mangroves grow, with their root-systems actually submerged in sea water. On the other hand certain regions, such as some desert areas around the Dead Sea, have a very high salt content but are practically dry throughout the year because of absence of rain and relatively high temperatures all the year round.

2. Temperature

The soil temperature is extremely variable and shows both diurnal and seasonal changes. The extent of the changes depends on the type of the soil (heavy or light, with or without leaf litter) and on the climatic conditions prevailing. Steep gradients of temperature with depth usually exist. These again will depend on the type of the soil. Soil texture and structure, the amount of water present in the soil, conditions for evaporation of water and plant cover, all play a part in determining soil temperature. Usually, the upper layers of the soil show wide fluctuations and as greater depths are reached, conditions become more constant throughout the year.

3. Gases

The composition of the gaseous phase in the soil can be variable. Oxygen, nitrogen and carbon dioxide are the three gases normally present. As the equilibrium between the air above and the gaseous phase in the soil is attained very slowly, appreciable differences between the phases may exist. The lighter the soil, the lower its organic matter content and the smaller the number of microorganisms in it, the greater will be the correspondence between composition of the air in the soil to that above it. In soils having a high organic content and containing many microorganisms the carbon dioxide content may be very much higher and the oxygen content much lower than in the air. In water-logged soils and especially in heavy soils the oxygen content of the gaseous phase may drop considerably below that normal in the atmosphere. This is also generally true for soils having appreciable vegetation. In such soils the roots of the plants will take up oxygen and produce carbon dioxide, again changing the balance of gases.

The volume of the gaseous phase differs greatly in different soils, being greatest in non-compact soils and least in heavy, compacted or water-logged soils. In addition to the three main gases, the soil may contain others, chiefly due to the activity of microorganisms and the absence of oxygen. Soils may contain methane, hydrogen sulphide, hydrogen, nitrous oxide and probably also small amounts of carbon monoxide, ethylene and ammonia.

4. Light

Light is abundant usually only on the surface of the soils. In light or sandy soils light penetrates a short distance into the soil, although its intensity falls off rapidly. In heavy soils light hardly penetrates at all. In cases where the soil is covered by water, light will penetrate considerable distances provided the water is clear.

Light intensity will fall off rapidly under a vegetation cover and its spectral composition is liable to change due to differential absorption and reflection under the canopy.

The change in the composition of light passing through a leaf canopy has been repeatedly documented. There is normally considerable absorption of red light, so that the R/FR ratio decreases. The nature of the canopy is a factor determining this ratio. In addition the R/FR ratio changes during the day, depending on the angle of elevation of the sun, and it also depends on climatic conditions, e.g. cloud cover.

5. Biotic Factors

Seeds in their natural habitat interact with other plants and with animals. The interaction with other plants may be due to inhibitors, stimulators or modification of the microhabitat. Animals may effect germination behaviour by seed softening in the digestive tract or due to distribution to other habitats. Man, using range management, or causing pollution or other technological changes in the environment, can be an important contributor to changes in the environment. Fires both accidental or planned can also affect germination behaviour.

Any two or all the factors mentioned can show marked interaction in their effect on germination. We will now attempt to see what part these various external factors can play in regulating germination.

III. Ecological Role of Environmental Factors

1. Moisture and Temperature

Probably the most crucial factor in determining germination of seeds in the soil is a suitable combination of temperature and moisture. As seeds are a means for propagation of the species, their germination may be expected at a time which will favour survival of the seedling.

The conditions favouring seedling survival differ in different climates. In temperate areas, with cold winters and summer rains, seed ripening and shed is frequently in the autumn. In areas with winter rains and hot dry summers, seed ripening and shed is usually at the beginning of the summer. In both cases the time of shedding of the seed does not coincide with conditions suitable for germination and seedling establishment. In fact, the seeds generally do not germinate even if moisture is present. This may be because moisture availability is not the only factor required. Some other factor such as an unsuitable temperature may prevent germination for some time after the seeds have been shed. The survival value of such a temperature requirement is obvious, both in the case of plants which shed their seeds in autumn, followed by a cold winter and of plants which shed their seed before the onset of a hot dry summer. Postponement of germination by some sort of dormancy, which may be due to a requirement of some special factor, will give the developing seedling a better chance for survival. Went (1957) cites examples of this type of behaviour for a number of plants growing in the Colorado desert. When soils from this desert were irrigated in the laboratory, it was the ambient temperature which determined whether summer or winter annuals germinated. At low temperatures (10°C) mainly winter annuals germinated, while at high temperatures (26 – 30°C) only the summer annuals germinated. At an intermediate temperature a third group of plants could be identified.

In Israel, in the Negev desert, there are no summer rains. The annuals are either winter annuals having a short life cycle, or biseasonal ones which germinate in winter and complete their life cycle by autumn (Gutterman, 1986). Most of the winter species do not have a special temperature requirement for their germination. Water is the limiting factor and they will germinate in summer if the fields are irrigated.

The desert regions of Western Australia have, like the Californian desert, both winter and summer rains. Observations made by Mott (1972) showed that following summer rains grasses predominated in the vegetation, while following winter rains dicotyledons were mainly found. The observed germination behaviour and seedling establishment could be reproduced in controlled experiments, when top soil containing the seeds was incubated in the laboratory under diffferent conditions of temperature. Here as in the Californian desert a combination of moisture and temperature conditions in the top soil determined germination behaviour and survival of the different species.

A requirement for a defined temperature is not an invariable rule. Often alternating temperatures are required which constitute an adaptation to certain climatic conditions. Juhren *et al.* (1953) investigated the germination of various grasses and especially of a number of species of *Poa*. They found that although all the species they studied could germinate and develop at moderate temperatures and with moderate diurnal alternations of temperatures, only one species, *Poa pratensis*, could stand diurnal changes of 36°C by day, and 20°C at night. This condition is one which would occasionally occur in its normal habitat. The seeds of *P. pratensis* did not germinate at all in the cold. Other species of *Poa, P. scabrella* and *P. bulbosa*, were apparently adapted to germination in the cold, day and night at 3.5°C, or day at 6°C and night at 3.5°C. These plants occur in regions where cold conditions do prevail.

Fluctuating temperatures can be an important environmental signal for seeds located deep in the soil or buried in the mud under water. Such conditions exist for example in the Scottish wet-lands, which are water-logged in winter. The water and mud serve as a kind of insulation for the seeds from changes in external conditions. In spring the water table falls, the soil is exposed and the temperature in its top layer begins to fluctuate. Seeds which require diurnal fluctuating temperatures will germinate in this season, under conditions which are favourable for seedling establishment. Seeds which require both temperature fluctuations and light will not germinate unless they are somehow brought to the surface. In certain environments changing temperatures may, for some species, act as indicators of the presence of gaps in the leaf canopy or leaf litter (Thompson and Grime, 1983). A similar requirement for temperature fluctuation has been reported for seeds of *Fimbristylis* and *Scirpus*, both weeds occurring in fields of low land rice (Pons and Schroder, 1986). In the case of *Fimbristylis*, fluctuating temperatures served as a kind of depth gauge and oxygen monitor, since this species cannot germinate in the absence of oxygen. In *Scirpus juncoides*, which germinates even under nitrogen, germination was repressed by oxygen concentrations above 10% and showed a slight increase between 0 and 8% oxygen. In this species the temperature fluctuation may have been the signal to ensure that germination would not occur under anaerobic conditions, when the seeds are buried very deep in the mud of the rice field. They do, however, germinate in the upper layers of the mud, where temperature fluctuations do occur and conditions are not completely anaerobic.

Avena fatua seems to germinate particularly well after frequent alterations between freezing and thawing, followed by periods of wetting and drying. Such conditions exist in certain periods of the year in Bavaria and were followed by massive infestations of this weed (Bachthaler, 1957).

The optimum temperature for germination may change during storage. For example in *Manihot esculentum*, which requires rather high temperatures for germination, it drops from 35° to 30°C during storage (Ellis and Roberts, 1979). The effect of temperature cannot be readily isolated from that of moisture. Other examples for special requirements of temperature are known. Seeds of plants from the Mojave and Sonora deserts germinate rapidly after treatment at 50°C, while lower temperatures result in long delays till full germination occurs (Capon and van Asdall, 1967), an apparent adaptation to high temperature prevailing in desert areas, prior to rainfall. The seeds of *Silene secundiflora* when freshly harvested germinate only at 7 – 16°C, but the temperature range broadens during after-ripening for 4 months. The temperature range of its germination and changes in it appear to ensure non-germination after shedding and germination in about September (Thompson, 1971a). This plant is restricted in its range of distribution.

Good correlation between geographical distribution and temperature requirements for germination have been observed in various species of Caryophyllaceae, but transposition of a given species, by man, from one geographic area to another did not result in changes in its germination behaviour and temperature requirements. This has been shown convincingly for *Agrostemma githago* (Thompson, 1970, 1971b). However, different populations of the same species of *Mimulus* did show adaptations in their germination requirement in accordance with temperature conditions prevailing in different areas (Vickery, 1967). Thus temperature alone can be an important ecological factor in seed germination.

The germination response of six species of the genus *Dioscorea* in Japan correlated well with their distribution. Species occurring in the colder climatic regions required only a short chilling and germinated rapidly and at higher temperature than the species from warmer climates, which required much longer chilling (Okagami and Kawai, 1982, 1983). This suggests that the chilling requirement is a feature of species occurring in a warmer climate, which appear to have a deeper dormancy. The temperature effect involved both dormancy release and a response of germination to temperature. High temperatures also were able to induce secondary dormancy, which could be released by chilling.

A complicated interaction between light, temperature and high humidity is characteristic of *Oldenlandia corymbosa* which grows in tropical environments. One line of this tropical plant produced seeds which germinated in the light at very high temperatures, 35 – 40°C. Such seeds were regarded as non-dormant. However, a second line of plants produced dormant seeds. These will not germinate in the light at high temperatures without prior exposure to lower ones, 25°C, and high relative humidities (Attims and Come, 1978).

The moisture tension under which different plant seeds will germinate can be different. Some species will germinate under a wide range of conditions of moisture tension, while others are much more specific in their requirements.

The seeds of willow will only germinate if they are shed in a very moist habitat. The seeds are characterized by a very rapid loss of viability. The growth of willows in very moist habitats may therefore be related to the property of the seeds.

Seeds of many plants will not normally germinate under water and frequently, if such seeds are kept under water for any length of time, their viability becomes impaired. Other species can withstand being placed under water and will even germinate under such conditions, for example seeds of *Scirpus*. Frequently, germination is greatly improved by increased aeration.

The number of seeds which germinate under water is comparatively small, but in some cases a period of submergence is beneficial. Seeds of *Saponaria officinalis* are dormant at maturity but lose their dormancy after submergence for 2 months under water, followed by exposure to air (Lubke and Cavers, 1969). This plant grows normally on gravel banks of rivers.

Instances are known where submergence of seeds in water, followed by exposure to air, promotes germination. Seeds of both *Heliotropium supinum* and *Mollugo hirta* will germinate only if they are buried in wet mud for a certain period of time and then exposed to air. These seeds have an additional requirement for germination, viz. low temperature for a period of time (Mall, 1954). It appears that continuous soaking increases the permeability of the seed coat. Both these plants occur in the drying pools and puddles which form following monsoon rains. Thus it might be possible that the conditions required constitute an adaptation to a very specific habitat. However, another plant, *Polygonum plebejum*, seems to occur in the same habitat and yet does not require presoaking for its germination.

Other species such as *Rorippa nasturtium* (water cress) and *Typha latifolia* germinate under water (Morinaga, 1926a,b). Again this requirement seems to be not for underwater conditions as such, nor for the presence of an excess of liquid water, but a requirement for the low oxygen tensions which are associated with this. *Typha* will normally germinate under water if given a light stimulus. The seeds will germinate not only under water, but also in light on moist filter paper. When the oxygen tension is lowered germination is much improved (Table 7.2). Sifton (1959) found that germination is not simply conditioned by light and oxygen tension but also by temperature. The general conclusion reached from this study was that the germination of *Typha* was determined by the amount of water which can be held by colloidal proteins of the aleurone grain.

TABLE 7.2. *Germination of Typha latifolia L. seeds under varying conditions (compiled from data of Sifton, 1959)*

	% Germination		
	Seeds immersed in water	Seeds on moist blotting paper In air	In 2% O_2
35°C Light	81	48	—
30°C Light	89	61	96
30°C Dark	4	—	—
25°C Light	86	44	—
20°C Light	69	37	—
15°C Light	24	20	—

An adaptation of germination and seedling survival under anaerobic conditions also seems to exist in some varieties of rice (Chapter 3), but in rice this is not an obligatory habitat or even one desirable for seedling development, while in *Typha* anaerobic conditions were preferred.

The seeds of wild rice, *Zizania aquatica* germinate in a habitat undergoing extreme temperature variations, which necessitates some degree of dormancy (Simpson, 1966). The mature seeds normally drop to the bottom of a lake bed, where they remain for long periods, partly under lake ice. Dormancy is lost under low oxygen tensions, and at low temperatures, i.e. under those conditions which prevail in the period between the shedding of the seeds and the time favourable for their germination, in spring. Drying the seeds in air leads to a rapid loss in viability.

The sensitivity of seeds to water tension, i.e. their ability to germinate at low (more negative) water potentials, is determined in some cases by the presence or absence of a mucilage cover (Harper and Benton, 1966). Seeds of *Lepidium sativa* which contain a large amount of mucilage are not sensitive to water tension, *Plantago major* shows some sensitivity, *Reseda alba* even more, while *Vicia faba* which is devoid of mucilage shows considerable sensitivity to water tension. Mucilages may function in a different way. The mucilagenous hairs of *Blepharis persica* actually serve to orientate the seed, when wetted, at an angle ensuring contact of the radicle with the soil (Gutterman *et al.*, 1967; Gutterman and Witztum, 1977). Excess moisture can prevent germination due to the creation of a diffusion barrier to oxygen by the imbibed mucilage. *Blepharis* is characterized by very rapid germination in wadi beds, where moisture conditions favourable for germination and survival are usually of very brief duration. In many species orientation of seeds towards the soil is determined by morphological features of the seed, for example in the Compositae (Sheldon, 1974).

Mucilage may have a role in assisting seedling establishment. The fruits of *Cavanillesia* contain a very large amount of mucilage adhering to the seeds. The seeds germinate equally well in the absence or presence of the mucilage (Garwood, 1985). However, seedling establishment was assisted by the presence of the mucilage, which may act as a buffer against dehydration of the seedling, since it retains water for long periods.

An entirely different problem exists in plants growing in arid regions, where the period during which germination and seedling establishment can take place is very short. In these plants, survival of the species is determined by mechanisms which ensure that germination occurs at a time when the seedling will be able to establish itself quickly, and to complete its life cycle while conditions are still suitable. Ideally germination should occur when moisture and temperature conditions favour both germination and seedling growth.

An examination of the effect of soil water potential in the germination of *Zygophyllum* gave a rather surprising result, in that the germinating rate was maximal at field capacity, ≈ -0.03 MPa. Under conditions of constant soil water potentials, the rate of germination was inhibited both at low soil water potentials, < -0.10 MPA, or at high ones, >0.04 MPa. However, in all cases the final germination percentages achieved were similar (Agami, 1986). When situations of soil dehydration were simulated in the laboratory, both high and low water potentials inhibited germination, as they did in the field.

The germination behaviour of *Calligonum comosum*, a desert shrub growing in the Sahara and similar habitats, appears to restrict the plant to such habitats as coarse sandy soils with low rainfall. The seeds of this plant will not germinate if in contact with liquid water for any length of time. The germination behaviour is complicated by the light sensitivity of the seeds. Germination occurs only in the dark when the seeds are buried in the sand (Koller, 1956).

Went (1953, 1957) ascribed the ability of plants to gauge the water status of the soil to the presence of inhibitors. He suggested that in certain desert seeds germination occurs if the rate of reformation of an inhibitor in the seeds, when they are moistened, is slower than the process of germination itself. Such a situation might occur only under very specific conditions of moisture and temperature. In seeds of *Pectis pappoza* germination occurs only after 25 mm of summer rains and not after an equal or greater amount of rain in the winter.

On the basis of these and many other instances of the behaviour of desert seeds, Went suggested that the seeds contain one or more inhibitors which separately or together act as a kind of rain gauge. This determines when germination will occur, as the seeds wil not germinate until they are washed out. Evidence for the presence of inhibitory substances in seeds or fruits of certain desert plants is available (Koller, 1955; Koller and Negbi, 1959). However, it is unfortunate that the nature of these inhibitors has never been ascertained. Moreover, quantitative changes in them have never been proved experimentally. Thus attractive as this theory is, it is still lacking final experimental proof.

Not all species exposed to uncertain conditions of water in the soil are able to gauge it. Seeds of *Salsola inermis* and *S. volkensii* germinate immediately after the first rain (Evenari *et al.*, 1980). Seeds of *Hybanthus prunifolius* germinated if sufficient rain fell (Augspurger, 1979). In both cases germination occurred irrespective of the subsequent chance of survival of the seedling. Attempts to find a mechanism in *Hybanthus* which regulates germination on exposure to uncertain and scattered rainfall were unsuccessful. Presumably other mechanisms exist to ensure survival in this shrub, such as the production of a large number of seeds.

Seeds of *Panicum antidotale, P. turgidum* and *Atriplex dimorphostegia* show a different type of adaptation to changes in water content (Koller, 1954). Seeds of these plants germinate to a high percentage in the laboratory if the seeds, together with their dispersal unit, are dried over calcium chloride before sowing. In their natural habitat the seeds are often exposed to hot dry weather particularly in the direct sun, before they germinate. The mechanism of predrying might thus constitute a special adaptation to their normal habitat. Predrying is certainly not essential for germination, as some of the seeds germinate even without it and the whole mechanism only points to a correlation between habitat and germination behaviour.

All these mechanisms, which either delay germination or limit it to a very special and often rarely occurring set of conditions, ensure that not all the seeds will germinate together. If, subsequent to germination, the seedlings are destroyed, the entire population will not be wiped out, as part of the seeds will still be viable, and remain in the soil, but will not germinate because of their very specific requirements for germination.

2. Gases

It is almost impossible to generalize about the composition of gases in the soil, in natural habitats. The gaseous composition depends very much on the nature of the soil, its physical structure, chemical composition, water content and the organisms living in it. It is usually accepted that the "normal" conditions for germination require the gaseous composition of air, i.e. 20% oxygen and very low content of carbon dioxide. However, sometimes other gases may be present in the soil, for example ethylene, which may affect germination.

As discussed previously, germination of many seeds is prevented when the oxygen concentration is lowered appreciably. However, the seeds of certain plants and particularly of aquatics such as *Typha* prefer anaerobic conditions for germination. Morinaga (1926a,b), when testing a variety of seeds, found that among 70 species at least 43 were capable of germinating under water, i.e. under almost anaerobic conditions; 18 out of these 43 species germinated equally in water and in air on moist filter paper, i.e. these species do not have absolute requirement for 20% oxygen and can germinate under very low oxygen pressures such as those prevailing under water. Only two species had a definite preference for anaerobic conditions.

No precise information of the internal atmosphere of the seed is available. The only attempt to measure this seems to have been made by Kidd (1914) in peas (Table 7.3). Kidd ground peas either in water or in barium hydroxide. Those ground with water were allowed to equilibrate with the air and then barium hydroxide was added. The difference in titer between the two treatments was taken as a measure of the carbon dioxide content of the seeds. This method must be regarded with considerable reserve with regard to the carbon dioxide content of the seeds, and moreover does not indicate the carbon dioxide oxygen ratio within the seed. All other conclusions on the gaseous atmosphere in the seeds are based on the response of seeds, with or without seed coat, to changes in the composition in the external atmosphere. Such responses need not always reflect a change in the internal atmosphere.

TABLE 7.3. *The CO_2 content of peas, Pisum sativum, while germinating (after Kidd, 1914)*

Time of germination (hr)	CO_2 content ml CO_2/100 g of seeds
Dry Seeds	145
18	64
25	41
39	43
64	39
97	16

Excised oat embryos can germinate under rather low oxygen pressures (Lecat, 1987). At a pressure of 10% oxygen, germination was 95% and even in the presence of only 3 – 4% oxygen almost 60% of the seeds germinated. Oxygen uptake by the glumules did not apparently interfere with respiration of the embryos, since their transient intensive oxygen uptake occurs much earlier, and is completed before the respiratory rise in oxygen uptake of the germinating embryo. These data suggest that even in cultivated crops there is no absolute requirement for 20% oxygen. The problems related to the permeability of isolated membranes from seeds has been discussed in Chapter 4.

Relatively little information is available about the direct effect of oxygen on germination. Proof of impermeability or partial permeability of seed coats to oxygen comes from the work of Marchaim *et al.* (1972) and Come (1970). Indirect evidence for impermeability to oxygen exists for the seed coat of the upper seed of *Xanthium*. A rise in the external oxygen tension enhances germination. In these seeds, therefore, the seed coat is not completely impermeable to oxygen, but partially permeable to it. However, as already discussed, under natural conditions oxygen concentrations do not rise above 20%. The ecological significance of these findings is not entirely clear. Wareing and coworkers suggested that the germination of *Xanthium* seeds seems to occur only when a certain minimal threshold of the internal oxygen concentration is reached. This results in the oxidation of an inhibitor present in the seeds. This inhibitor apparently prevents germination. Nevertheless, although there is a good correlation between external oxygen concentration and germination, it is not absolutely certain that germination is the direct result of destruction of the inhibitor (Wareing and Foda, 1957; Porter and Wareing, 1974). Under natural conditions the necessary threshold value is attained only very slowly due to the low permeability of the seed coat to oxygen. If the seed coat is punctured or damaged in some other way, these processes will be greatly accelerated. If the seed coat is completely impermeable to oxygen, then only damage can enable germination (Crocker, 1906; Shull, 1911).

The work of Edwards (1969) discussed in Chapter 4, describing a similar situation, indicates a role of oxygen in controlling the concentration of a germination inhibitor in seeds of charlock (*Sinapis arvensis*). This could be a more general phenomenon by which the internal oxygen concentration controls germination. Come (1970) has indicated that in apple seeds a low oxygen concentration induces secondary dormancy, possibly because of the interaction of oxygen with germination-inhibiting substances in the seed. Examples therefore exist which show correlations between external oxygen concentration and germination. This kind of results can provide explanations for the mechanism by which oxygen concentration can affect germination. In general it can be stated that low oxygen concentrations are usually accompanied by either an excess of water or lack of light or both. In all probability the role of oxygen as an ecological factor must be sought under those conditions under which one of these additional factors is also of importance (see also section III.1).

Kidd (1914a,b) concluded from his experiments that internal carbon dioxide accumulation, above a certain level, is an important factor in inhibiting germination. Germination will occur only when this level falls. According to Kidd such a drop was observed in peas about 18 hr after the seeds were placed in water (Table 7.3). This idea that the internal carbon dioxide concentration may possibly regulate germination requires further verification. Thornton (1943-45) showed that a large number of seeds, particularly of crop plants, germinated in carbon dioxide concentrations as high as 40 – 80%, provided 20% oxygen was also present. He concluded that, at least in the case of the seeds used, carbon dioxide did not inhibit germination in the presence of oxygen. Kidd also found that increasing oxygen concentrations and increasing temperatures decreased the inhibition caused by carbon dioxide, which was confirmed by Thornton.

Under natural conditions a situation is rarely met where carbon dioxide concentrations rise and yet the oxygen content remains high. Thus, except for the species which require very low oxygen concentrations, the conclusions reached by Kidd for *Brassica alba* seem to correspond more closely to events occurring naturally than the examples quoted by Thornton. In *Brassica* seeds the carbon dioxide was assumed to decrease the permeability of the seed coat, especially towards carbon dioxide, presumably causing an accumulation of carbon dioxide and an increase in the CO_2/O_2 ratio. For *Cucurbita* (Chapter 4) it has been shown that the seed membrane, which controls gaseous diffusion, is more permeable to carbon dioxide than to oxygen, also resulting in changes in the CO_2/O_2 ratio.

In some seeds small amounts of carbon dioxide can promote germination. The dormancy-breaking action of carbon dioxide on *Medicago* and *Trifolium* was mentioned in Chapter 4. In these plants the response to carbon dioxide may have resulted from a process of selection by agricultural practice rather than from natural selection. It seems possible that the sensitivity to small amounts of carbon dioxide causes rapid germination and that plants whose seeds germinate rapidly are selected for, as this is often a desirable property in agricultural crops.

A suggested difference between weed seeds and the seeds of crop plants is the ability of the former to survive while buried in soils for long periods. Under these conditions the seeds are liable to be exposed to raised carbon dioxide and lowered oxygen concentrations while imbibed. Weed seeds are not damaged by burial and germinate rapidly when removed from the soil, as for example after ploughing. However, some weed seeds lose their viability quite rapidly if buried. For example *Scandix pecten, Linaria minor, Bartsia odontites* and *Polygonum aviculare* showed a reduction of 90% in their viability in a 2-year period of burial (Brenchley and Warington, 1933). The seeds of many crops seem to suffer during burial in the soil. In the seed burial experiments of Duvel (1905) it was found that *Avena fatua* and *Lactuca scariola* survived under conditions where *A. sativa* and *L. sativa* lost their viability. Similar differences were observed between seeds from wild and cultivated varieties of *Helianthus annuus.*

From the above it is clear that no definite conclusion can be drawn about the role of the two gases, carbon dioxide and oxygen in regulating germination. It seems probable that the internal concentration of the gases in the seed is the determining factor. It is possible that the CO_2/O_2 ratio is the important factor rather than the absolute concentration of either of them.

Another gas which may be present in the soil is ethylene, which is known to affect seed germination (Chapter 4). Tree roots can, under conditions with a favourable water regime, produce ethylene some of which may be retained in the soil (Stumpff and Johnson, 1987). Ethylene production decreased as the water potential of the soil decreased, but increased again 4 – 5-fold following irrigation of the soil. The ability of sunflower seeds to synthesize ethylene seems to correlate with their inability to germinate under certain specific conditions (Corbineau *et al.*, 1988). Changes in the ethylene content of the soil, caused by environmental conditions such as changes in water potential, may have a role in regulating seed germination.

However, there are few reliable detailed data on the composition of the soil atmosphere. Furthermore it is unlikely that ethylene will accumulate to any extent in the soil, since it diffuses rapidly and also can react with soil constituents. Other gases released by bacterial and fungal fermentation processes might also be present in soils, but there is no direct evidence to indicate whether they accumulate and whether they have any effect on germination.

3. Light

Light is one of the environmental factors which may regulate germination. Seeds requiring light will not germinate when they are buried under soil or leaf litter, but will germinate when they fall on the soil surface. As already discussed, with regard to their response to light, seeds can be divided into groups, viz. (1) those which require light, (2) those which are inhibited by light (Table 7.4 and see also Table 3.8) and (3) those which are indifferent to light. It would therefore be expected that plants of species which require light for germination would distribute their seeds on

TABLE 7.4. *Classification of some seeds according to their light requirement. A — Seeds whose germination is stimulated by light. B — Seeds whose germination is retarded by light*

A	B
Arceuthobium oxicedri	*Bromus sp.*
Daucus carota	*Datura stramonium*
Elatine alsinastrum	*Lycopersicum esculentum*
Ficus aurea	Liliaceae, various species
Ficus elastic	*Nigella sp.*
Gloxinia hybrida	*Phacelia sp.*
Gramineae, various species	*Primula spectabilis*
Gesneriaceae, various species	
Lactuca saliva	
Lobelia cardinalis	
Lobelia inflata	
Loranthus europaeus	
Lythrum ringens	
Mimulus ringens	
Nicotiana tabacum	
Nicotiana affinis	
Oenothera biennis	
Primula obconica	
Phoradendron flavescens	
Raymondia pirenaica	
Rumex crispus	
Verbascum thapsus	

the surface of the soil or in its uppermost layers. On the other hand species whose germination is inhibited by light, no matter where the seeds land, will only germinate if they are covered by a certain amount of soil. However, this is a gross oversimplification. Frequently germination of seeds which are inhibited by prolonged illumination is stimulated by short illumination, for example in *Amaranthus blitoides* and *Atriplex dimorphostegia*. Furthermore, seeds of the same genus may be light-requiring or light-inhibited, for example *Primula obconica*, which is light-stimulated, and *Primula spectabilis* which is light inhibited. In more extreme cases different varieties of the same species may be light-requiring or light-indifferent, for example *Lactuca sativa*. A requirement for short, as opposed to continuous, illumination may be advantageous in certain situations, for example if the seeds are entirely uncovered they will fail to germinate as continuous light inhibits them, but if the seeds are only partially covered, they may be exposed briefly to light at certain periods of the day. Under these conditions they will germinate. In many seeds light sensitivity is confined to only one well-defined region, for example in *Phacelia* sensitivity is localized at the chalazal and micropylar ends; in lettuce only the micropylar end of the seeds is light-sensitive. Such sensitivity could result in germination of the seed occurring only in certain orientations in the soil. There is not enough evidence, however, to prove that localization of sensitivity has ecological importance.

Light sensitivity is often induced in seeds. The seeds of *Spergula arvensis* and *Stellaria media* do not require light for germination when they are fresh. After burial in the soil a light requirement is induced (Wesson and Wareing, 1969). The ecological importance of this induced sensitivity may be that it serves as a signal showing that the seeds have again reached the soil surface. The induction of a light requirement has also been reported for *Capsella bursa pastoris* and *Senecio vulgaris*.

Light and temperature frequently interact. This interaction may make seeds sensitive to light at certain temperatures but not at others. Some of the possible combinations of light and temperature are shown in Table 7.5.

TABLE 7.5. *Classification of light – temperature interactions of some seeds (compiled from data quoted by Stiles, 1950 and Koller, 1955)*

Plant	Temperature	Light			Dark		
		Low	High	Alternating	Low	High	Alternating
Veronica longifolia		+	–	–	–	+	+
Epilobium hirsutum		+	–	–	–	+	+
Poa pratensis		+	–	–	–	–	+
Rumex crispus		+	–	–	–	–	+
Apium graveolens		+	–	–	–	–	+
Oryzopsis miliacea		+	–	–	–	–	+
Gesneriaceae species		+	–	+	–	–	–
Ranunculus sceleratus		–	–	+	–	–	–
Chloris ciliata		×	+	–	–	–	–
Amaranthus retroflexus		×	+	–	–	–	–
A. lividus		×	–	–	–	–	–
A. caudatus		×	+	–	–	–	–

\+ signifies germination is promoted by combination of treatments.
– signifies no promotion.
× germination inhibited by this set of conditions.

These interactions are very complex and difficult to interpret. Such varying requirements for conditions for germination may permit adaptation to life in different habitats. The first six species appearing in Table 7.5 will germinate in the light only at relatively low temperatures, which prevail for example in the Mediterranean areas with winter rains. In areas with summer rain and high or fluctuating temperatures they may germinate when buried in the soil. The six additional species in Table 7.5 are light-requiring summer plants, which will germinate either on the surface or when slightly covered with soil. They will not germinate when either deeply buried in the soil or at low temperatures. The above interpretation of the response of these species is based on their germination behaviour under laboratory conditions. Such combinations of a number of factors, created in the laboratory, may not in fact exist in nature. The observed germination behaviour of the seeds may be no more than residual genetic properties which no longer have any direct survival value and which are retained as long as they have no harmful effect. Stoutjesdyk (1972) suggested that such features, which are revealed in laboratory experiments, may be part of a pool of unutilized traits on which natural selection may act in the future.

Light requirement is frequently associated with small seeds, which are supposed to contain rather small amounts of reserve materials, so that it might be advantageous for them to germinate under conditions where photosynthesis occurs very soon after germination. But seed size is no measure of the amount of storage materials present relative to the requirements of the seedling.

In some cases, the light requirement may be related to seedling establishment. Seeds of *Narthecium ossifragum* are light sensitive, but the light requirement is such that 10% of the intensity of day light was sufficient to induce germination. However, subsequent growth of the seedling requires much more light. The germination behaviour of this species is complicated by the interaction of light and moisture requirements. Germination requires a high water table, but excess water causes both a loss of viability of the seeds and damping of the seedlings (Summerfield, 1973). The preferred combination of favourable light, moisture, temperature and suitable substrate for germination and seedling establishment apparently is rare in nature. This plant grows characteristically in mire. The main mechanism in the species to ensure survival is the production of a very large number of seeds to counterbalance the high rate of failure in seedling establishment.

The light response of seeds is mediated by phytochrome. It has been attempted to relate the R/FR mechanism in regulating germination to its possible role in given ecological situations. The distribution of red and far-red light of incident radiation changes during the day, enrichment in far-red occurring at dawn and dusk, i.e. at lower angles of elevation of the sun. Moreover the spectral composition of radiation changes as it penetrates through a leaf canopy, red light being selectively absorbed. There is a several fold increase in the relative far-red irradiance compared to red as light passes through a corn canopy (Sinclair and Lemon, 1973). Passage of light through a leaf canopy of wheat resulted in similar changes. The magnitude of these effects depended on both the solar elevation and the length of passage through the leaf canopy. Thus the ratio R/FR decreased from 1.2 in the open air to 0.2 under a canopy when the leaf area index was 5, i.e. the canopy was very dense (Holmes and Smith, 1977a,b,c). Direct attempts to investigate the significance of such changes were made by Stoutjesdyk (1972). His experiments show that in a number of species germination was inhibited when the seeds were sown under a canopy of *Crataegus monogyna*. This applies both to seeds whose germination is stimulated by light, such as *Betula pubescens* and *Epilobium hirsutum*, and those whose germination in inhibited by light, such as *Bromus tectorum*. He sowed seeds either under the *Crataegus* canopy or just outside it, but shielded from direct sunlight and also ran suitable controls in the dark. He also examined the spectral composition of the light reaching the seeds. The light under the canopy was enriched in FR while that outside the canopy contains less FR and relatively more blue than direct sunlight. Seeds of different species in the same genus, for example *Sagina*, responded differently. The results suggest that it is possible that the ratio R/FR of light under natural conditions could affect germination and that when the FR content of the radiation is high, germination inhibition may result. Presumably inhibition of germination will depend not only on the R/FR ratio of the radiation but also on the amount of phytochrome in the seeds and their sensitivity to the P_R/P_{FR} ratio in them. Thus when considering the ecological role of light it is no longer sufficient to consider whether seeds are exposed to light, but the spectral composition of the light reaching them as well as the phytochrome ratio in the seeds must be taken into account.

Different plants growing in the same habitat may adapt different strategies with regard to their light response. Both *Geum urbanum* and *Cirsium palustre* occupy coppices covered by ash (*Fraxinus*) which are felled at regular intervals. Before felling *Geum* predominates, while after felling *Cirsium*, which is a shade-avoiding plant, is the predominating species (Pons, 1983, 1984). *Geum* is not light sensitive and its primary dormancy is broken by stratification. The seeds germinate independently of the presence of the tree layer. In contrast *Cirsium*, which is light sensitive, is able to store and maintain a light stimulus if kept at low temperatures. The seeds of *Cirsium* are inhibited by FR light, as it exists under the leaf canopy. They germinate after felling, when exposed to higher temperatures, as well as to light.

The complexity of light requirement can be illustrated by the behaviour of two species colonizing bare soil, *Verbascum* and *Oenothera*. The former has a very low requirement for light for its germination which is rapidly saturated. In contrast *Oenothera* has a requirement for prolonged illumination and higher light intensities. Both are inhibited by light filtered through a leaf canopy but to different extents (Table 7.6).

TABLE 7.6. *Response of Verbascum thapsus and Oenothera biennis to light filtered through leaf canopies of different species (from data of Gross, 1985)*

Type of canopy	Radiation level relative to that giving maximal germination	Germination as % of maximum in full light	
		Verbascum	*Oenothera*
Livistonia	0·395	23	0
Diffenbachia	0·344	66	0
Anthurium	0·256	24	0
Coccolobis	0·18	59	4

The function of R/FR response of seeds in their natural habitat is much more complex than was asssumed at first. No single ecological explanation will apply, and each case must be studied in detail. Even in a given habitat the signals used to sense optimal condition for germinating can differ greatly.

Light, as sun radiation, affects various other external factors in the habitat and interacts with them in their effect on germination. A good example is provided by the behaviour of seeds of plants on different slopes of a hill. In the Negev highlands of Israel, two species of *Helianthus* grow on the northern and southern slopes of the hills. The north facing slopes receive less radiation, have lower soil temperatures, lower soil salinity and higher moisture content than the south facing slopes. The seeds of *H. vesicarium*, which grow on the northern slopes, have a narrower range of temperature for germination and are more sensitive to salinity than the seeds of *H. ventosum*, which grow on the southern slopes (Fig. 7.1) (Gutterman and Agami, 1987). On such slopes sunlight may also affect the seeds while they are still

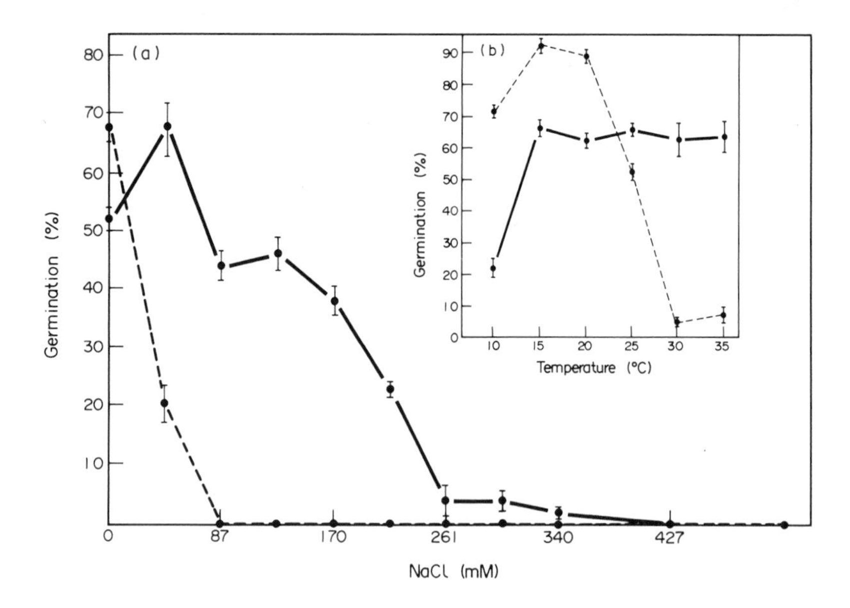

FIG. 7.1. Germination of seeds of *Helianthus vesicarium* (– – –) and *H. ventosum* (———) when exposed to different salinities (a) and to different temperatures (b) (from data of Gutterman and Agami, 1987).

on the mother plant. On the two sides of the hill the intensity of the light may be quite different and result in different conditions for seed maturation which may affect their subsequent germination behaviour.

4. Soil Conditions

As previously mentioned, the physical and chemical properties of the soil determine the microhabitat of the seed. Physical properties will determine the water holding capacity and aeration of the soil as well as its hydraulic conductivity, which in turn determines the rate of flow of water to the seed. Particle size of the soil will influence contact between the seed and its substrate, and through it, the amount and rate of water taken up by the seed (Hadas, 1977; Hadas and Russo, 1974). Soil composition affects the solute content of soil water, which affects germination in several ways. The total soluble solids can affect germination through osmolarity and through ionic strength. In addition toxic effects due to certain ions present in the soil solution may occur. However, metal ions generally regarded as phytotoxic, such as A1, Cd, Cu, Zn and Pb had little or no effect on the germination of a number of tree species at pH between 3 and 5. The presence of such ions in the soil solution does not suppress germination at the concentrations found in mountain ecosystems (Scherbatskoy, Klein and Badger, 1987), although they may affect seedling establishment. The nitrogenous compounds frequently present in soils affect germination of seeds, as some of them are known to break dormancy, for example of *Sinapis arvense*. The germination response in the soil is determined by a combination of factors in the microhabitat such as light and temperature (Goudey *et al.*, 1987), as well as by soil structure and soil composition. Some of the environmental factors can be modified artificially and thus affect germination.

Sometimes it is desirable, for practical purposes, to modify some environmental factor artificially in order to cause a sudden uniform germination of weed seeds so that their seedlings can then be eradicated before cultivated crops are sown in the field.

Under field conditions the moisture content changes continuously, especially in the surface layer of the soil where it is depleted due to evaporation. As a result the germinating seed has to compete for water with the atmospheric evaporative demand. This competition is severe under dry-farming conditions and probably even more severe under natural conditions.

A number of different definitions have been suggested to describe optimal soil-water relationships in the microenvironment of a germinating seed. One definition assumes that a critical soil water potential exists, which is suitable for germination. This critical value is characteristic for each species (Hunter and Erickson, 1952). Another definition is based on the seed – soil contact. It was shown that when seed – soil contact was poor, the germination percentage was reduced even at high water potentials of the soil (Collis-George and Hector, 1966). Apparently the soil – water matrix potential is the important factor. An attempt to analyse the actual conditions adjacent to the germinating seed was made by Hadas (1970) and Hadas and Stibbe (1973). They propose a model, describing the water relations between a single seed and its immediate environment as a dynamic process. Their model considers the water uptake of the germinating seed as a function of the changing factors of the soil, such as water content, water potential, water diffusivity and water flow towards the seed. From their work it appears that the germinating seed in its microhabitat in the soil is exposed to a much more severe water stress than is measured by conventional methods in the bulk of the soil (Chapter 3).

The high salt content of soils, especially of sodium chloride, can inhibit germination, primarily due to osmotic effects. In such saline environments the development of the seedling is severely delayed. Although the effect of sodium is primarily osmotic, toxic effects on the hydrated seeds have been reported, for example in barley (Bliss *et al.*, 1986).

Germination and initial seedling growth are the stages of development most sensitive to salinity, irrespective of the salt tolerance of the mature plant. Two cultivars of tomato (*Lycopersicum esculentum*), differing considerably in their salt tolerance, responded almost identically to salinity during germination. They were able to endure prolonged exposure to high, germination inhibiting, levels of salinity (460 mM) and germinated normally when the salt was removed. On the other hand a wild species of tomato, *L. cheesmanii*, which grows naturally on the coast a little above sea level, was severely damaged by exposure to such high salinity (Kurth *et al.*, 1986). Some plants can develop or possess resistance to salt. The special vegetation found in saline habitats is frequently termed halophytic. Many of the plants so classified are distinguished by a high salt tolerance at various stages of their development, including germination, rather than by a positive salt requirement. But some plants show a definite requirement for a certain salinity and these germinate better in the presence of low concentrations of salt, and their subsequent development is also better. This is true for *Atriplex halimus* for example (Fig. 7.2). Even in this plant the tolerance to sodium chloride during growth and development of the seedling is 10 to 100 times greater than during germination. However, in *Atriplex* salinity improved growth only under conditions of high evaporative demand of the atmosphere; under humid conditions any addition of salt retarded growth (Gale *et al.*, 1970). It is possible that the response of germination to salinity is dependent on some additional factor.

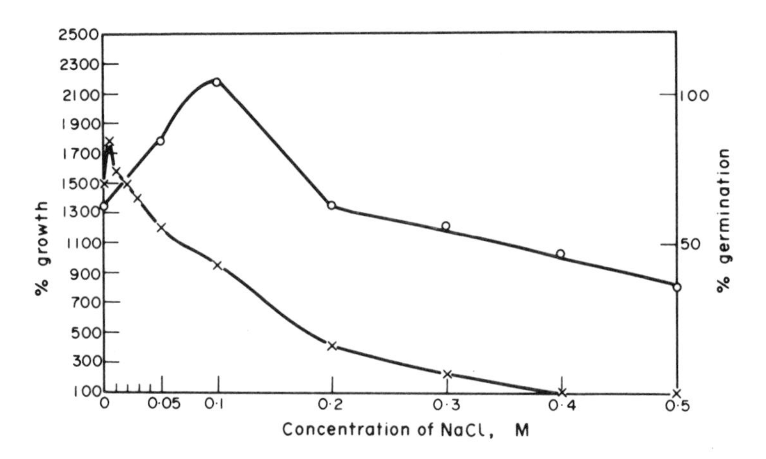

FIG. 7.2. Effect of NaCl concentration on growth and germination of *Atriplex halimus* (Poljakoff-Mayber, unpublished) ×——× % germination, o——o growth, final dry weight as % of initial dry weight.

In a different species of *Atriplex, A. triangularis*, with polymorphic seeds, an interaction between salinity and temperature was observed (Khan and Ungar, 1984). The larger, less dormant seeds germinated early in the growing season when part of the salt was washed out by rain and the temperatures fluctuated between 5° and 25°C. The small seeds, which were more dormant, had a more precise temperature requirement, were much less salt tolerant and with temperature alternations between 5° and 25°C germinated only 5% in the presence of 1.5% NaCl, while the large ones germinated 60%. Most of the small seeds did not germinate in the year of their maturation, but remained dormant in the soil, while most of the large ones germinated.

A study of seeds of a number of coastal species showed that they can be separated into three categories: those from the salt marshes; those from the dunes and those from intermediary areas. The germination of some species of plants from a salt marsh were stimulated by salt. This effect was particularly noticeable when seeds were exposed to sea water and then placed in fresh water, for example *Limonium vulgare*. This species survived 21 weeks immersion in sea water. Seeds of plants from dunes were more adversely affected by salt than those growing nearer the sea (Woodell, 1985). Apparently the group of plants from salt marshes which are exposed to sea water have distinct germination responses to salt determined by the length of exposure and the final conditions under which they germinate.

In many cases in saline habitats, germination occurs when the salt content of the habitat has reached its lowest level, e.g. towards the end of or after the rainy period. However, some plants can germinate under conditions of very high salinity, for example *Arthrocnemum halocnemoides*. Its seeds are still able to germinate in solutions containing 20 g/l NaCl, although the germination percentage under such conditions was half that in water (Malcolm, 1974). The seeds for these experiments were taken from plants growing in saline water-logged areas of Western Australia. The water table there is 1 – 1.5 m deep, and the ground water contains between 14 and 55 g/l soluble salts. It appears that these seeds are adapted to germinate and grow under highly saline conditions.

The ionic composition of the soil water also frequently affects germination. In most areas of the world salinity problems are due to excess Na^+ with Cl^- or SO_4^{2-} as the counter ions, but cases are known where the Ca^{2+} content may also cause problems. Seeds of *Hypericum perforatum* do not germinate in soils containing more than traces of calcium, whereas other seeds seem to be indifferent to the calcium content of the soil. Calcium has been reported in germination, as in other cases of salinity damage, to reverse toxicity caused by NaCl.

Avena fatua germinates well in sandy loam and loess clay which also favour the growth of the seedling. In peaty soils germination is good but growth poor. Calcereous soils were found to be least suitable for its germination (Bachthaler, 1957).

Salt frequently accumulates in the organs of plants growing in saline habitats. As a result the fruits or the dispersal units of such plants often contain salt, although the seeds themselves are usually devoid of NaCl. In such cases germination is to some extent controlled by the salt content of the fruit, and the salt serves as an inhibitor of germination. In other seeds more specific inhibitors, not acting through osmotic effects, are thought to be present.

In a few plants occurring in a saline habitat vivipary is observed. The best-known case is probably that of the mangroves, which grow for most of the year directly in sea water. It seems that during the process of germination these plants are much more sensitive to salt and are more easily damaged by it than the growing seedling as is the case in many plants. In the mangroves vivipary may be regarded as a means of evading the unfavourable environment during germination.

Salinity is one of the most important problems endangering the soils of cultivated and natural habitats. A lot of research is directed towards understanding the effect of salinity on the plant in its different stages of development. But even in non-saline soils many factors can change plant behaviour. When seeds are buried in the soil their germination behaviour may be affected in various ways. For example, induction of cyclical changes in dormancy following burial have been reported (Karssen, 1980/81). Seeds of *Polygonum persicaria, Chenopodium bonus-henricus, Sisymbrium officinale* and *Senecio vulgaris* all had germination characteristics consistent with the behaviour of summer annuals. They had maximum germination ability in spring or early summer. If they failed to germinate, they remained buried and entered secondary dormancy, which was lost only in the following spring. It is not yet certain just what factors in the soil are responsible for such cyclical changes, but temperature, soil moisture and perhaps nitrate content of the soil contributed to them. Thus, here is an instance of seeds whose germination is controlled in such a way that they will not germinate in the wrong season and this mechanism persisted over a number of seasons.

It should be remembered that soils are quite heterogeneous and germination and seedling establishment may be determined by microheterogeneity in the soil. In germination studies there is often a discrepancy between the germination percentage achieved under laboratory conditions and that achieved, for the same stock of seeds, in the field. It is very difficult to reproduce in the laboratory the constantly changing conditions which occur in the field. For this reason extrapolation of the results of laboratory experiments to field conditions should be carried out with great caution.

5. Germination Inhibitors

Germination inhibitors have been mentioned earlier and one of the roles ascribed to them was as sensors of external conditions, especially of rainfall. Germination inhibitors are apparently also present in fruits. The observation that many seeds do not germinate within the fruit led to the idea that fruits may fulfil an ecological role in seed dispersal, as germination will not occur unless the seeds are liberated from the fruit. This will occur if the fruit decomposes, is broken or damaged or if it is eaten by animals and the seeds subsequently excreted. In all these cases the dispersal of the seeds is increased and therefore the chances that the seeds will reach areas removed from the parent plant are improved. However, since fruits contain many compounds which are toxic or unpleasant in taste, such as polyphenols and their oxidation products formed by the action of polyphenol oxidases (Mayer and Harel, 1981), it has been proposed that fruits have a role as protectants of the seeds from danger of predators (Herrera, 1982).

Germination inhibitors have been found in many plants. Although in many cases they have not been identified, nevertheless correlation between inhibitory power of extracts and the germination behaviour of the seeds has been found. In fleshy fruits inhibition may be correlated with sugar content. In other instances salt content in the dispersal unit has been found to be the cause of germination inhibition. In the case of *Zygophyllum dumosum* the dispersal unit was found to contain salt in sufficient amounts to cause inhibition, but, in addition, weak inhibitory substances were also present which accentuated the inhibitory action of the salt (Lerner *et al.*, 1959).

In a few instances specific substances have been found to cause inhibition of germination. Among the inhibitors identified in fleshy fruits are parasorbic acid in *Sorbus aucuparia* (Kuhn *et al.*, 1943), and ferulic acid in tomatoes (Akkerman and Veldstra, 1947). However, the inhibitor alone could not account for the inhibition observed and it is probably not the only inhibitor present. In lemon, strawberry and apricot, Varga (1957a,b) showed the presence of a mixture of organic acids which increased in amount as the fruit ripened and could together act as an inhibitory factor. In dry fruits the only inhibitor identified with any certainty is coumarin, found in fruits of *Trigonella arabica* (Lerner *et al.*, 1959). Even in *Trigonella* probably other, additional, weak inhibitory substances are also present. Phenolics and coumarin and its derivatives are often reported as almost universally present inhibitors, which if present in husks, coats or fruits etc. can also act as germination inhibitors (Van Sumere *et al.*, 1972). In the case of *Fraxinus excelsior* the presence of an inhibitor in the embryo itself has been shown by Villiers and Wareing (1960). In most of these cases the role of the inhibitors may be supposed to be dispersion of seed germination over a period of time. Phenolic substances, which originate in leaf litter, have often been invoked as being germination regulatory substances. However, a thorough discussion of their possible role tends to throw doubt on their role as allelopathic agents (Williams and Hoagland, 1982).

It must be remembered that the presumed functions of inhibitors in fruits are by no means finally proven, and in fact they are very difficult to prove unequivocally. It is possible to interpret the observed facts differently (as already discussed). It is to be hoped that different approaches will lead to new lines of research into the probable biological and ecological function of inhibitors. One such study, accompanied by field observations, has been carried out on fruits of *Ammi majus* (Friedman *et al.*, 1982). Inhibitors of the furanocoumarin group are compartmentalized in the fruit between its inner and outer envelopes. The inner envelope prevents the inhibitor from reaching the embryo, thus avoiding autotoxicity. The outer envelopes permit the slow exit of the inhibitor into the surrounding environment. The high density of the fruits of *Ammi* near the mother plant limits the germination of other plants in the vicinity. The presence of the inhibitor in the fruit also affords protection against predators. More detailed examinations of extracts from fruits and seeds have shown that these contain a mixture of substances, some of which inhibit while others stimulate germination. Others are active in affecting growth. In natural environments substances affecting germination may be present and excreted or leached out also from other parts of the plant. Recently a germination stimulator of the parasite *Striga* was isolated from the exudate of its host, *Sorghum* (Chapter 4). This substance which is active as a germination stimulator of *Striga* at 10^{-7}M, was also isolated from a non-host (Chang *et al.*, 1986). Probably more examples of this kind will be found if the interaction between plants by allelopathy is studied more extensively. From the foregoing discussion it seems likely that germination is not simply controlled by inhibitors but that the interaction of both the promoting and inhibitory substances regulate it, as in the case of *Fraxinus* (Villiers and Wareing, 1960; see also Chapter 6).

6. Biotic Factors

Biotic factors in a given habitat are very variable. They include the interaction between plant parts of the same species, the interaction between different plant species and the role of animals and man in affecting germination.

Many plant organs other than fruits and seeds contain inhibitors of germination. It has frequently been observed that leaves or leaf litter contain compounds which can inhibit germination of a number of seeds. The accumulation of leaf litter under trees could have a regulating effect on germination. If species differ in their sensitivity to inhibitors, then those plants whose seeds are most sensitive to the inhibitor will not germinate in the immediate vicinity of plants whose leaves or straw contain such inhibitors. The presence of inhibitors in leaf litter might affect the distribution of certain plants, favouring some species and preventing the occurrence of others. It is often observed that some plants are entirely absent in the immediate vicinity of certain other species. For example, in Israel the soil surrounding *Eucalyptus* trees is very poorly vegetated. It was suggested that the inhibitors present in the soil or leaf litter might be responsible for this. Yardeni and Evenari (1952) showed that leaves of

Eucalyptus contained germination inhibitors. However, later experiments showed that although a very strong inhibitor could be isolated from the leaves, the soil from *Eucalyptus* groves did not inhibit germination of either wheat or lettuce (Lerner and Evenari, 1961). Thus the inhibitor in the leaves is either washed out or destroyed and no longer exercises a biological function. However, it is possible that very small amounts of inhibitor, together with the mechanical effects caused by the leaf litter, may be responsible for prevention of germination. For example, Dinoor (1959) showed that the leaf litter under Valonia oak constituted a mechanical obstacle to seedling establishment. When germination occurred on the leaf litter, the roots failed to reach the soil and if it occurred underneath the litter, the shoot failed to emerge. Similar effects may occur in *Eucalyptus* groves or in other habitats covered with leaf or straw litter.

Inhibitors are present in the litter of a number of plants. In rice straw, Koves and Varga (1958) showed the presence of a number of phenolic compounds which have inhibitory properties and they suggest that these compounds may have some biological function. Beech litter also contains inhibitors. In this case inhibitors are absent in the litter immediately after the leaves are shed, but appear after it has been exposed to one winter (Winter and Bublitz, 1953). This could be related to the relatively brief viability of beech seeds, which germinate shortly after shedding, the seedlings establishing themselves before the inhibitor develops in the litter. Among the species containing a variety of compounds inhibitory to germination is *Sorghum bicolor* (Lehle and Putnam, 1983; Pasaniuk *et al.*, 1986) and the litter of the plant may therefore have some allelopathic role; but a real biological effect has never been demonstrated.

The caution with which results of experiments with extracts of leaf litter should be regarded is demonstrated by the case of leaf litter of *Backousia* (Cannon *et al.*, 1962). *Araucaria* fails to regenerate in regions where there is a large amount of leaf litter from *Backhousia*. Extracts of the leaf litter do in fact inhibit germination of *Araucaria*, the active compound being dehydroangustione. However, field experiments showed that on the leaf litter itself *Araucaria* germinates well. Watering of the litter increases germination. Thus, in this case the litter was acting in a different way, perhaps causing seedling death, rather than by inhibiting germination.

Other causes for the absence of plants in the vicinity of certain trees may be the competition for water and light, which may have a more decisive effect than inhibitors. If competition is very strong, then small amounts of inhibitors may also play some part in regulating plant distribution.

The presence of fungi, particularly in decomposing leaf litter can complicate the interpretation of germination studies due to the possible liberation of volatile stimulating compounds by the litter decomposing fungi. Volatile substances produced by fungi may affect germination of many seeds. The response of *Rumex* and of *Sorghum* are examples. *Rumex* germination was stimulated by 5 – 10 ppm of nonanenitrile, 2-nonanone and even low concentrations of octylthiocyanate. *Sorghum halepense* germination was stimulated by low concentrations of methyl salicylate (French, 1985).

Competition apparently is the crucial factor in establishment of *Juncus* species among the sward of grasses. The *Juncus* can germinate and establish itself well when the soil is very moist. However, if the soil dries out a little, then *Juncus* no longer germinates readily. Under these circumstances other plants germinate and establish a cover which prevents light reaching the *Juncus* seeds, which require light for their germination. The *Juncus* is subsequently effectively prevented from germinating (Lazenby, 1956). Such *Juncus* seeds retain their light sensitivity, even if buried, for many years.

Many members of the Cruciferae, such as *Brassica* and *Sinapis,* contain complexes of mustard oils in their fruit as well as in other parts of the plant. In *Sinapis arvensis* the fruits contain mustard oils, but the seeds which are shed germinate readily. The upper part of the fruit does not open readily and retains its single seed. This seed only germinates later, when the mustard oils have been washed out of the fruit. An interesting example of inhibitory action by other parts of *Brassica* is provided by the case of *B. nigra*. In the forties certain areas of California, which were damaged by fire, were subsequently resown with *Brassica nigra*. In the areas so treated many species failed to re-establish themselves (Went *et al.*, 1952). Apparently, *Brassica* leaf litter contains an inhibitor which prevented the germination of certain plants. The presence of a water-soluble inhibitor in leaves of *Brassica nigra* was in fact proved (Scherzer, 1954).

In the aerial parts of *Echium plantagineum* two distinct inhibitors were demonstrated which are of special interest because they affected different species of plants quite differently (Ballard and Grant-Lipp, 1959).

The leaves of *Encelia farinosa* contain a powerful inhibitor of both germination and growth. A mulch of leaves of *Encelia* can inhibit the germination of a number of plants (Bonner, 1950). However, the suggestion that this accounts for the absence of plants very near to *Encelia* shrubs has been disputed by Muller (1953). He claims that *Franseria* plants growing in a similar habitat also contain powerful inhibitors, yet its immediate environment is populated by other plants. Muller ascribes the differences between the ground cover in the immediate vicinity of these two shrubs as being due to differences in the growth habit and in the formation of leaf litter. Under *Encelia* open ground species develop readily.

Another instance of a plant containing a germination inhibitor is *Artemisia absinthium*, the inhibitor apparently being the alkaloid absinthin. A number of plants sown in the vicinity of *Artemisia* fail to germinate or to develop. For example, *Levisticum officinale* is killed up to a distance of one metre. Some plants are more resistant than others, *Senecio, Lathyrus clymena* and *Linum austriacum* being extremely sensitive, while *Stellaria* and *Datura* are quite resistant (Funke, 1943). The ecological importance is again in doubt, as seedlings even of sensitive plants, which survive for a year near *Artemisia absinthium*, subsequently develop quite normally, without further inhibition. The decomposing leaf litter of *Celtis laevigata* did significantly inhibit the germination of a number of species (Lodhi and Rice, 1971). An interesting mechanism has been suggested for the seeds of *Coumarouna odorata* (Valio, 1973). Seeds of this plant have a high coumarin content and their germination is quite insensitive to exogenous coumarin. After germination the radicle excretes coumarin and this can then apparently inhibit the germination of other seeds in the vicinity. It remains to be proved that this effect occurs under natural conditions in the soil as well as in the laboratory.

Seeds of *Heracleum laciniatum* contain inhibitors which delay the germination of the seeds, for example lettuce and *Salix*, and which particularly powerfully inhibit root elongation. Among the active compounds are furano coumarins. Whether these inhibitors, which leak out quite easily, have a role in preventing germination of other seeds near the *Heracleum* seeds is unclear (Juntilla, 1975, 1976).

Seeds of *Coffea* contain large amounts of caffeine in their endosperm, which may cause self intoxication of the root cap during germination, by inhibition of mitosis (Friedman and Waller, 1983). The plant apparently avoids autointoxication by a relatively rapid elongation of the hypocotyl, whereby the meristematic root tip is pushed away from the caffeine-containing endosperm so that it can resume cell division. During germination the caffeine is transported from the endosperm to the cotyledons. The caffeine appears to be sequestered from sites where mitosis occurs. The mature plant and senescing seeds can also liberate caffeine, which may damage roots. In old plantations degeneration of plants may be ascribed to the toxic effect of caffeine on the roots. This compound is also presumed to protect seeds from predators. *Coffea* provides an example of the complex allelopathic effects of a donor plant, which secretes an inhibitory substance, which on the one hand protects it from competition and on the other hand causes damage by autotoxicity. How these opposing effects are balanced and whether ecological benefits predominate still remains unclear (Friedman and Waller, 1985).

Many examples are found in botanical literature of the effect of plants on each other. Many of these are based on visual observation and most experiments have not been sufficiently rigorous to permit interpretation. Although it appears that plants do excrete substances, or that substances leach out from them, which can affect other plants or seeds in their environment, the magnitude of the effects and their biological importance is still not clear. This whole problem has been reviewed by Evenari (1965); see also Rice (1983).

The seeds of certain parasitic plants present a special ecological problem. Both *Striga* and *Orobanche* can develop only if the seeds germinate very near to plants which they can parasitize (see Chapter 4 and previous section). The germination of these parasitic plants normally requires the presence of a stimulator, but they can be induced to germinate artificially. It appears likely that the stimulator which is required by these seeds is excreted or leached out from the roots of very many plants, and that it constitutes a normal metabolite in such plants. The requirement of *Striga* and *Orobanche* for such a factor in their germination would ensure that they germinate only under conditions where the chance to parasitize a suitable host is good (Musselman, 1980). A compound stimulating the germination of *Striga* has been isolated from cotton, which is not a host for the parasite (see previous section). It has been chemically identified and named strigol. Its structure is shown in Fig. 4.5, VII. This compound is very active and stimulates germination to 50% at 10^{-11}M, but appears to be rather non-specific. Recently it has been shown that exudates from *Sorghum*, which is a host for *Striga*, contain a hydroquinone derivative, which may be the host specific stimulant for *Striga* germination (Chang *et al.*, 1986). The structure of this compound is shown in Fig. 4.5, IX (Chapter 4). It is rather unstable and decomposed readily. This decomposition occurs a short distance from the host root and might be the mechanism which assures *Striga* germination near the host root.

The seeds of parasitic plants such as *Striga* germinate near their host, and soon the development of the radicle is stopped and a haustorium formed to attach to and invade the host. It has now been shown that the haustorium-inducing factor for *Striga* is host cell derived and is 2,6-dimethoxy – *p.* – benzoquinone. This compound apparently is produced by the oxidative attack of a laccase-like enzyme from the parasitic root which oxidizes compounds on the host root surface (Chang and Lynn, 1986).

In the case of *Viscum album* and other *Viscum* species nothing is known about special germination requirements, the sticky nature of the fruit merely increasing the chances of it adhering to trees after being carried by birds. This increases the likelihood of the seeds reaching a suitable host. The chemical composition of the sticky material, viscin, has recently been described (Gedalovich *et al.*, 1988a). Viscin is a mucilage located in special cells in the inner layer of the ovary. The mucilage is made up of highly branched polysaccharides, mainly xylans but also containing arabinans, galactomannans, xyloglucans and rhamnogalacturonans, as well as neutral sugars and some protein. The viscin of the Viscaceae and Loranthaceae differ in composition, the latter containing less neutral sugars and more protein than the former (Gedalovich *et al.*, 1988a,b). The subject of germination of parasitic angiosperm seeds has been reviewed by Brown (1965), Edwards (1972) and by Worsham (1987).

Extracts or diffusates from other plants have been known to have germination-inhibiting properties, e.g. from *Coridothymus*, an aromatic shrub (Katz *et al.* 1987) and *Ambrosius* (Dalrymple and Rogers, 1983).

Fire is an environmental factor but it may be regarded at least in some senses as a biotic factor because it changes the flora and fauna in the environment in which the seed eventually germinates. Fire can remove vegetation and so improve light and aeration and at the same time remove competition for space, light, water and nutrients between the seedlings which are establishing themselves and the existing plants. Went *et al.* (1952) suggested that in addition fire can destroy accumulated inhibitors present in the soil cover and uppermost layers of the soil, again removing a possible cause of failure of seeds to germinate, or of the seedlings to establish themselves. The charred remains of chapparal vegetation appeared actually to stimulate the germination of *Emmeranthe penduliflora* (Wicklow, 1977). It appears that during charring promotory substances are formed. In the case of chapparal herbs, oligosaccharide type molecules have been postulated as the active compounds stimulating germination, which arise during charring. Even heating of wood to 175°C produced active compounds, which may arise from hemicelluloses or xylans (Keeley and Pizzorna, 1986). Another important effect of fire may be the subsequent loosening of hard seed coats of seeds in the soil by heating and charring. This will make them more permeable to water and gases and will reduce the resistance to the pressure of the elongating rootlet.

Many animals can change the balance of different plants in a given area by grazing, by distributing the seeds, by the excretion of seeds in new habitats different from those in which the fruit was eaten and by other means. Very few attempts have been made to study quantitatively the effect of passage through the digestive tract on seed germination and distribution. Cottontail rabbits commonly eat seeds of *Polygonum*. While the seeds of two species, *P. lapanthifolium* and *P. persicaria* germinated after secretion, the larger seed of *P. pennsylvanicum* were destroyed in the digestive tract. The retention in the digestive tract seems to aid seed distribution (Staniforth and Cavers, 1977). Passage of seeds of *Najas* through the digestive tract of mallard ducks resulted in the destruction of 70% of the seeds. However, the remainder had much improved germination. The remaining viable seeds may be carried for very long distances by the ducks, since they remain in the digestive tract for more than 10 hr (Agami and Waisel, 1986).

Goats grazing in the Middle East denuded the area of forests and as a result caused the establishment of a different vegetation containing many annuals and shrubs. The removal of the upper stratum of plants may be advantageous to annual plants, by reducing competition for light due to the removal of shrubs. Insects of various kinds which collect seeds may have certain local effects, for example, ants accumulating certain kinds of seeds in or near their heaps. Ants may play an important role in the destruction of seeds, particularly those seeds which are used for forage. They may collect seeds near their nests, some of which still may be able to germinate (Carroll and Janzen, 1973). Some species of seeds have a special appendage containing oil droplets, the elaiosome, which makes the seeds attractive to ants (Berg, 1972). The oil-containing appendage is utilized by the ants, but the seed itself is not eaten. In this case the ants act as a dispersal agent.

Seeds of *Calathea* are collected by ponerine ants, which are attracted by the aril. The aril is removed in the nest, and the seeds are scattered. They germinate better without the aril (Horvitz and Beattie, 1980). This may contribute to seed distribution and improved germination. Probably other cases of ant involvement in seed distribution and modified germination are known, but only a few have been studied.

Man has exercised a profound effect on the distribution of plants in certain areas. Agricultural practice has always been designed to cause the establishment of certain plants to the entire exclusion of others. The disastrous results which this policy may have was shown in the centre of the United States which was converted into an almost barren area for many years, the so-called “dustbowl”. Jungle clearance in South America for temporary agriculture, without subsequent reafforestation, is also resulting in man-made deserts. The methods formerly used to control vegetation were relatively simple, consisting of ploughing and weed eradication. However, modern usage of herbicides has greatly increased the means at the disposal of man to alter or control vegetation in all stages of development, and consequently the danger of major changes over widespread areas has also increased.

7. Seedling Establishment

Root growth ensuring water supply, seedling vigour in piercing the soil surface and the ability to begin photosynthesis are some of the factors which ensure the establishment of seedlings from small seeds. The fact that most small seeds have as their main storage materials fats, which are of high caloric value, may also be important. Small, fat-containing dicotyledonous seeds have epigeal germination and green cotyledons.

In most cases, the germinating seed must first of all establish anchorage by the root in the soil and ensure commencement of water and solute absorption, which are prerequisites for the growth of the shoot.

The question of the pressure that must be developed by the rootlet to pierce the soil was discussed by Pfeffer (1893). Some empirical data were collected over the years, but no attempt to evaluate the relevant soil parameters was made. A more precise attempt to explain certain features of growth of pea and wheat roots was made by Barley *et al.* (1965). In these experiments, soil density, matric potential and apparent soil cohesion were varied, and root growth followed. Soil cores of different densities and matric potentials were prepared and placed in the experimental pots to form a resistance to the advance of the roots. This resistance delayed further root growth by at least 24 hr and induced an increase in root diameter. Production of lateral roots was reduced by the resistance of the cores. Eventually the roots did penetrate the cores and grew through them. The data show that soil strength does have an important effect on the ability of the roots to penetrate it, especially through clods or finely structured layers. Fine metal probes and penetrometers were developed to study the problem more precisely (Farrell and Greacen, 1966). It seeems that the maximal pressure that can be developed by the growing root is close to 10 bars, a value that was already suggested by Pfeffer (1893).

It is root elongation which generates the forces needed to pierce and penetrate the soil or the artificial cores in the experimental systems. However, in some cases not only the root but also the plumule is pushed into the soil. This phenomenon is known as *cryptogeal germination* and was described by Jackson (1974). A case of cryptogeal germination has been described by Clarkson and Clifford (1987) in the fruits of *Jedda multicaulis*, an Australian woody dicotyledon, which grows in a very limited area in an open Eucalyptus forest on the northern shore of Australia (Clarkson and Clifford, 1987). In this plant the seed germinates within the fruit. The rootlet, together with the fused petioles of the cotyledons, penetrate the soil where they form a kind of pocket in which the plumule is protected. When the underground structure has developed a root system, the plumule emerges above the ground. Cryptogeal germination apparently provides the seedling with efficient protection so that even if the part of the young seedling above the ground is damaged, the buried part of the stem can regenerate vegetatively.

The diffusion of solutes towards the growing root in a fairly dry soil may apparently result in concentrations, at the root surface, 10 times higher than the average concentration in the soil solution (Passioura and Frere, 1967). This may create osmotic problems in water absorption.

When anchorage is accomplished and water and solute absorption is ensured, there is still a problem of piercing the soil surface and raising the plumule. Natural rainfall, or irrigation, cause the breakdown and rearrangement of the structural units of the soil surface. Due to water flow and splash, and due to sedimentation of the disturbed soil particles, the surface tends to be covered by a continuous layer of closely packed particles. These compact layers are often described as "seals". Besides this effect they decrease the infiltration rate and gaseous exchange of the soil; they also form a considerable mechanical impedance to seedling emergence (Arndt, 1965a,b). In some cases a single seedling is incapable of piercing the seal and only when a few seedlings germinate simultaneously are they capable of lifting the seal and breaking it, thus creating pathways for emergence above ground. Arndt (1965b) suggests a model to permit the estimation of the axial force which can be developed by the seedling in order to penetrate the seal. The seedling can be regarded as a column supporting a load. The initial load-bearing potential depends to a large extent on the cross-section at the root – stem junction. This potential decreases with the growth of the stem, or the hypocotyl, as the buckling tendency of a loaded column increases with its length, independently of the cross-sectional area. As the length of the stem increases, due to deep burial of the seed, the combined effects of the reduced cross-sectional area of the stem and increased tendency for buckling tend to reduce the effective axial force of the seedling. From this point of view a shallow position position of the seeds in the soil may be therefore advantageous to seedling establishment.

The hypocotyl frequently is the organ which thrusts through the soil. In tomato seedlings, the anatomy of the hypocotyl changes. As the compression of soil increases, the hypocotyl thickens and seedling emergence is slowed down (Liptay and Geier, 1983). At soil shear strength between 0 – 9.3 kPa seedling emergence of wheat was markedly affected, although germination was only slightly delayed, by 1 – 12 hr, perhaps due to a reduced rate of water uptake. At a soil shear strength of 9.3 kPa seedling emergence was reduced to 7% and coleoptile and root length were adversely affected (Collis-George and Yoganathan, 1985a). When soil shear strength was increased above 30 kPa, germination and water uptake were significantly reduced, e.g. at 52 kPa germination was reduced to 8.3% but emergence was zero at 28 kPa. The shear strength was measured before water uptake by the seeds occurred and depended on the nature of the soil and its compressibility (Collis-George and Yoganathan, 1985b).

The seeds of the various species contain different amounts of storage material and therefore have different needs for the onset of independent photosynthesis and production of protein and other plant components. Most cotyledons are leaf-like and especially in the epigeal species they become green and apparently capable of photosynthesis.

The different epigeal species differ in the ability of their cotyledons for expansion and photosynthesis. Lovell and Moore (1970) studied the cotyledons of 11 species ranging from hypogeal, through non-expanding epigeal to species whose cotyledons develop a very considerable photosynthetic area. The cotyledons of the hypogeal *Pisum sativum* and *Phaseolus multiflorus* did not expand, lost weight with time and survived for a very short period. They produced almost no chlorophyll when exposed to light, had no stomata and practically no capacity for carbon dioxide fixation. They merely served as food reserve for the growing embryo.

In epigeal, photosynthesizing cotyledons, the products of $^{14}CO_2$ fixation were not transported from them until they reached their maximal size. Chlorophyll production in cotyledons attached to the axis was greatly enhanced compared to detached ones (Moore *et al.*, 1972). It also seems that the embryonic axis has some differential effect on the development of the photosynthetic systems in the cotyledons (Moore and Lovell, 1970). The growth rate of seedlings was highest in those species in which the cotyledons showed most expansion and the highest capacity for carbon dioxide fixation (Lovell and Moore, 1971). In such seedlings the development of the first leaves was delayed.

Competition between individuals of the same species is an additional important factor in seeding establishment. One of the strategies for survival is the production of very many seeds. However, only a small proportion of them survive and complete their life cycle. This is especially marked in desert environments. In such desert areas, the rare combination of sufficient rain, at the right time with a suitable temperature, causes the appearance of flushes of vegetation. Large numbers of seeds germinate, forming dense covers usually near the mother plant. These seedlings compete with each other for water and light. In counts made in the winter of

1964/65 near Avdath experimental station in the Negev desert, 7623 seedlings of *Salsola inermis* were noted in an area of 16 m^2, of which only 284 became established and completed their life cycle. In the winter of 1963/64 in the same area, 2250 seedlings of *Diplotaxis* emerged but only 709 survived till the next year. The survivors subsequently prevented the establishment and survival of 1000 new seedlings which emerged in the following year (Evenari and Gutterman, 1976; Gutterman, 1983). In the competition between seedlings of different species, the various substances leaching from fruits and seeds, which delay or prevent the germination, but somehow avoid autotoxicity, may be a way of providing more living space (see above).

From the foregoing discussion of the ecology of germination it is clear that much remains to be discovered in this respect. There is an urgent need for controlled experiments and for rigorous comparisons between the ecological behaviour of seeds in the laboratory and in their natural habitat. Germination is a key process in determining plant distribution and a study of its ecology will add greatly to our understanding of plant ecology in its wider aspects.

Bibliography

Agami M. (1986) *Physiol. Plant.* **67**, 305.
Agami M. and Waisel Y. (1986) *Oecologia* **68**, 473.
Akkerman A. M. and Veldstra H. (1947) *Rec. Trav. Chim. Pays-Bas* **66**, 441.
Arndt W. (1965a) *Austr. J. Soil Res.* **3**, 45.
Arndt W. (1965b) *Aust. J. Soil Res.* **3**, 55.
Attims Y. and Come D. (1978) *C. R. Acad. Sc. Paris* **286D**, 1669.
Augspurger C. K. (1979) *Oecologia* **44**, 53.
Bachthaler G. (1957) *Z. Acker und Pflanzenbau* **103**, 128.
Ballard L. A. T. and Grant-Lipp A. E. (1959) *Aust. J. Biol. Sci.* **12**, 342.
Barley K., Farrell D. A. and Greacen E. L. (1965) *Aust. J. Soil Res.* **3**, 69.
Berg Y. (1972) *Am. J. Bot.* **59**, 109.
Bliss R. D., Platt-Aloia K. A. and Thompson W. W. (1986) *Plant Cell Environment* **9**, 727.
Bonner J. (1950) *Bot. Rev.* **16**, 51.
Brenchley W. E. and Warington K. (1933) *J. Ecol.* **21**, 103.
Brown R. (1965) in *Handbuch der Pflanzen Physiologie* **15**, 925, Springer-Verlag, Berlin.
Cannon J. R., Corbett N. H., Haydock K. P., Tracey J. G. and Webb L. J. (1962) *Aust. J. Bot.* **10**, 119.
Capon B. and van Asdall W. (1967) *Ecology* **48**, 305.
Carroll C. R. and Janzen D. H. 1973) *Ann. Rev. Ecol. Syst.* **4**, 231.
Cavers P. B. and Harper J. L. (1966) *J. Ecol.* **54**, 367.
Cavers P. B. and Harper J. L. (1967) *J. Ecol.* **55**, 73.
Chang M. and Lynn D. G. (1986) *J. Chem. Ecol.* **12**, 562.
Chang M., Netzly O. H., Butler L. G. and Lynn D. G. (1986) *J. Am. Chem. Soc.* **108**, 7858.
Clarkson J. R. and Clifford H. T. (1987) *Aust. J. Bot.* **5**, 715.
Cohen D. (1966) *J. Theoret. Biol.* **12**, 119.
Cohen D. (1986) *J. Ecol.* **56**, 219.
Cohen D. and Levin S. A. (1987) in *Mathematical Topics, Population Biology and Neurosciences*, Lectures in Biomathematics p. 110, vol. 71 (Eds E. Teramoto and M. Yomaguti).
Collis-George N. and Hector R. (1966) *Aust. J. Soil. Res.* **4**, 145.
Collis-George N. and Yoganathan P. (1985a) *Austr. J. Soil Res.* **23**, 577.
Collis-George N. and Yoganathan P. (1985b) *Austr. J. Soil Res.* **23**, 589.
Come D. (1970) *Les Obstacles à la Germination*, p. 162, Masson & Cie, Paris.
Cook C. E., Whichard L. P., Wall M. E., Egley G. H., Coggan P., Luhan B. A. and McPahil A. T. (1972) *J. Am. Chem. Soc.* **94**, 6198.
Corbineau F., Rudnicki R. M. and Come D. (1988) *Physiol. Plant.* (In press).
Crocker W. (1906) *Bot. Gaz.* **42**, 265.
Dalrymple, R. L. and Rogers, J. L. (1983) *J. Chem. Ecol.* **9**, 1073.
Dinoor A. (1959) M. Sc. (Agr.) Thesis, Rehovot (in Hebrew).

Duvel J. W. T. (1905) U.S.F.A. *Bureau of Plant Industry*, Bull. No. 83.
Edwards M. M. (1969) *J. Exp. Bot.* **20**, 876.
Edwards W.G. H. (1972) in *Phytochemical Ecology*, p. 235 (Ed. J. B. Harborne), Academic Press, London.
Ellis R. H. and Roberts E. H. (1979) *Ann. Bot.* **44**, 677.
Evenari M. (1965) in *Handbuch der Pflanzen Physiologie*, p. 691, vol. 16, Springer-Verlag, Berlin.
Evenari M. and Gutterman Y. (1976) in *Etude de Biologie Vegetal*, p. 57. (Ed. R. Jacques) CNRS Gif sur Yvette.
Evenari M., Shannon L. and Tadmore N. (1980) *The Negev, the Challenge of a Desert*, Bialik Publishing Co., Jerusalem (in Hebrew).
Farrell D. A. and Greacen E. L. (1966) *Aust. J. Soil Res.* **4**, 1.
Forsyth C. and Brown N. A. C. (1982) *New Phytol.* **90**, 151.
French R. C. (1985) *Ann. Rev. Phytopathol.* **23**, 173.
Friedman J. and Waller G.R. (1983) *J. Chem. Ecol.* **9**, 1099.
Friedman J. and Waller G. R. (1985) *Trends Biochem. Sci.* **10**, 47.
Friedman J., Rushkin E. and Waller G. R. (1982) *J. Chem. Ecol.* **8**, 55.
Funke G. L. (1943) *Blume* **5**, 281.
Gale J., Naaman R. and Poljakoff-Mayber A. (1970) *Aust. J. Biol. Sci.* **23**, 947.
Garwood N. (1985) *Am. J. Bot.* **72**, 1095.
Gedalovich E., Kuijt J. and Carpita N. C. (1988a) *Physiol. Mol. Plant Pathol.* **32**, 61.
Gedalovich E., Delmer D. and Kuijt J. (1988b) *Ann. Bot.* (In press).
Goudey J. S., Saini H. S. and Spenser M. S. (1987) *Can. J. Bot.* **65**, 849.
Gross K. L. (1985) *New Phytologist* **170**, 531.
Gutterman Y. (1980/81) *Israel J. Bot.* **29**, 105.
Gutterman Y. (1983) in *Ecology and Environmental Quality*, p. 1 (Ed. H. I. Shuval), Balaban, ISS, Rehovot.
Gutterman Y. (1986) in *Environmental Quality and Ecosystem Stability* vol. 3A/B, p. 135 (Eds Z. Dubinski and Y. Steinberger), Bar Ilan Univer. Press.
Gutterman Y. and Agami M. (1987) *J. Arid Environ.* **12**, 215.
Gutterman Y. and Heydecker W. (1973) *Ann. Bot.* **37**, 1049.
Gutterman Y. and Porath D. (1975) *Oecologia* **18**, 37.
Gutterman Y. and Witztum A. (1977) *Bot. Gaz.* **138**, 29.
Gutterman Y., Witztum A. and Evenari M. (1967) *Isr. J. Bot.* **16**, 213.
Hadas A. (1970) *Isr. J. Agr. Res.* **20**, 3.
Hadas A. (1977) *J. Exp. Bot.* **28**, 977.
Hadas A. and Stibbe E. (1973) in *Physical Aspects of Soil in Eco-systems*, p. 97 (Ed. A. Hadas, D. Swartzendruber, P. E. Rijtema, M. Fuchs and B. Yaron), Springer-Verlag, Berlin.
Hadas A. and Russo D. (1974) *Agrom. J.* **65**, 643.
Harper J. L. and Benton R. A. (1966) *J. Ecol.* **54**, 151.
Harper J. L., Lovell P. H. and Moore K. G. (1970) *Ann. Rev. Ecol. Syst.* **1**, 327.
Herrera C. M. (1982) *American Naturalist* **120**, 218.
Hilton J. R. (1985) *J. Exp. Bot.* **36**, 974.
Holmes M. G. and Smith H. (1977a) *Photochem. Photobiol.* **25**, 533.
Holmes M. G. and Smith H. (1977b) *Photochem. Photobiol.* **25**, 539.
Holmes M. G. and Smith H. (1977c) *Photochem. Photobiol.* **25**, 551.
Horvitz C. C. and Beattie A. J. (1980) *Am. J. Bot.* **67**, 321.
Hunter J. R. and Erickson A. F. (1952) *Agr. J.* **44**, 107.
Jackson G. (1974) *New Phytol.* **73**, 771.
Juhren M., Hiesen W. H. and Went F. W. (1953) *Ecology* **34**, 288.
Juntilla O. (1973) *Physiol. Plant.* **29**, 264.
Juntilla O. (1975) *Physiol. Plant.* **33**, 22.
Juntilla O. (1976) *Physiol. Plant.* **36**, 374.
Karssen C. M. (1970) *Acta Bot. Neerl.* **19**, 81.
Karssen C. M. (1980/81) *Isr. J. Bot.* **29**, 65.
Katz D. A., Sneh, B. and Friedman, J. (1987) *Plant and Soil* **98**, 53.
Keeley S. C. and Pizzorna M. (1986) *Am. J. Bot.* **73**, 1284.
Khan M. A. and Ungar I. A. (1984) *Am. J. Bot.* **71**, 481.
Khan M. A. and Ungar I. A. (1986) *Bot. Gaz.* **147**, 148.
Kidd F. (1914a) *Proc. R. Soc. B.* **87**, 408.
Kidd F. (1914b) *Proc. R. Soc. B.* **87**, 609.

Kigel J., Ofir M. and Koller D. (1977) *J. Exp. Bot.* **28**, 1125.
Kigel J., Gibly A. and Negbi M. (1979) *J. Exp. Bot.* **30**, 997.
Koller D. (1954) Ph.D. Thesis, Jerusalem (in Hebrew).
Koller D. (1955) *Bull. Res. Council, Israel* **5D**, 85.
Koller D. (1956) *Ecology* **37**, 430.
Koller D. (1972) in *Seed Biology*, vol. 2, p. 1 (Ed. T. T. Kozlowski), Academic Press, London.
Koller D. and Negbi M. (1959) *Ecology* **40**, 20.
Koves E. and Varga M. (1958) *Acta Biol. Szeged.* **4**, 13.
Kuhn R., Jerchel D., Moewus F. and Moeller E. F. (1943) *Naturwissenschaften* **31**, 468.
Kurth E., Jensen A. and Epstein E. (1986) *Plant Cell Environ.* **9**, 667.
Lazenby A. (1956) *Herbage Abstracts* **26**, 71.
Lecat S. (1987) Ph.D. Thesis Univ. Pierre and Marie Curie, Paris.
Lehle F. R. and Putnam H. R. (1983) *J. Chem. Ecol.* **9**, 1223.
Lerner H.R. and Evenari M. (1961) *Physiol. Plant.* **14**, 229.
Lerner H. R., Mayer A. M. and Evenari M. (1959) *Physiol. Plant.* **12**, 245.
Liptay A. and Geier T. (1983) *Ann. Bot.* **51**, 409.
Lodhi M. AK. and Rice E. L. (1971) *Bull. Torrey Bot. Club* **98**, 83.
Lovell P. M. and Moore K. G. (1970) *J. Exp. Bot.* **21**, 1017.
Lovell P. M. and Moore K. G. (1971) *J. Exp. Bot.* **22**, 153.
Lubke M. A. and Cavers P. B. (1969) *Can. J. Bot.* **47**, 529.
McComb J. A. and Andrews R. (1974) *Aust. J. Exp. Agr. Animal Husbandry* **14**, 68.
Malcolm C. V. (1974) *J. R. Soc. West Australia* **47**, 73.
Mall L. P. (1954) *Proc. Nat. Acad. Sci. India* **24**, Sec. B., 197.
Marchaim U., Birk Y., Dovrat A. and Berman T. (1972) *J. Exp. Bot.* **23**, 302.
Mayer A. M. and Harel E. (1981) *Phytochem Soc. Sump. Ser.* **19**, 161.
Moore K. G. and Lovell P. H. (1970) *Planta* **93**, 284.
Moore K. G., Bentley K. and Lovell P. H. (1972) *J. Exp. Bot.* **23**, 432.
Morinaga T. (1926a) *Am. J. Bot.* **13**, 126.
Morinaga T. (1926b) *Am. J. Bot.* **13**, 159.
Mott J. J. (1972) *J. Ecol.* **60**, 293.
Muller C. H. (1953) *Am. J. Bot.* **40**, 53.
Musselman L. J. (1980) *Ann. Rev. Phytopathol.* **18**, 463.
Okagami N. and Kawai M. (1982) *Bot. Mag. Tokyo* **95**, 151.
Okagami N. and Kawai M. (1983) *Plant Cell Physiol.* **24**, 509.
Pasaniuk O., Bills D. D. and Leather G. R. (1986) *J. Chem. Ecol.* **12**, 1553.
Passioura J. B. and Frere M. H. (1967) *Aust. J. Soil Res.* **5**, 149.
Pfeffer E. (1893) *Abh. Sachs. Akad. Wiss.* **20**, 233.
Pons T. L. (1983) *Plant Cell Environ.* **6**, 385.
Pons T. L. (1984) *Plant Cell Environ.* **7**, 263.
Pons T. L. and Schroder H. F. J. M. (1986) *Oecologia* **68**, 315.
Porter N. G. and Wareing P. F. (1974) *J. Exp. Bot.* **25**, 583.
Quinn J. A. (1977) *Am. Midland Naturalist* **97**, 484.
Rice E. E. (1983) *Allelopathy*, 2nd Edn, Academic Press, New York.
Scherbatskoy T., Klein R. M. and Badger G. F. (1987) *Environ. Exp. Bot.* **27**, 157.
Scherzer R. (1954) M.Sc. Thesis, Jerusalem (in Hebrew).
Sheldon J. C. (1974) *J. Ecol.* **62**, 47.
Shull C. A. (1911) *Bot. Gaz.* **52**, 455.
Sifton H. B. (1959) *Can. J. Bot.* **37**, 719.
Silvertown J. W. (1984) *Am. Naturalist* **124**, 1.
Simpson G. M. (1966) *Can. J.Bot.* **44**, 1.
Sinclair T. R. and Lemon R. (1973) *Sol. Energy* **15**, 89.
Staniforth R. J. and Cavers P. B. (1977) *J. Appl. Ecol.* **14**, 261.
Stiles W. (1950) *Introduction to the Principles of Plant Physiology*, Methuen, London.
Stoutjesdyk Ph. (1972) *Acta Bot. Neerl.* **21**, 185.
Stumpff N. J. and Johnson J. D. (1987) *Physiol. Plant.* **69**, 167.

Summerfield R. J. (1973) *J. Ecol.* **61**, 387.
Thomas T. H., Biddington N. L. and O'Toole D. F. (1979) *Physiol. Plant.* **45**, 492.
Thompson K. (1987) *New Phytol.* **66**, (Supplement) 23.
Thompson K. and Grime J. R. (1983) *J. Appl. Ecol.* **20**, 141.
Thompson P. A. (1970) *Ann. Bot.* **34**, 427.
Thompson P. A. (1971a) *Physiol. Plant.* **23**, 734.
Thompson P. A. (1971b) *J. Ecol.* **58**, 699.
Thorton N. C. (1943-45) *Contr. Boyce Thompson Inst.* **13**, 357.
Valio F. M. (1973) *J. Exp. Bot.* **24**, 442.
Van Sumere C. F., Cottenie J., de Greef J. and Kint J. (1972) *Recent Adv. Phytochem.* **4**, 165.
Varga M. (1957a) *Acta Biolog. Szeged.* **3**, 213.
Varga M. (1957b) *Acta Biolog. Szeged.* **3**, 225.
Vickery R. K. (1967) *Ecology* **48**, 649.
Villiers T. A. and Wareing P. F. (1960) *Nature, Lond.* **185**, 112.
Wareing P. F. and Foda H. A. (1957) *Physiol. Plant.* **10**, 266.
Went F. W. (1953) *Desert Research Proceedings, Int. Symp.* pp. 230-240, Jerusalem Special Publication No. 2 Res. Council, Israel.
Went F. W. (1957) *Experimental Control of Plant Growth* pp. 248-251, Chronica Botanica Co., Waltham, Mass.
Went F. W., Juhren G. and Juhren M. C. (1952) *Ecology* **33**, 351.
Wesson G. and Wareing P. F. (1969) *J. Exp. Bot.* **20**, 414.
Wicklow D. T. (1977) *Ecology* **58**, 201.
Williams R. D. and Hoagland R. E. (1982) *Weed Sci.* **30**, 206.
Winter A.G. and Bublitz W. (1953) *Naturwissenschaften* **40**, 416.
Woodell S. R. J. (1985) *Vegetatio* **61**, 223.
Worsham A. D. (1987) in *Parasitic Weeds in Agriculture*, vol. 1, p. 45 (Ed. L. J. Musselman), CRC Press, Boca Raton.
Yardeni D. and Evenari M. (1952) *Phyton* **2**, 11.

8

Seeds and Other Methods of Propagation and their Use by Man

In the previous chapters we have considered seeds as the major means by which reproduction and continuation of the species, and its distribution in time and space are achieved. Seed longevity, dormancy and viability were discussed. Another aspect of the role of seeds is their use by man in agriculture and horticulture, their use as a source of food and feed for man and domesticated animals and their use as a source of various raw materials for industry. In the following an overview is given of some of the practical aspects of seed use and technology and on tissue culture as a way of propagation.

I. Seeds as Food

The most important seeds grown for food and feed as a source of carbohydrates are the cereals, while the legumes are the chief source of protein. In addition various species belonging to different families are used as a source for lipids, for example oils and waxes. Over 1500 million metric tonnes of cereals are grown each year, as compared to 150 million tonnes of legumes. Large amounts of seeds are allocated for sowing every year, although the bulk of the seeds are used for consumption. Even for vegetable crops or ornamental plants, the amount of seeds required is very considerable. We will not concern ourselves with the production process, i.e. the growing of seeds, the conditions needed for flowering and for maximizing seed formation (see Chapters 3, 4 and 7; and Seeds, 1961; Hebblethwaite, 1980).

The large amounts of seeds produced have to be stored, sometimes for long periods. The storage requirements for seeds destined for reproduction are not exactly the same as that for storage of seeds as a source of food. Prolonged viability is essential in the first case, but of less importance in the second.

Seeds stored in bulk may be attacked by rodents, insects, mites, fungi and bacteria. The extent of the infection is determined both by the moisture content of the seeds and by the temperature of the seeds in bulk.

Seeds for food are usually stored in large silos, warehouses or even on ships. In all these cases control of humidity and temperature is difficult and expensive, requiring aeration, which also serves as a means of cooling. Although the metabolism of the stored seeds is very low, it nevertheless can cause a rise in temperature of the seeds in bulk. Bacterial and fungal infection also add to this temperature rise. Bacterial and fungal spores are frequently seed borne, either on the surface of the seeds or between the seed coat and the embryo (Halloin, 1987). They are very difficult to get rid of. If temperature and humidity are not controlled and infections are present, the increases in temperature due to the metabolism of bacteria and fungi may cause severe increases in temperature, leading in extreme cases to charring of the seeds and even to fires. In less severe cases it may lead to changes in the properties of seed proteins, e.g. they become unsuitable for bread baking (Duffus and Slaughter, 1980).

A severe danger during the prolonged storage of seeds as food is the production of mycotoxins. The chief toxin-producing species are *Aspergillus* and *Fusarium*, but other fungi can also occasionally produce toxins (Stoloff, 1976). The toxins persist in the infected seeds long after the causative organism has been eliminated. The toxins in maize have been reported to persist for 12 yr (Duffus and Slaughter, 1980). Some of the toxins, e.g. aflatoxins produced by *Aspergillus flavus* and *A. parasitica*, are potent carcinogens. Severe outbreaks of cirrhosis of the liver were reported in infants, in India, which could be traced to *Aspergillus flavus* infection in groundnuts (Robinson, 1967). During the Second World War, in the Soviet Union, an outbreak of alimentary toxic aleukia was traced to a toxin produced by *Fusarium* infection of wheat grains wintering under snow (Joffe, 1986).

In the case of seeds used for certain industries, e.g. brewing or milling of different varieties of wheat, special attention must be paid in order to maintain specific properties. For brewing, high percentage germination is essential, while for milling water content must be higher for hard than for soft wheats (Duffus and Slaughter, 1980). Since it is very difficult and costly to create specific storage conditions for a large bulk of seeds, storage losses are very high.

II. Seeds for Propagation in Agriculture

In order to meet the needs of the growers, seeds produced in large quantities must meet certain national and international criteria. Among the most important are viability, vigour, purity and uniformity.

Germination must fall into well-defined percentage ranges. Germination should be as uniform as possible and occur within certain time limits. Failure to meet these standards can lead to severe losses because the normal pattern of agricultural practices may be disturbed. As a result losses in yield can occur. As well as the requirement for germination, vigour is also necessary, i.e. the ability of the seedling to emerge, develop and become established quickly under field conditions. The concept of vigour is important, but often difficult to define accurately (Heydecker, 1972; Roberts, 1973, see also Chapter 3). No clear markers exist to identify vigour in the dry seeds, but large seeds are more vigorous than smaller ones of the same cultivar. Despite this observation, size as such cannot be used as a marker for vigour.

When supplied to the grower all the seeds, in a given lot, must belong to the same species, uncontaminated by other species and especially not by seeds of weeds often present in the same field with the crop plant. Since modern agronomy often uses cultivars of species having certain especially desirable properties, such as disease resistance, the seeds provided must be of the pure specific cultivar. The purity of seeds of such cultivars is of enormous economic importance, especially when expensive hybrids are involved. Examination of the purity of cultivars often involves quite sophisticated techniques, for example characterization of storage protein using electrophoretic techniques.

The seeds provided to the grower must be as uniform in size as possible. This is becoming increasingly important when seeds are sown mechanically and now are frequently coated, in a pelleting process, which changes the size of the unit sown. It is obviously desirable that the seeds should be as free as possible from bacterial, fungal or viral pathogens, as well as of predators, such as insects or their eggs or larvae. Storage conditions similar to those of seeds for food are also required. A critical factor determining the longevity of seeds during storage, their behaviour during processing and their susceptibility to spoilage is their moisture content. The problem of water content and its effect on longevity has already been discussed (Chapter 3). However, the moisture content of the seeds immediately prior to harvest is also of great importance and affects storage.

It is often necessary to harvest crops before their moisture content is suitable for storage and at a stage of maturity unfavourable for mechanical handling. This can result in severe damage both during harvest itself and during subsequent treatment. Therefore harvest at the correct stage of maturity and water content is desirable, climatic factors allowing (Carver, 1980).

Thus, in seed production, quality comprises a variety of factors from purity of lines to uniformity in seed shape, size, water content and maturity, as well as high vigour and germinability.

Since in the economy of the 20th century seeds are very often moved from one country to another, it may be necessary to quarantine seeds prior to permitting their use for sowing, in order to ensure that no pathogens or pests are introduced with them. This problem is even more serious when new species are introduced from one country into another, especially in the case of wild species, which are to be cultivated. The danger of introducing an entirely new, previously unrecognized, pathogen which may damage other plants is quite real. Most countries therefore have special facilities for quarantining seeds before they are released for use.

Quarantine may therefore be considered as a further aspect of control of seed quality. Although quarantine does not exist for the bulk import of food grains and legumes, health control is also important.

The seed industry is concerned not only with quality control during seed production, but also with all the steps involved in harvesting and subsequent processing.

The simplest form of harvesting seeds is by hand. The crop is examined for the maturity of the seeds by visual inspection, i.e. are they sufficiently dry? The plant or plant parts are then cut for harvest. In more sophisticated systems the plants are harvested mechanically. In both cases, manual and mechanical harvesting, the seeds must be separated from the other parts of the plant by some process of threshing. Threshing is essentially some kind of beating process which ensures, in cereals for example, that the seeds are separated from the ears and from the rest of the plant. In wild species, seed dispersal often occurs without separation of the seed from the fruit or ear, which abscind together, with the seeds still in them. Such dispersal of spikelets of the ear of wild Gramineae, or the rapid splitting of the pod of legumes or the siliqua of Cruciferae assist in distributing the seeds in space (Chapter 7). During the development of modern varieties of cereals, by breeding, natural abscission of the ear is a trait which has been selected against. It is desirable that all the seeds remain in the spikelets, or in legumes in the pod, until they are harvested and not scattered (McWilliam, 1980). In the most advanced harvesting procedures, all the steps are carried out by one machine, a combined harvester of some kind.

Following threshing the seeds are moved in bulk along some kind of conveyor belt, they are separated from contaminating species and dirt removed in order to prepare seeds for storage.

During harvesting and threshing, the seeds can easily be damaged which may impair their subsequent ability to germinate. A crucial factor at this stage is the water content of the seeds. The nature of harvesting and threshing devices is very important. They must be designed to fulfil their function as gently as possible. Some of these aspects are considered by Thompson (1979), and in an FAO Agricultural Development Paper (1975).

After threshing and removal of contaminants it is often necessary to dry seeds to a certain desired moisture content which will determine their longevity during storage (see above). Whether drying is necessary is determined by climatic conditions prevailing during harvest. The simplest form of drying involves spreading the seeds out and allowing them to dry in the air in the field, weather permitting. In modern seed technology mechnical dryers are used, for example fans blow air through the seeds or conveyor belts move the seeds over a source of air, or various combinations of such devices (McLean, 1980). Whatever the method used, the rate of drying is extremely important. Too rapid drying may lead to damage, while if drying is too slow, the chance of infection by pathogens is greatly increased.

During drying the seed undergoes metabolic changes, many of which are critical for the subsequent ability of the seed to germinate. Drying of pea seeds while they are still synthesizing DNA leads to poor viability (Sen and Osborne, personal communication). Drying of seeds at a certain stage of development may induce dormancy, while at other stages, at a different water content dormancy is not induced.

The next step is storage. Whether the seeds are to be stored for food or for sowing, they must be protected from damage. Protection against damage by rodents, birds and insects is achieved by keeping out the pests by various means, for example sealing the storage containers. Protection against fungi and bacteria is more difficult since some of the pathogens are seed borne, and may be located under the seed coat. Consequently seeds are often treated with fungicides and/or insecticides prior to and during storage. Such treatments are limited or impossible in the case of food grains.

The most favourable storage conditions for most seeds are very low humidity and low temperatures. Seed viability is best maintained if the water content of the seeds is also low, but conditions differ for different seeds (Chapter 3). Modern technology has developed various means to preserve seeds better and for longer periods than obtained in silos or warehouses. One of the most suitable techniques for preservation of small amounts of seeds for germplasm is cryopreservation (see later).

The last step in processing involves preparation for marketing, either before or after storage. This involves the separation of seeds according to their size, shape or length, by their specific gravity (Brandenburg, 1977), their surface properties, colour, affinity for certain liquids or even their ability to bounce (Linnett, 1977). Each of these, or combinations of any of them, is characteristic of a given seed and will eventually ensure uniformity of the product. At this stage also mechanical damage to seed can occur and must be minimalized. Although seeds are dry and hard, they can nevertheless easily be broken and damaged. Particularly sensitive are the seed coat and the embryo or the embryonic axis. Damage to the embryonic axis depends to a large extent on its location in the seed and whether the embryo is deeply buried in the enveloping tissues.

Processing may also involve blending of different lots of seeds to give a desired moisture content or achieve the guaranteed germination percentage. If 80% germination is guaranteed, seeds with 95% germination might be mixed with an equal amount of seeds germinating at only 65%. However, commercially marketed seeds usually have germination of 90% or more.

The treatment with insecticides or fungicides, if not carried out previously, is done during the final stages of processing. At this stage the seeds may also be pelleted with pesticides, herbicides active against weed species, nutrients or with mixtures of growth substances designed to give the seedling a head start during development, (Heydecker and Coolbear, 1977). The latter treatment is sometimes referred to as envigoration.

Finally the seed industry packages the seeds. This may be in large sacks or containers for wheat or rice seeds or small carefully sealed aluminium foil envelopes containing one gram of minute *Petunia* seeds. The processing, storing and marketing of seeds are of great importance in modern agriculture and horticulture. At each stage through which the seed passes its germination behaviour, vigour and seed physiology may be affected or altered. The relevance of the physiology and other aspects of germination behaviour seems therefore obvious.

The extensive attempts to improve crop varieties by breeding, selection and other techniques, to adopt the crop products to long-distance transportation, durability and longer shelf life causes continuous changes in the varieties currently in use. As a result older varieties disappear, and with them various traits which may be important in breeding at some future time. To preserve the gene pool, seed storage laboratories have been designed. Thus in 1958 a National Seed Storage Laboratory was opened in the University of Colorado — Fort Collins and another one in Japan (Buss, 1979; Timothy and Goodman, 1979). Prolonged seed storage is important not only for preserving seeds of cultivated crops, but also for those of wild plants, which often have traits which may be valuable for breeding, for example tolerance to adverse external conditions or resistance to disease. The seed storage laboratories provide free germplasm to research institutes all over the world. Since the amounts of seeds stored are relatively small, the preferred method is cryopreservation. With time the scope of this task is broadening and germplasm is being preserved not only in the form of seeds but also as tissue cultures and somatic embryos.

III. Somatic Embryogenesis and Organogenesis as a Means of Propagation

Many tissues of plants can be induced to initiate and develop embryos under suitable conditions. Such embryogenesis from somatic tissue rather than from generative cells has been termed somatic embryogenesis. Somatic embryos can be made to develop into normal intact plants. It is also possible to induce formation of embryos from the culture of generative cells, which again can be made to form adult fertile plants. A special form of embryogenesis is that of the haploid or dihaploid embryos obtained from anther culture. Immature zygotic embryos can also be cultured successfully to produce, eventually, fertile adult plants (Raghavan, 1980). In soybeans an average of two somatic embryos may be produced per cotyledon (Parrott *et al.*, 1988). The nucellus of some plants also develops into secondary embryos. This phenomenon of polyembryony is rather common in nature in the Citrus family (Button and Kochba, 1977). Such nucellar embryos are found in addition to the normal zygotic embryos. Other parts within the embryo sac are also capable of developing into embryos.

The phenomenon of somatic embryo formation, from tissue cultures, was apparently first described by Levin (1951) and by Steward (1958) for *Daucus carota*. It is now regarded as a general phenomenon, not restricted to a few plant families (Ammirato, 1983). Although embryos can be derived from many kinds of cells, some apparently are embryogenic while others are not, or at least it is very difficult to induce them to form embryos. There appear to be some differences at the biochemical level between embryogenic and non-embryogenic cells. It is by no means certain that these differences are the cause of their different behaviour (Carlberg, 1987). There is a distinction between embryos derived from predetermined embryogenic cells, termed embryogenesis, and those derived from callus via organogenesis (Thorpe, 1980). There is some difference in origin and treatment of propagules derived from somatic embryos, obtained from embryogenic cells, predetermined indirectly and propagules obtained by organogenesis. In the former the various stages of embryo formation can be observed. In the latter different media, used sequentially, are generally needed to induce organogenesis, for example in a callus (Evans *et al.*, 1983). The callus is exposed first to a medium to induce shoot formation and then to another to induce root formation but more than two media may be used. The process of producing many plantlets in this way is usually referred to as clonal propagation (Vasil and Vasil, 1980).

Success in growing somatic embryos and getting them to form plantlets suitable for propagation is usually dependent on a variety of factors, similar to those known to affect cell callus and suspension cultures. Among the important factors are: the nature of the explant; from which part of the plant it is derived; its size; the culture medium and the ratio between the growth regulators in the medium. The ratio of the regulators and the sequence of their application is especially important in determining the successful development of the explant into an embryo and then into an adult plant. As mentioned above, frequently, the ratio between the growth regulators must be changed during different stages of the development of the explant, in order to channel the development in the desired direction. Other factors contributing to success are the solidity and viscosity of the culture medium, gaseous diffusion and the number and density of embryos developing in the culture vessel. As in other plant tissue cultures, considerable genetic variability may arise in somatic embryo cultures, the so called *soma-clonal variability* exists in them as in other tissue cultures. This variability is an important source for selection of desirable properties, which may be used in breeding (Jones, 1985). The same general principles apply to both forms of propagation, by embryogenesis and by organogenesis (Evans *et al.*, 1981).

One of the important features of somatic embryos no matter how they are derived, is that they can be produced in a relatively small space, in large amounts and can subsequently be sown or planted out (Hussey, 1983). Often this procedure can save a great deal of time, particularly when normal propagation is slow, for example in orchids, or it is necessary to establish new varieties rapidly or speed up propagation when there is a sudden large demand for a given cultivar, as happens with ornamental plants (Evans *et al.*, 1981). This form of propagation also has the advantage that virus free cultures can sometimes be established readily and maintained in a seed or gene bank. However, the considerable genetic variability in plants propagated by somatic embryos is in most cases a distinct disadvantage (Evans and Sharp, 1988).

The potential use of these culture methods is enhanced by the possibility of being used for genetic manipulation. It is not intended to discuss this topic here but it should be mentioned that a variety of techniques exist for the exchange of genetic material between cells of plants other than by the normal sexual process. These include somatic hybridization, transfer of DNA (either directly or by using various vectors such as plasmids or viruses) and protoplast fusion. All of these techniques ultimately depend on the ability to establish stable clones which can be propagated and which will survive under normal conditions for growing plants in the field or which can be adapted to survive such conditions.

The use of embryogenesis and organogenesis in order to propagate certain kinds of plant under special conditions is finding increasing use in agricultural and horticultural practice. Some of its potential advantages have been briefly mentioned. It is now possible to cryopreserve cells or propagules obtained by various techniques. Such preservation extends the potential use of these means of propagation as it lengthens the life span of the embryos and permits their storage (James, 1983; Withers, 1983; Finkle and Ulrich, 1983; Kartha, 1985).

In many cases *in vitro* embryo culture is used as a means of propagation, particularly in cases where embryo abortion is common and seedless fruits are produced, for example stone fruit or grapes. In practice excised ovules, 30 – 40 days after anthesis, are cultured *in vitro*. Although success of the cultures is not 100%, it is a technique which has proved to be useful (Spiegel-Roy *et al.*, 1985).

IV. Cryopreservation

Conservation of genetic traits of plants can be done efficiently by seed storage. However, recalcitrant seeds cannot be stored for long periods. Although not strictly speaking recalcitrant, soybean crops present great problems of preservation during storage. Numerous methods have been devised and tried out to achieve long-term preservation of genetic material for future use.

Almost all these methods are based on attempts to lower metabolism, without damaging the ability of the seed, the embryo or the tissue culture, to resume growth and development when desirable. These attempts were reviewed in detail by Withers (1987). The most successful and best advanced method of storage is that of cryopreservation.

In essence this method involves rapid cooling of biological material to very low temperatures, and storage under conditions which will preserve the ability to resume growth when brought back to normal temperatures. Most biological material will not endure such conditions unless previously treated with cryoprotectants, compounds which protect it during deep freezing.

Actively growing plant tissues appear to be more resistant to freezing than quiescent ones, with some exceptions such as seeds or dormant buds. Therefore in most cases a "pre-growth" period is allowed before freezing. This may be considered as the first step in a cryopreservation protocol.

During the next step, cryoprotection, the tissue is exposed to a complex mixture of highly hypertonic solutions. Among the most frequently used compounds are dimethylsulphoxide mixed with sugars, sugar alcohols or high molecular weight polymers such as polyethyleneglycol 6000. Different ratios of the compounds and different lengths of exposure are used for different tissues, the treatment being rather empirical.

The third step is the actual cooling, usually in liquid nitrogen. Optimal rates of cooling must be determined empirically for each tissue (Kartha, 1985; Withers, 1986). Subsequent storage is at very low temperatures, below −100°C.

A further critical stage in cryopreservation is the return of the tissue to normal temperatures. Warming should be such as to avoid recrystallization of intracellular ice. Generally rapid warming at 40°C is recommended. The thawed material is transferred to the recovery stage for resumption of growth and development. Each tissue appears to respond differently to the cryoprotectants and to the rates of cooling and rewarming. There are as yet no hard rules or recipes applicable to all tissues. Preservation of somatic embryos appears to be particularly difficult (Withers, 1979).

Cryopreservation is a relatively new technique and more research will hopefully improve it and lead to widespread use.

Finally a word of caution is needed. The procedures briefly discussed in this chapter hold out considerable hope for the future. However, it is perfectly clear that they will not now, nor probably in the future, become a substitute for the propagation of plants by seeds. The seed is still the ideal way for storing packages containing genetic information for long periods, for producing new plants and for normal routine agricultural and horticultural procedures. It should be mentioned that there is reason for concern about the possible implications of using what are essentially unconventional means of plant propagation. Whether such means will have any long-term effects cannot be predicted at this stage, but all of us should be alert to the problem and give it due consideration.

Bibliography

Ammirato V. (1983) in *Handbook of Cell Culture* vol. 1, p. 82 (Eds D. A. Evans, W. R. Sharp, P. V. Ammirato and Y. Yamada), Macmillan, London.

Brandenburg N. R. (1977) *Seed Sci. Technol.* **5**, 173.

Buss L. N. (1979) in *The Plant Seed: Development, Preservation and Germination*, p. 145 (Eds I. Rubinstein, R. L. Phillips, C. E. Green and B. G. Gengenbach), Academic Press, New York.

Button J. and Kochba J. (1977) in *Plant Cell Tissue and Organ Culture*, p. 70 (Eds J. Reinert and P. S. Bajaj), Springer-Verlag, Berlin.

Carlberg I. (1987) Ph.D. Thesis, Uppsala University, *Acta Universitatis Upsaliensis* **72**, Uppsala, Sweden.

Carver M. (1980) in *Seed Production*, p. 259 (Ed. P. D. Hebblethwaite), Butterworths, London.

Duffus C. and Slaughter C. (1980) *Seeds and their Uses,* John Wiley and Sons, Chichester.

Evans D. S. and Sharp W. R. (1988) in *International Ass. Plant Tissue Culture Newsletter*, No. 54, p. 2 (Ed. Ad. J. Kool).

Evans D. A., Sharp W. R. and Bravo J. E. (1983) in *Handbook of Plant Cell Culture*, vol. 2, p. 47 (Eds D. A. Evans, W. R. Sharp, P. V. Ammirato and Y. Yamada), Macmillan, London.

Evans D. A., Sharp W. R. and Flick C. E. (1981) in *Plant Tissue Culture*, p. 45 (Ed. T. A. Thorpe), Academic Press, New York.

F.A.O. Agricultural Development Paper No. 98 (1975) *Cereal Technology,* F.A.O., Rome.
Finkle B. and Ulrich J. (1983) in *Handbook of Plant Cell Culture*, vol. 2, p. 806 (Eds D. A. Evans, W. R. Sharp, P. V. Ammirata and Y. Yamada), Macmillan, London.
Halloin J. M. (1986) in *Physiology of Seed Deterioration*, p. 89 (Eds M. B. McDonald Jr and C. J. Nelson) Special Publication No. 11, Crop Science Society of America Inc., Madison, Wisconsin.
Hebblethwaite P. D. (1980) *Seed Production*, Butterworths, London.
Heydecker W. (1972) in *Viability of Seeds*, p. 209 (Ed. E. H. Roberts), Chapman and Hall, London.
Heydecker W. and Coolbear P. (1977) *Seed Sci. Technol.* **5**, 353.
Hussey G. (1983) in *Plant Biotechnology*, p. 111 (Eds S. H. Mantell and H. Smith), *Soc. Exp. Biol. Seminar Ser.* **18**, Cambridge University Press, Cambridge.
James E. (1983) in *Plant Biotechnology*, p. 163 (Eds S. H. Mantell and H. Smith) *Soc. Exp. Biol. Seminar Ser.* **18**, Cambridge University Press, Cambridge.
Joffe A. Z. (1986) *Fusarium Species; Their Biology and Toxicology*, John Wiley and Sons, Chichester.
Jones M. G. K. (1985) in *Plant Products of the New Technology*, p. 215 (Eds K. Fuller and J. P. Gallon), *Ann. Proc. Phytochem. Soc. Europe*, vol. 26, Clarendon Press, Oxford.
Kartha K. K. (Ed.) (1985) *Cryopreservation of Plant Cells and Organs,* CRC Press, Boca Raton.
Levin M. (1951) *Am. J. Bot.* **38**, 132.
Linnett B. (1977) *Seed Sci. Technol.* **5**, 194.
McLean K. A. (1980) *Drying and Storing Combinable Crops*, Farming Press Ltd.
McWilliam J. R. (1980) in *Seed Production*, p. 51 (Ed. P. D. Hebblethwaite), Butterworths, London.
Parrott W. A., Williams E. G., Hildebrand D. F. and Collins G. B. (1988) in *Newsletter of The Intern. Assoc. Plant Tissue Culture* No. 54, p. 10 (Ed. Ad. J. Kool).
Raghavan V. (1980) in *Perspectives in Plant Cell and Tissue Culture*, (Ed. I. K. Vasil), *Intern. Rev. Cytology* Suppl. 11B, p. 209, Academic Press, New York.
Roberts E. H. (1973) *Seed Sci. Technol.* **1**, 529.
Robinson P. (1967) *Clin. Pediatrics* **6**, 57.
Seeds (1961) *The Yearbook of Agriculture* (Ed. A. Stefferand), U.S. Department of Agriculture, Washington.
Spiegel-Roy P., Sahar N., Baron J. and Levi U. (1985) *J. Am. Hort. Soc.* **119**, 109.
Steward F. C., Mapes M. O. and Mears K. (1958) *Am. J. Bot.* **45**, 705.
Stoloff L. (1976) in *Mycotoxins and other Fungal Related Food Problems*, p. 23, *Adv. Chem Series* **149**, Am. Chem. Soc. Washington, D.C.
Thompson J. R. (1979) *An Introduction to Seed Technology*, L. Hill.
Thorpe T. A. (1980) in *Perspectives in Plant Cell and Tissue Culture*, (Ed. I. K. Vasil) *Internat. Rev. Cytology* vol. 11A, p. 71, Academic Press, New York.
Timothy D. H. and Goodman M. M. (1979) in *The Plant Seed: Development, Preservation and Germination*, p. 171 (Eds I. Rubinstein, R. L. Phillips, C. E. Green and B. G. Gengenbach), Academic Press, New York.
Vasil I. K. and Vasil V. (1980) in *Perspectives in Plant Cell and Tissue Culture*, (Ed. I. K. Vasil) *Internat. Rev. Cytology*, vol. 11A, p. 145, Academic Press, New York.
Withers L. A. (1979) *Plant Physiol.* **63**, 460.
Withers L. A. (1983) in *Plant Biotechnology*, p. 187 (Eds S. H. Mantell and H. Smith) *Soc. Exp. Biol. Seminar Series,* **18**, Cambridge University Press, Cambridge.
Withers L. A. (1986) in *Plant Cell Culture Technology*, p. 96 (Ed. M. M. Yeoman), Blackwells, Oxford.
Withers L. A. (1987) in *Oxford Survey of Plant Molecular and Cell Biology*, p. 221 (Ed. B. J. Miflin), Oxford University Press, Oxford.

Plant Index

Author Index

Subject Index